# Iron Nanomaterials for Water and Soil Treatment

# Iron Nanomaterials for Water and Soil Treatment

edited by

**Marta I. Litter | Natalia Quici | Martín Meichtry**

Pan Stanford Publishing

*Published by*

Pan Stanford Publishing Pte. Ltd.
Penthouse Level, Suntec Tower 3
8 Temasek Boulevard
Singapore 038988

Email: editorial@panstanford.com
Web: www.panstanford.com

**British Library Cataloguing-in-Publication Data**
A catalogue record for this book is available from the British Library.

**Iron Nanomaterials for Water and Soil Treatment**



ISBN 978-981-4774-67-3 (Hardcover)
ISBN 978-981-4669-49-8 (eBook)

# Contents

# Preface

Nanotechnology has a great potential for providing efficient, cost-effective and environmentally acceptable solutions to face the increasing requirements on quality and quantity of fresh water for industrial, agricultural, or human use. Iron nanomaterials, either zerovalent iron, iron oxides, or those prepared from natural extracts and iron salts, present key physicochemical properties that make them particularly attractive as contaminant removal agents for water and soil cleaning. The large surface area of these nanoparticles imparts high sorption capacity to them, along with the ability to be functionalized for the enhancement of their affinity and selectivity. However, one of the most important properties is the outstanding capacity to act as redox-active materials, transforming the pollutants to less noxious chemical species by either oxidation or reduction, such as reduction of Cr(VI) to Cr(III) and dehalogenation of hydrocarbons.

This book focuses on the methods of preparation of iron nanomaterials that can carry out contaminant removal processes and the use of these nanoparticles for cleaning waters and soils. It carefully explains the different aspects of the synthesis and characterization of iron nanoparticles and methods to evaluate their ability to remove contaminants, along with practical deployment. It overviews the advantages and disadvantages of using iron-based nanomaterials and presents a vision for the future of this nanotechnology.

Although there are a number of books published on the subject of nanomaterials, not too many of them are especially devoted to iron materials, which are rather of low cost, are nontoxic, and can be prepared easily and envisaged to be used in a large variety of applications. The literature has scarce reviews on preparation of iron nanoparticles from natural sources and lacks emphasis on the different processes, such as adsorption, redox pathways, and ionic exchange, taking place in the removal of different pollutants. Reports and mechanisms on soil treatment are not commonly found in the literature. This book opens a multidisciplinary scope for engineers and scientists and introduces undergraduate and postgraduate students to a new world of nanomaterials useful in treatment of waters and soils.

We thank the publisher for presenting the possibility to write a book on this exciting and innovative topic. We also warmly appreciate the contribution of all colleagues who participated in the making of this book and promptly resolved issues that surfaced during the various stages of its development.

We are particularly grateful to Stanford Chong and Shivani Sharma for their extreme help during the elaboration of this book.

**Marta I. Litter**

**Natalia Quici**

**Martín Meichtry**

# Chapter 1

# The Story and Future of Nanoparticulated Iron Materials

**Marta I. Litter**
*Gerencia Química, Comisión Nacional de Energía Atómica, 1650 San Martín, Prov. de Buenos Aires, Argentina*
marta.litter@gmail.com

## 1.1 Introduction

Due to the demand for fresh water, which has tremendously increased [1], it is necessary to process huge amounts of wastewater effluents containing different types of pollutants. The requirements for clean and safe water, continuously increasing due to the increase in population, draughts, and contamination of water sources, are more and more stringent. For this purpose, new innovative treatment processes, together with improvement in the conventional ones, are essential to sustain the demand for clean water. The conventional technologies, although well understood and deployed, are generally costly and sometimes ineffective to reach environmental regulations.

Effluent slurries coming from various industrial processes together with domestic wastes are still present in the ground

*Iron Nanomaterials for Water and Soil Treatment*
Edited by Marta I. Litter, Natalia Quici, and Martín Meichtry

ISBN 978-981-4774-67-3 (Hardcover), 978-981-4669-49-8 (eBook)
www.panstanford.com

and sometimes contain high amounts of heavy metals and dense nonaqueous phase liquids (DNAPLs), for example, chlorinated solvents, because they have been improperly disposed in the past. These wastes migrated into the ground. For decades, it was not recognized that they could persist in the subsurface, potentially contaminating drinking-water sources. Both heavy metals and DNAPLs are particularly persistent contaminants and are common contaminants at industrialized sites. The existing technologies (steam and density modified displacement, stabilization/solidification, and in situ redox treatment) cannot achieve the regulations for cleanup of contaminated aquifers.

Recently, nanotechnology has emerged as a very important tool for the removal of several pollutants from wastewaters and soils, with considerable potential benefits even with the possibility of use to obtain safe drinking water [2, 3]. Nanoparticles are small atomic clusters in the range of 1–100 nm with size-dependent properties. On the other hand, nanotechnology can be defined as a new area of science and engineering that involves the discovery and ability to manipulate, control, and apply nanoscale material in a useful way. The name "nano" comes from the size of the molecules, which is measured in nanometers or one billionth of a meter (i.e., $1\ nm = 1 \times 10^{-9}\ m$) [4].

The use of zerovalent iron nanoparticles (nZVI) or nanostructured iron oxides (nFeOx) is a rather new technology for the treatment of contaminated land and water (surface and groundwater). This technology can be widely adopted as a rapid, highly effective, and low-cost alternative to conventional remediation technologies. However, several steps should be first envisaged to solve some remaining problems for a wide application, such as the lack of instability of free nanoparticles, or difficulties for injection in soil to achieve good contact of the nanoparticles with the subsurface groundwater.

The unique properties of nanoparticles, i.e., the large surface-area-to-volume ratio and the high surface energy lead to a significantly improved reactivity with regard to contaminants [5]. Use of nanoparticles is convenient for in situ applications. Particularly, iron nanoparticles have proved to be very effective for the transformation and detoxification of a wide variety of contaminants, such as chlorinated organic solvents, pesticides, PCBs, arsenic and heavy metals, among others.

The use of nanometals for subsurface remediation of chlorinated compound and heavy metal contaminated sites has received significant attention due to the ability of nanometals to rapidly transform contaminants in controlled laboratory experiments. Nanometals used for these purposes include nanoiron and nanozinc, but nZVI is most commonly used. In addition, other metals such as palladium or nickel are usually added to increase the reduction rate, forming a bimetallic nanometal. The reactivity and the availability of the existing technology to precisely design and synthesize nanometals make these materials particularly attractive for the remediation of subsurface contaminants.

## 1.2 Use of Zerovalent Iron for Environmental Purposes

The use of iron in its elemental state, $Fe^0$ or ZVI, has been recognized as a potential tool for the removal of contaminants due to the exceptional properties of the material: ZVI is abundant (the fourth most abundant element in the earth's crust), nontoxic, cheap, easy to produce, and the process requires little maintenance. ZVI is an effective reductant but offers also good properties to degrade or oxidize other pollutants, can be combined with $H_2O_2$ to produce Fenton-type processes, and can be used to remove Cr(VI) and other heavy metals, halogenated organics, nitroaromatic, metalloids such as As, nitrate, dyes, phenolic compounds, etc. The use of the technology has enormously increased in the last years.

In the 1980s, ZVI began to be used for environmental purposes in the form of powders or filings taking profit of the iron corrosion chemistry. In 1982, Gould published the first study on the environmental use of ZVI, describing the kinetics of reduction of hexavalent chromium by metallic iron [6]. In 1991, the first field application of ZVI in permeable reactive barrier (PRB) technology for in situ remediation of groundwater contaminated by trichloroethylene (TCE) and pentachloroethylene (PCE) was performed at the Canadian Forces Base, Borden, Ontario; removal of about 95 and 91% for TCE and PCE, respectively, was found [7]. Since then, the use of iron as a reactive material in PRB led to considerable R&D [8, 9]. Zerovalent iron PRB systems may remove chlorinated organics by reductive dechlorination, whereas metals, metalloids, and radionuclides may be removed via reductive precipitation, surface adsorption or complexation, or coprecipitation with Fe oxyhydroxides formed on the ZVI surfaces. This work was followed by studies on Cr, As, halogenated hydrocarbons, etc. with microsized materials, for example, references [10–17]. Due to the success of the $Fe^0$-based PRB implementation, ZVI technology gained much attention. Several papers detailed studies on the removal of organic and inorganic compounds by ZVI for hazardous waste removal [e.g., 18–22], which indicated a great potential use of the material.

Concerning the involved reactions, Matheson and Tratnyek [15] related the chemistry of the iron corrosion with the dehalogenation pathways of halogenated compounds, and for the first time, they showed that increasing the clean surface area of iron greatly increased the degradation rate of, for example, carbon tetrachloride. An important study in 1996 described the kinetics of dehalogenation promoted by ZVI using an extended set of various iron powders; the reaction rates depended strongly on the reactive surface area of the materials [23]. Hereafter, ZVI has been extensively applied in the remediation/treatment of groundwater and wastewater contaminated with halogenated organics [23–26], nitroaromatics [22, 27], dyes [28, 29], phenolic compounds [30], heavy metals [31, 32], arsenate and arsenite [33–35], chromate [36, 37], nitrate [36, 38, 39], bromate [40], selenite [41], selenate [42, 43], and uranyl [44, 45]. In the last two decades, the number of publications increased exponentially [46].

Environmental applications of metallic iron have been enthusiastically accepted by many users and regulatory agencies, largely due to the low costs and the absence of any known toxicity induced by the use of iron. Metallic iron in the form of iron microparticles (granular or powdered, >0.1 mm in diameter) has been used in packed bed reactors and PRBs intersecting the contamination plume [47, 48].

Although ZVI has been extensively applied for the remediation/treatment of groundwater and wastewater contaminated with various organic and inorganic pollutants, serious limitations appeared for the correct deployment of the technology. These limitations included low reactivity due to the intrinsic passive layer, narrow working pH, reactivity loss with time due to the precipitation of metal hydroxides and metal carbonates, low selectivity for the target contaminant, especially under toxic conditions, limited efficacy for treatment of some refractory contaminants, and passivity of ZVI arising from certain contaminants [49]. In this last work, studies on ZVI technologies were divided into three categories: (1) studies to overcome the limitations of ZVI; (2) studies on the use of ZVI to sequester hazardous contaminants; and (3) application in real practice to tackle the derived problems.

Some review papers focused on microsized ZVI (mZVI) or ZVI with larger particle size [46, 50, 51]. Various criticisms about the technology arose [52, 53].

Actually, the efficiency of the ZVI technology was improved with various actions such as the pretreatment to remove the passive layer, the addition of a second metal, generally a noble one, to produce ZVI-based bimetals, the coupling of ZVI with other adsorptive materials, and the recovery of the reactivity of aged ZVI. However, the most important advance was the fabrication of nanosized ZVI to increase the surface area of the material, leading to an exceptional reactivity.

## 1.3 Development of Nanostructured Zerovalent Iron

The investigations about the use of nZVI for groundwater remediation began in the 1990s, and the prepared materials showed a tremendous promise in the environmental sector due to its high reactivity and effectiveness.

The first report on the preparation of metal nanoparticles was published by Schlesinger et al. in 1953 [54], who reduced transition metal ions, including Fe(II) with borohydride with the objective of hydrogen production; this technique was later confirmed by Brown and Brown [55]. In the 1990s, the nanosize of the products was ascertained [56, 57] and the term "nanoscale Fe powder" began to be used.

The small size makes the deployment of the nanoparticles very flexible due to their conceptually high mobility through porous media and their potential for injection at almost any location and depth in terrestrial groundwater systems. Nanoparticulated iron materials have been proven to be highly effective for the removal of a wide range of pollutants in water, including chlorinated organics, inorganic anions, heavy metals and metalloids, including Pb, Cr, Cu, As, Ni, Zn, Cd, and Ag, and radionuclides such as radioisotopes of Ba, $TcO_4$, and U [5 and references therein]. In 1997, preparation of nanoscale iron particles by reducing Fe(II)/Fe(III) ions of different salts in water ($FeCl_2$, $FeSO_4$) with sodium borohydride was reported; the authors pointed out that nZVI might be directly injected into the contaminated subsurface for in situ dechlorination in the subsurface and remediation of contaminated soil and groundwater [58, 59].

Over the last 25 years, the technologies based on iron nanomaterials for remediation of soil and groundwater have undertaken different issues, comprising the comparison with the use of iron (macro or micro) powder/filings, non-stabilized and stabilized nZVI, improvement in the mobility of nanoparticles, and injection procedures. Reviews on the subject appeared in last times, treating different scientific topics on the technology, for example, synthesis, properties, and applications [60–62], nZVI aqueous corrosion, manufacture, enhancement of reactivity, stability and subsurface mobility [63], nZVI applications for groundwater remediation in Europe [64], nZVI/bimetallic nanomaterials [65, 66], benefits and risks of nZVI for in situ remediation [67, 68], and general overview [5, 69–71]. Zhao et al. [71] have described the evolution of ZVI-based cleanup technologies, and Fig. 1.1., taken from this reference, provides an overview of the major milestones on the environmental uses of the technology.

According to Fig. 1.1, the following stages can be considered for the development of nZVI-based remediation technology: (1) synthesis, characterization, and testing of bare or non-stabilized ZVI, which are essentially ZVI aggregates with some primary particles in the nanoscale, (2) synthesis, characterization, and testing of stabilized nZVI for reductive dechlorination and reductive immobilization of redox active metals and radionuclides, (3) application of stabilized nZVI for environmental remediation, and (4) studies on environmental fate, transport and toxicity of bare and stabilized nZVI. Examples of potential applications of nanoscale iron particles for in situ remediation are depicted in Fig. 1.2., taken from Ref. [5]. As it can be observed, this includes transformation of fertilizers such as nitrate, dechlorination of organic solvents such as carbon tetrachloride or tetrachloroethylene, degradation of pesticides such as Lindane or DDT, and immobilization of heavy metals or arsenic.

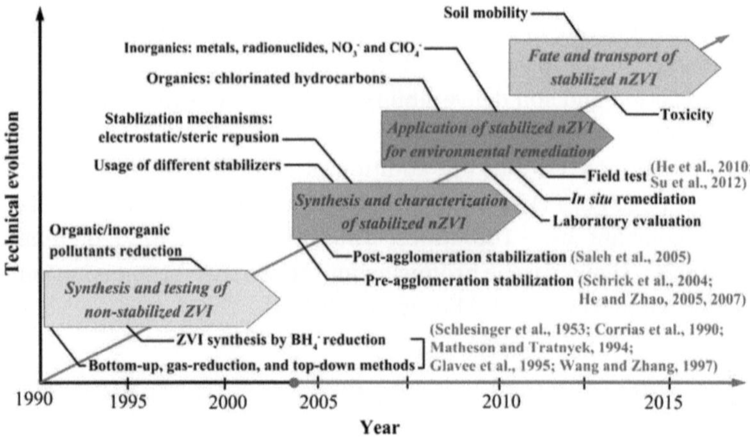


**Figure 1.1** Development and technical evolution of stabilized nZVI technologies. Reprinted from Ref. [71], Copyright 2016, with permission from Elsevier.

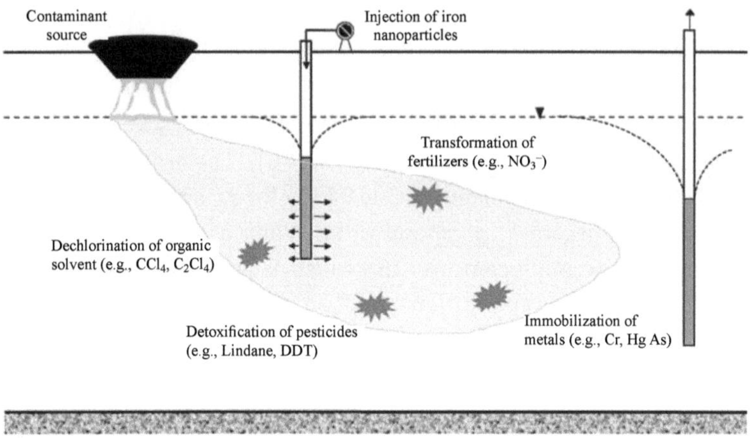


**Figure 1.2** Nanoscale iron particles for in situ remediation. Reprinted by permission from Springer Nature: Kluwer Academic Publishers, *Journal of Nanoparticle Research*, Ref. [5], Copyright 2003.

Wang and Zhang [58] demonstrated that non-stabilized synthetic ZVI particles were effective for reductive dechlorination and noted that the freshly prepared nZVI particles were much more reactive than commercial iron powders. Due to the importance of chlorinated hydrocarbons, these findings were judged of extreme relevance for decontamination of soils. However, it was needed to investigate the influence of the properties of the nZVI materials such as particle aggregation and soil deliverability, aspects that began to be studied [72–74].

It was noticed that the addition of a small fraction of a noble metal (e.g., Pd, Pt) to nZVI increased the surface-area-normalized rate constant by ~100 times. Thus, the use of nZVI/bimetallic materials showed to be promising for the subsurface remediation of chlorinated solvents and heavy metals [59].

## 1.4 Field Studies on Application of Stabilized nZVI for in situ Remediation

In recent years, a range of inexpensive, iron-based, water cleanup technologies have been developed to address the major problem of arsenic contamination in groundwater used for drinking [51]. Quinn et al. [75] highlighted the potential of nZVI to treat dissolved chlorinated solvents in situ, and to remediate zones with DNAPLs in aquifers. Since then, several field experiments were developed, typically injecting nanoparticles directly as a slurry (nanofluid) into the subsurface environment (e.g., [5, 26, 76]). The first field injection test of nZVI was reported in 2001 [26] for in situ degradation of chlorinated organics in groundwater. When a non-stabilized nZVI suspension was injected into the subsurface, it was observed that nZVI aggregates rapidly deposited on the well screen, while the remaining finer nZVI could travel from a few inches to a few feet before becoming immobilized in the soil matrix. The trend was then to investigate more effective stabilization techniques and more deliverable nZVI. The utility of iron nanoparticles in removing or stabilizing metallic and metalloid contaminants has also been demonstrated in a variety of soil and water media. Examples of reducing As [77, 78], Cu(II) [79], and Cr(VI) [80] mobility in groundwater or soils have been reported. A rapid transfer of nano-iron-based remediation technologies from laboratory to field-scale application and full-scale commercial applications of nZVI in land and groundwater remediation have rapidly developed [3]. In 2009, the technology was tested at the pilot- or field-scale at more than 58 sites [81], which increased to around 70 projects worldwide at the pilot or full scale in 2015 [82, 83]. Various remediation efficiencies have been observed in field trials in the USA and in Europe [75, 84–88].

## References

1. WHO (1997). *Water Pollution Control: A Guide to the Use of Water Quality Management Principles*, Helmer, R. and Hespanhol, I. (eds.) (E&FN Spon, London, UK).
2. Sharma, Y. C., Srivastava, V., Singh, V. K., Kaul, S. N., and Weng, C. H. (2009). Nano-adsorbents for the removal of metallic pollutants from water and wastewater, *Environ. Technol.*, **30**, pp. 583–609.
3. Tratnyek, P. G. and Johnson, R. L. (2006). Nanotechnologies for environmental cleanup, *Nano Today*, **1**, pp. 44–48.
4. Maclurcan, D. and Radywyl, N. (eds.) (2011). *Nanotechnology and Global Sustainability, Perspectives in Nanotechnology* (CRC Press).
5. Zhang, W. X. (2003). Nanoscale iron particles for environmental remediation: An overview, *J. Nanoparticle Res.*, **5**, pp. 323–332.
6. Gould, J. (1982). The kinetics of hexavalent chromium reduction by metallic iron, *Water Res.*, **16**, pp. 871–877.
7. O'Hannesin, S. F. and Gillham, R. W. (1992). A permeable reaction wall for in situ degradation of halogenated organic compounds. *The 45th Canadian Geotechnical Society Conference*, 25–28 October 1992, Toronto, Ontario.
8. Blowes, D. W. and Ptacek, C. J. (1994). System for treating contaminated groundwater. US Patent number US5362394.
9. Litter, M. I., Cortina, J. L., Fiúza, A. M. A., Silva, A., and Tsakiroglou, C. (2014) In-situ technologies for groundwater treatment: The case of arsenic, in *In-Situ Remediation of Arsenic-Contaminated Sites*, Holländer, H., Bundschuh, J., and Ma, L. (eds.) (CRC Press/Balkema, Taylor & Francis Group, London), pp. 208–252.
10. Balko, B. A. and Tratnyek, P. G. (1998). Photoeffects on the reduction of carbon tetrachloride by zero-valent iron, *J. Phys. Chem. B*, **102**, pp. 1459–1465.

11. Cantrell, K. J., Kaplan, D. I., and Wietsma, T. W. (1995). Zero-valent iron for the in situ remediation of selected metals in groundwater, *J. Hazard. Mater.*, **42**, pp. 201–212.

12. Domga, R., Togue-Kamga, F., Noubactep, C., and Tchatchueng, J.-B. (2015). Discussing porosity loss of $Fe^0$ packed water filters at ground level, *Chem. Eng. J.*, **263**, pp. 127–134.

13. Gillham, R. W. (1993). Cleaning halogenated contaminants from groundwater. U.S. Patent No. 5, 266, 213.

14. Orth, W. S. and Gillham, R. W. (1996). Dechlorination of trichloroethene in aqueous solution using Fe(0), *Environ. Sci. Technol.*, **30**, pp. 66–71.

15. Matheson, L. J. and Tratnyek, P. G. (1994). Reductive dehalogenation of chlorinated methanes by iron metal, *Environ. Sci. Technol.*, **28**, pp. 2045–2053.

16. Senzaki, T. and Kumagai, Y. (1988). Removal of chlorinated organic compounds from wastewater by reduction process: Treatment of 1,1,2,2-tetrachloroethane with iron powder, *Kogyo Yosui*, **357**, pp. 2–7.

17. Senzaki, T. and Kumagai, Y. (1989). Removal of chlorinated organic compounds from wastewater by reduction process: II. Treatment of trichloroethylene with iron powder, *Kogyo Yosui*, **369**, pp. 10–25.

18. Khudenko, B. M. (1991). Feasibility evaluation of a novel method for destruction of organics, *Water Sci. Technol.*, **23**, pp. 1873–1881.

19. Khudenko, B. M. (1985). Mechanism and kinetics of cementation processes, *Water Sci. Technol.*, **17**, pp. 719–731.

20. Senzaki, T. (1991). Removal of chlorinated organic compounds from wastewater by reduction process: III Treatment of trichloroethylene with iron powder, *Kogyo Yosui*, **391**, pp. 29–35.

21. Gillham, R. W. and O'Hannesin, S. F. (1994). Enhanced degradation of halogenated aliphatics by zero-valent iron, *Ground Water*, **32**, pp. 958–967.

22. Agrawal, A. and Tratnyek, P. G. (1995). Reduction of nitro aromatic compounds by zerovalent iron metal, *Environ. Sci. Technol.*, **30**, pp. 153–160.

23. Johnson, T. L., Scherer, M. M., and Tratnyek, P. G. (1996). Kinetics of halogenated organic compound degradation by iron metal, *Environ. Sci. Technol.*, **30**, pp. 2634–2640.

24. Arnold, W. A. and Roberts, A. L. (2000). Pathways and kinetics of chlorinated ethylene and chlorinated acetylene reaction with Fe(0) particles, *Environ. Sci. Technol.*, **34**, pp. 1794–1805.

25. Roberts, A. L., Totten, L. A., Arnold, W. A., Burris, D. R., and Campbell, T. J. (1996). Reductive elimination of chlorinated ethylenes by zero-valent metals, *Environ. Sci. Technol.*, **30**, pp. 2654–2659.

26. Elliott, D. W. and Zhang, W. X. (2001). Field assessment of nanoscale bimetallic particles for groundwater treatment, *Environ. Sci. Technol.*, **35**, pp. 4922–4926.

27. Keum, Y. S. and Li, Q. X. (2004). Reduction of nitroaromatic pesticides with zero-valent iron, *Chemosphere*, **54**, pp. 255–263.

28. Nam, S. and Tratnyek, P. G. (2000). Reduction of azo dyes with zerovalent iron, *Water Res.*, **34**, pp. 1837–1845.

29. Pereira, W. S. and Freire, R. S. (2006). Azo dye degradation by recycled waste zero-valent iron powder, *J. Braz. Chem. Soc.*, **17**, pp. 832–838.

30. Morales, J., Hutcheson, R., and Cheng, I. F. (2002). Dechlorination of chlorinated phenols by catalyzed and uncatalyzed Fe(0) and Mg(0) particles, *J. Hazard. Mater.*, **90**, pp. 97–108.

31. Rangsivek, R. and Jekel, M. R. (2005). Removal of dissolved metals by zero-valent iron (ZVI): Kinetics, equilibria, processes and implications for stormwater runoff treatment, *Water Res.*, **39**, pp. 4153–4163.

32. Shokes, T. E. and Moller, G. (1999). Removal of dissolved heavy metals from acid rock drainage using iron metal, *Environ. Sci. Technol.*, **33**, pp. 282–287.

33. Neumann, A., Kaegi, R., Voegelin, A., Hussam, A., Munir, A .K. M., and Hug, S. J. (2013). Arsenic removal with composite iron matrix filters in Bangladesh: A field and laboratory study, *Environ. Sci. Technol.*, **47**, pp. 4544–4554.

34. Su, C. M. and Puls, R. W. (2001). Arsenate and arsenite removal by zerovalent iron: Effects of phosphate, silicate, carbonate, borate, sulfate, chromate, molybdate, and nitrate, relative to chloride, *Environ. Sci. Technol.*, **35**, pp. 4562–4568.

35. Su, C. M. and Puls, R. W. (2001). Arsenate and arsenite removal by zerovalent iron: Kinetics, redox transformation, and implications for in situ groundwater remediation, *Environ. Sci. Technol.*, **35**, pp. 1487–1492.

36. Alowitz, M. J. and Scherer, M. M. (2002). Kinetics of nitrate, nitrite, and Cr(VI) reduction by iron metal, *Environ. Sci. Technol.*, **36**, pp. 299–306.

37. Puls, R. W., Paul, C. J., and Powell, R. M. (1999). The application of in situ permeable reactive (zero-valent iron) barrier technology for the remediation of chromate-contaminated groundwater: A field test, *Appl. Geochem.*, **14**, pp. 989–1000.

38. Chi, I., Zhang, S. T., Lu, X., Dong, L. H., and Yao, S. L. (2004). Chemical reduction of nitrate by metallic iron, *J. Water Supply Res. Technol.*, **53**, pp. 37–41.

39. Huang, C. P., Wang, H. W., and Chiu, P. C. (1998). Nitrate reduction by metallic iron, *Water Res.*, **32**, pp. 2257–2264.

40. Xie, L. and Shang, C. (2007). The effects of operational parameters and common anions on the reactivity of zero-valent iron in bromate reduction, *Chemosphere*, **66**, pp. 1652–1659.

41. Liang, L. P., Yang, W. J., Guan, X. H., Li, J., Xu, Z. J., Wu, J., Huang, Y. Y., and Zhang, X. G. Z. (2013). Kinetics and mechanisms of pH-dependent selenite removal by zero-valent iron, *Water Res.*, **47**, pp. 5846–5855.

42. Klas, S. and Kirk, D. W. (2013). Understanding the positive effects of low pH and limited aeration on selenate removal from water by elemental iron, *Sep. Purif. Technol.*, **116**, pp. 222–229.

43. Yoon, I. H., Kim, K. W., Bang, S., and Kim, M. G. (2011). Reduction and adsorption mechanisms of selenate by zero-valent iron and related iron corrosion, *Appl. Catal. B: Environ.*, **104**, pp. 185–192.

44. Fiedor, J. N., Bostick, W. D., Jarabek, R. J., and Farrell, J. (1998). Understanding the mechanism of uranium removal from groundwater by zero-valent iron using X-ray photoelectron spectroscopy, *Environ. Sci. Technol.*, **32**, pp. 1466–1473.

45. Gu, B., Liang, L., Dickey, M. J., Yin, X., and Dai, S. (1998). Reductive precipitation of uranium(VI) by zero-valent iron, *Environ. Sci. Technol.*, **32**, pp. 3366–3373.

46. Fu, F. L., Dionysiou, D. D., and Liu, H. (2014). The use of zero-valent iron for groundwater remediation and wastewater treatment: A review, *J. Hazard. Mater.*, **267**, pp. 194–205.

47. Blowes, D. W., Ptacek, C. J., Benner, S. G., McRae, Che, W. T., Bennett, T. A., and Puls, R. W. (2000). Treatment of inorganic contaminants using permeable reactive barriers, *J. Contam. Hydrol.*, **45**, pp. 123–137.

48. Litter, M. I., Morgada, M. E., and Bundschuh, J. (2010). Possible treatments for arsenic removal in Latin american waters for human consumption, *Environ. Pollut.*, **158**, pp. 1105–1118.

49. Guan, X., Sun, Y, Qin, H., Li, J., Lo, I. M., He, D., and Dong, H. (2015). Review. The limitations of applying zero-valent iron technology in contaminants sequestration and the corresponding countermeasures: The development in zero-valent iron technology in the last two decades (1994–2014), *Water. Res.*, **75**, pp. 224–248.

50. Noubactep, C. (2008). A critical review on the process of contaminant removal in $Fe^0$-$H_2O$ systems, *Environ. Technol.*, **29**, pp. 909–920.

51. Cundy, A. B., Hopkinson, L., and Whitby, R. L. D. (2008). Use of iron-based technologies in contaminated land and groundwater remediation: A review, *Sci. Total Environ.*, **400**, pp. 42–51.

52. Noubactep, C. (2015). Designing metallic iron packed-beds for water treatment: A critical review, *Clean Soil, Air, Water*, **43**, pp. 1–11.

53. Noubactep, C. (2016). Commentary. No scientific debate in the zero-valent iron literature, *Clean Soil, Air, Water*, **43**, pp. 1–3.

54. Schlesinger, H. I., Brown, H. C., Finholt, A. E., Gilbreath, J. R., Hoekstra, H. R., and Hyde, E. K. (1953). New developments in the chemistry of diborane and of the borohydrides. 9. Sodium borohydride, its hydrolysis and its use as a reducing agent and in the generation of hydrogen, *J. Am. Chem. Soc.*, **75**, pp. 215–219.

55. Brown, H. C. and Brown, C. A. (1962). A simple preparation of highly active platinum metal catalysts for catalytic hydrogenation, *J. Am. Chem. Soc.*, **84**, pp. 1494–1495.

56. Corrias, A., Ennas, G., Licheri, G., Marongiu, G., and Paschina, G. (1990). Amorphous metallic powders prepared by chemical-reduction of metal-ions with potassium borohydride in aqueous-solution, *Chem. Mater.*, **2**, pp. 363–366.

57. Glavee, G. N., Klabunde, K. J., Sorensen, C. M., and Hadjipanayis, G. C. (1995). Chemistry of borohydride reduction of iron(II) and iron(III) ions in aqueous and nonaqueous media. Formation of nanoscale Fe, FeB, and $Fe_2B$ powders, *Inorg. Chem.*, **34**, pp. 28–35.

58. Wang, C. B. and Zhang, W. X. (1997). Synthesizing nanoscale iron particles for rapid and complete dechlorination of TCE and PCBs, *Environ. Sci. Technol.*, **31**, pp. 2154–2156.

59. Zhang, W. X., Wang, C. B., and Lien, H. L. (1998). Treatment of chlorinated organic contaminants with nanoscale bimetallic particles, *Catal. Today*, **40**, pp. 387–395.

60. Kharisov, B. I., Rasika Dias, H. V., Kharissova, O. V., Jiménez-Pérez, V. M., Pérez, B. O., and Flores, B. M. (2012). Iron-containing nanomaterials: Synthesis, properties, and environmental applications, *RSC Adv.*, **2**, pp. 9325–9358.

61. Kharisov, B. I., Kharissova, O. V., Rasika Dias, H. V., Ortiz Méndez, U., Gómez de la Fuente, I., Peña, Y., and Vázquez Dimas, A. (2016). Iron-based nanomaterials in the catalysis, in *Advanced Catalytic Materials:*

*Photocatalysis and Other Current Trends*, Norena, L. E. and Wang, J.-A. (eds)(InTechOpen), http://dx.doi.org/10.5772/61862.

62. Sun, Y.-P., Li, X.-Q., Cao, J., Zhang, W.-X., and Wang, H. P. (2006). Characterization of zero-valent iron nanoparticles, *Adv. Colloid Interf. Sci.*, **120**, pp. 47–56.
63. Crane, R. A. and Scott, T. B. (2012). Nanoscale zero-valent iron: Future prospects for an emerging water treatment technology, *J. Hazard. Mater.*, **211–212**, pp. 112–125.
64. Mueller, N. C., Braun, J., Bruns, J., Černík, M., Rissing, P., Rickerby D., and Nowack, B. (2012). Application of nanoscale zero-valent iron (nZVI) for groundwater remediation in Europe, *Environ. Sci. Pollut. Res.*, **19**, pp. 550–558.
65. O'Carroll, D., Sleep, B., Krol, M., Boparai, H., and Kocur, C. (2013). Nanoscale zero-valent iron and bimetallic particles for contaminated site remediation, *Adv. Water Resour.*, **51**, pp. 104–122.
66. Zou, Y., Wang, X., Khan, A., Wang, P., Liu, Y., Alsaedi, A., Hayat, T., and Wang, X. (2016). Environmental remediation and application of nanoscale zerovalent iron and its composites for the removal of heavy metal ions: A review, *Environ. Sci. Technol.*, **50**, pp. 7290–7304.
67. Grieger, K. D., Fjordbøge, A., Hartmann, N. B., Eriksson, E., Bjerg, P. L., and Baun, A. (2010). Environmental benefits and risks of zero-valent iron nanoparticles (nZVI) for in situ remediation: Risk mitigation or trade-off? *J. Contam. Hydrol.*, **118**, pp. 165–183.
68. Burton, A. (2009). Hit or miss? Benefits and risks of using nanoparticles for in situ remediation, *Environ. Health Perspect.*, **117**, p. A552.
69. Tesh, S. J. and Scott, T. B. (2014). Nano-composites for water remediation: A review, *Adv. Mater.*, **26**, pp. 6056–6068.
70. Stefaniuk, M., Oleszczuk, P., and Ok, Y. S. (2016). Review on nano zerovalent iron (nZVI): From synthesis to environmental applications, *Chem. Eng. J.*, **287**, pp. 618–632.
71. Zhao, X., Liu, W., Cai, Z., Han, B., Qian, T., and Zhao, D. (2016). An overview of preparation and applications of stabilized zero-valent iron nanoparticles for soil and groundwater remediation, *Water Res.*, **100**, pp. 245–266.
72. Schrick, B., Hydutsky, B. W., Blough, J. L., and Mallouk, T. E. (2004). Delivery vehicles for zerovalent metal nanoparticles in soil and groundwater, *Chem. Mater.*, **16**, pp. 2187–2193.
73. He, F. and Zhao, D. (2005). Preparation and characterization of a new class of starch-stabilized bimetallic nanoparticles for degradation

of chlorinated hydrocarbons in water, *Environ. Sci. Technol.*, **39**, pp. 3314–3320.

74. He, F., Zhao, D., Liu, J., and Roberts, C. B. (2007). Stabilization of Fe–Pd nanoparticles with sodium carboxymethyl cellulose for enhanced transport and dechlorination of trichloroethylene in soil and groundwater, *Ind. Eng. Chem. Res.*, **46**, pp. 29–34.

75. Quinn, J., Geiger, C., Clausen, C., Brooks, K., Coon, C., O'Hara, S., Krug, T., Major, D., Yoon, W. S., Gavaskar, A., and Holdsworth, T. (2005). Field demonstration of DNAPL dehalogenation using emulsified zero-valent iron, *Environ. Sci. Technol.*, **39**, pp. 1309–1318.

76. ITRC. (2005). *Permeable Reactive Barriers: Lessons Learned/New Directions.* Report # PRB-4. Washington DC: Interstate Technology and Regulatory Council, Permeable Reactive Barriers Team.

77. Kanel, S. R., Manning, B., Charlet, L., and Choi, H. (2005). Removal of arsenic (III) from groundwater by nanoscale zero-valent iron, *Environ. Sci. Technol.*, **39**, pp. 1291–1298.

78. Kanel, S. R., Greneche, J. M., and Choi, H. (2006). Arsenic(V) removal from groundwater using nano-scale zero-valent iron as a colloidal reactive barrier material, *Environ. Sci. Technol.*, **40**, pp. 2045–2050.

79. Liu, R. Q. and Zhao, D. Y. (2007). In situ immobilization of Cu(II) in soils using a new class of iron phosphate nanoparticles, *Chemosphere*, **68**, pp. 1867–1876.

80. Xu, Y. H. and Zhao, D. Y. (2007). Reductive immobilization of chromate in water and soil using stabilized iron nanoparticles, *Water Res.*, **41**, pp. 2101–2108.

81. Karn, B., Kuiken, T., and Otto, M. (2009). Nanotechnology and in situ remediation: A review of the benefits and potential risks, *Environ. Health Perspect.*, **117**, pp. 1823–1831.

82. Bardos, P., Bone, B., Daly, P., Elliott, D., Jones, S., Lowry, G. V., and Merly, C. (2015). *A Risk/ Benefit Appraisal for the Application of Nano-scale Zero-valent Iron (nZVI) for the Remediation of Contaminated Sites.* Report for European Union Seventh Framework Programme.

83. Chowdhury, A. I. A., Krol, M. M., Kocur, C. M., Boparai, H. K., Weber, K. P., Sleep, B. E., and O'Carroll, D. M. (2015). nZVI injection into variably saturated soils: Field and modeling study, *J. Contam. Hydrol.*, **183**, pp. 16–28.

84. O'Hara, S., Krug, T., Quinn, J., Clausen, C., and Geiger, C. (2006). Field and laboratory evaluation of the treatment of DNAPL source zones using emulsified zero-valent iron, *Remediation*, **16**, pp. 35–56.

85. Gavaskar, A., Tatar, L., and Condit, W. (2005). *Cost and Performance Report: Nanoscale Zero-valent Iron Technologies for Source Remediation.* Contract report CR-05-007-ENV, Port Hueneme, California 93043-4370: Naval Facilities Engineering Command, Engineering Service Center.

86. He, F., Zhao, D., and Paul, C. (2010). Field assessment of carboxymethyl cellulose stabilized iron nanoparticles for in situ destruction of chlorinated solvents in source zones, *Water Res.*, **44**, pp. 2360–2370.

87. Bennett, P., He, F., Zhao, D., Aiken, B., and Feldman, L. (2010). In situ testing of metallic iron nanoparticle mobility and reactivity in a shallow granular aquifer, *J. Contam. Hydrol.*, **1116**, pp. 35–46.

88. Velimirovic, M., Tosco, T., Uyttebroek, M., Luna, M., Gastone, F., De Boer, C., Klaas, N., Sapion, H., Eisenmann, H., Larsson, P. O., Braun, J., Sethi, R., and Bastiaens, L. (2014). Field assessment of guar gum stabilized microscale zerovalent iron particles for in-situ remediation of 1,1,1-trichloroethane, *J. Contam. Hydrol.*, **164**, pp. 88–99.

## Chapter 2

# Zerovalent Iron Nanoparticle Composites for Water Treatment: An Overview

**Sarah J. Tesh, Huw Pullin, and Thomas B. Scott**
*University of Bristol, H. H. Wills Physics Laboratory, Tyndall Avenue*
*Bristol, BS8 1TL, United Kingdom*
sarah.tesh@bristol.ac.uk

## 2.1 Introduction

The use of reactive nanoparticles (NPs) for industrial applications and their incorporation into domestic consumer products is continually increasing. For water and environmental remediation, the use of zerovalent iron nanoparticles (nZVI) has been most prolific for a variety of reasons, including efficacy, cost, and perceived environmental compatibility. This chapter examines the state-of-the-art use of nZVI in land remediation and water treatment, specifically the development of "fixed" nanocomposites.

### 2.2.1 Nanoscale Zerovalent Iron

The use of zerovalent metals has been proven effective for the remediation of contaminated groundwater [1, 2]. Iron, above all

*Iron Nanomaterials for Water and Soil Treatment*
Edited by Marta I. Litter, Natalia Quici, and Martín Meichtry

ISBN 978-981-4774-67-3 (Hardcover), 978-981-4669-49-8 (eBook)
www.panstanford.com

other metals, has gained most attention in this field, being relatively inexpensive [3], and viewed as more environmentally benign than other metals, as oxidized iron forms are ubiquitous in nature—in soils, rocks, lakes, rivers, on the sea floor, in air, and in organisms [4].

Due to their size, nanomaterials exhibit different physical, chemical, and biological characteristics when compared to their bulk counterparts [5]. Nanomaterials have a larger surface-area-to-volume ratio and, consequently, a higher density of surface reaction sites per unit mass. Furthermore, surface free-energy is observed to be greater for nanomaterials than for micro- or macroscale counterparts. Nanomaterials, therefore, display a notably higher reactivity for surface-mediated processes.

#### 2.2.1.1 Drawbacks of nZVI use

When within the optimum size range, nZVI potentially represents a more efficient alternative to current materials used for water treatment [6]. nZVI has been thoroughly studied for remediation purposes (see Refs. [5] and [7] and all references therein), and it has been shown to remove a large number of contaminants, and at much greater reaction rates than bulk iron.

However, despite the unquestionable effectiveness of nZVI, and the simplicity of delivery (nZVI can be injected into the ground as a dry powder or slurry to directly treat contaminated groundwater), there are multiple disadvantages of using "free" NPs for remediation. It is well recognized that the dispersion of nZVI in an aquifer will be limited by multiple processes: geochemical/geophysical characteristics, mineral sorption, microbiological activity, and aggregation and formation of voluminous corrosion products. These variables are difficult to predict and would require unique tailoring of nZVI for each situation [8, 9]. Further, changes in the groundwater system (natural or otherwise) may also cause contaminants adsorbed to the surface of nZVI to become remobilized and surface adaptations to be reversed or become redundant [10], consequences that become inevitable when considering the operational difficulty of reclaiming nZVI after injection.

Multiple studies have developed methods for avoiding these problems by adapting the nZVI particles to limit or negate interparticle attractions. Surfactants [11] or polymers [12] can be added to nZVI surfaces to encourage steric hindrance or prevent electrostatic attraction, or nZVI can be incorporated into other mobile structures such as carbon [13], mesoporous silica [14], and colloidal clays [15]. However, these treatments are an additional cost, and some may be regarded as possible contaminants [16].

Furthermore, there is relatively little known about the long-term ecotoxicological effects of freely dispersed nZVI or other NPs in the environment [17], as the same properties that provide the remediative qualities could also make them hazardous for living organisms. If, in the future, nZVI is classed as toxic or an unacceptable risk, any deployed nZVI may have to be remediated itself.

Due to some of the uncertainties outlined above, the UK government is taking a precautionary approach for introducing engineered NPs into the environment [16, 18] and has called for further research into nanotoxicology and NP behavior in subsurface environmental systems.

## 2.2 Nanocomposites: The Alternative?

### 2.2.1 Static Nanocomposites

To avoid the limitations outlined above, it would be highly advantageous to develop a remediation method that utilizes the reactivity of nZVI while avoiding the release of "free" NPs into the environment. One possible route is to develop a "nanocomposite" where NPs are anchored to, and complement, a micro- or macroscale support material; in this way, they can be safely recovered. Nanocomposites can be utilized for fixed bed reactors, filter columns, permeable reactive barriers, and domestic filters. There are three general categories of static nanocomposites: (a) membranes/mats, (b) beads, and (c) three dimensional porous structures (Fig. 2.1), which will be discussed in the following sections.

#### 2.2.1.1 Membranes and mats

Membranes, with pore sizes of 50–200 nm in open structures [19], traditionally act as size-exclusion-based filters physically preventing harmful microbes or particles from passing through. However, the functionality of the membranes can be enhanced by modifying the pores with reactive functional groups/NPs that are readily accessible for reaction with aqueous contaminants.

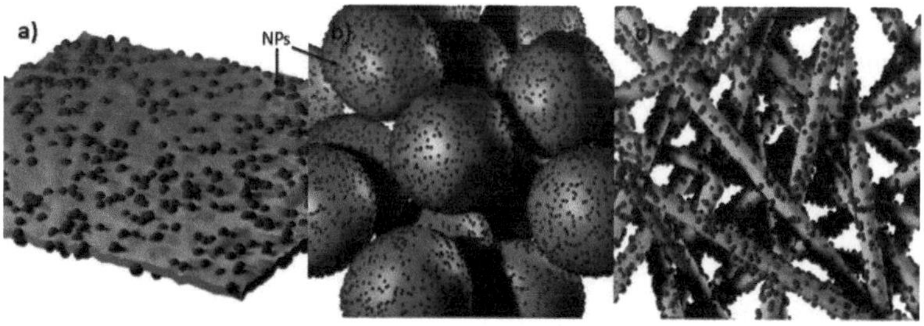


**Figure 2.1** A schematic diagram illustrating the three types of static nanocomposite: (a) membrane or mat, (b) beads, and (c) porous 3D structures.

The main bulk structure of the membrane is commonly made from polymers such as poly(vinyl alcohol) (PVA) [20], poly(acrylic acid) (PAA) [21], polyethersulfone (PES) [22], and chitosan [23]. Membrane fabrication methods include phase inversion [22], solution casting [24], and thermal-grafting polymerization [19]. However, the most popular process is electrospinning [22], where fibers are produced by electrostatic repulsion. The nZVI can then be incorporated by, for example, submerging the fiber membrane in an aqueous solution of iron salt, causing iron ions to complex with the fibers. After rinsing off excess salt, the ions are then chemically reduced, often using sodium borohydride ($NaBH_4$), to form the nZVI in situ.

Using this popular fabrication technique, Horzum et al. (2013) [23] created a chitosan fiber membrane functionalized with nZVI, and successfully removed $As^{3+}$ and $As^{5+}$ (25.0 mg/L) from synthetic solutions albeit under very specific conditions (pH 3.0 and low concentrations). A more successful nanocomposite, from cross-linked water-stable electrospun PAA/PVA, was used as a nanoreactor to complex $Fe^{3+}$ with the free PAA carboxyl residues for subsequent reductive formation of nZVI [20] and bimetallic FePdNP [25]. The resulting polymer nanofiber mats were stable, highly porous, reusable, and able to rapidly remediate waters spiked with various aqueous contaminants, including metal, TCE, and dyes. Daraei et al. [22] developed a novel nanocomposite membrane (using the phase inversion method) using a PES membrane matrix with polyaniline-magnetite (PANI-$Fe_3O_4$) NPs, which removed 85% of 20 mg/L $Cu^{2+}$ in 2 h. The membrane proved to be effective after regeneration with EDTA (for four cycles). Parshetti and Doong [26] functionalized two polymer membranes, polyvinylidene fluoride (PVDF) and nylon 66, with polyethylene glycol by trapping ferrous and nickel ions via dip-coating and thermal-grafting polymerization. The ions were subsequently reduced using $NaBH_4$ to create core–shell Fe/Ni NPs uniformly distributed over the membrane surface. The two membranes reduced 100% of TCE within 25 min, and the composite maintained a high reactivity after 10 days, during which 16 cycles of TCE injection occurred.

Although these example studies were shown to be conceptually successful within a laboratory under synthetic conditions, there are some limitations that need to be addressed. One disadvantage of membrane nanocomposites is that they can often only tolerate a slow flow rate (less than 1 L/min), to allow sufficient residence time for remediation reactions to occur, and low water pressure, to prevent the rupture of the membrane structure and the environmental release of NPs. Furthermore, to be suitable for real-world applications, there is still the hurdle of up-scaling and coping with complex environmental water chemistry. In the studies above, tests were performed in pure Milli-Q, which is not a realistic representation of a real-world water system. Detailed tests analyzing the effect of composite ageing (corrosion, dissolution, degradation, etc.) and clogging in real and complex environmental water samples are needed to address this limitation.

#### 2.2.1.2 Beads

Bulk nanocomposites can also be in the form of micro-/macroscale beads or particulates. Beads may be utilized in static treatment systems such as filter columns (operated in a manner akin to ion exchange columns) and permeable reactive barriers. The key advantage of these spherical nanocomposites is that they can be transplanted into existing technology and infrastructure, providing few barriers to operational uptake. Noubactep [27] has demonstrated the effectiveness of introducing iron into commonplace sand filter column systems. In this way, the performance is significantly improved, and the filter column naturally clogs when no reactive iron is left due to the generation of voluminous corrosion products from iron oxidation.

Commonly in nanocomposite beads, the NPs are incorporated into spherical polymer structures and resins, such as chelating resins [28] and ion exchange resins [29]; a popular example in the literature is the biopolymer alginate [30]. Bezbaruah et al. studied the entrapment of nZVI within calcium (Ca)-alginate beads [31]. The beads were approximately 5 mm in size (Fig. 2.2) and contained NPs with an average size of 35 nm and a range of 10–100 nm. The performance of these beads was proved to be as effective as "free" nZVI for the removal of nitrate (20–100 mg/L) [32], TCE (1–40 mg/L) [33], and $As^{5+}$ (1–10 mg/L) [34]. Alongside this work, Kim et al. (2010) [35] developed another nZVI-alginate bead composite, with nZVI created in situ by reducing previously emplaced ferrous ions within the beads. These beads proved effective at removing 99.8% of TCE within 4 h from a synthetic solution.

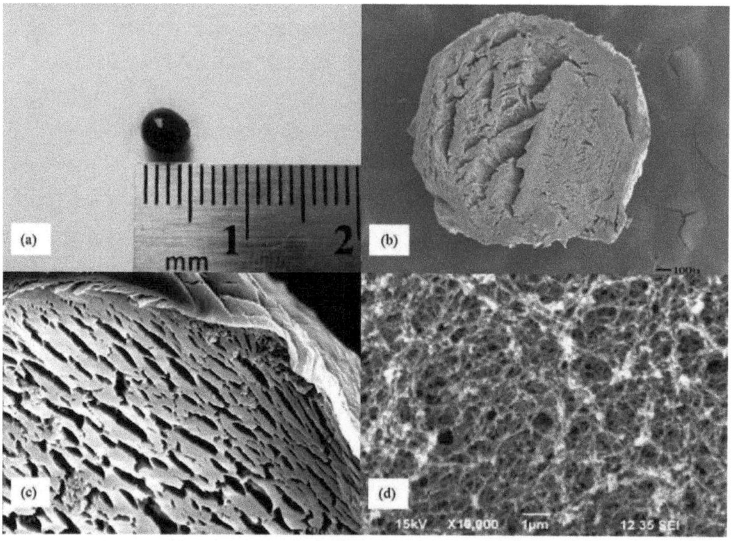


**Figure 2.2** (a) Photograph and (b) SEM image of an alginate bead containing nZVI; (c, d) SEM images of cross-sectioned bead. Reprinted from Ref. [31], Copyright 2009, with permission from Elsevier.

As another alternative material, ion exchange resins are also very much used for nanocomposite beads [29]. One particular example reached commercial availability in 2004 [34] under the name of $ArsenX_{np}$. Since 1997, American and Indian researchers have been working on units to remove arsenic from water in West Bengal, India. Initially, activated alumina was the employed adsorbent material, but to improve the performance, $ArsenX_{np}$ was introduced alongside. The ArsenX$_{np}$ beads (diameter 300–1200 μm) contained hydrated ferric oxide NPs within a macroporous anion exchange resin [36] designed to have a high selectivity for sorption of arsenic oxyanions. Besides the remediation properties of the material, the success of the product was also due to its sustainability. The unit (Fig. 2.3) is attached to hand-pump driven wells, is gravity fed, and requires no electricity or pH adjustment. After more than 20,000 bed volumes, when an arsenic breakthrough of 50 μg/L occurs, the product is taken to a central regeneration facility and all waste is removed and safely stored.

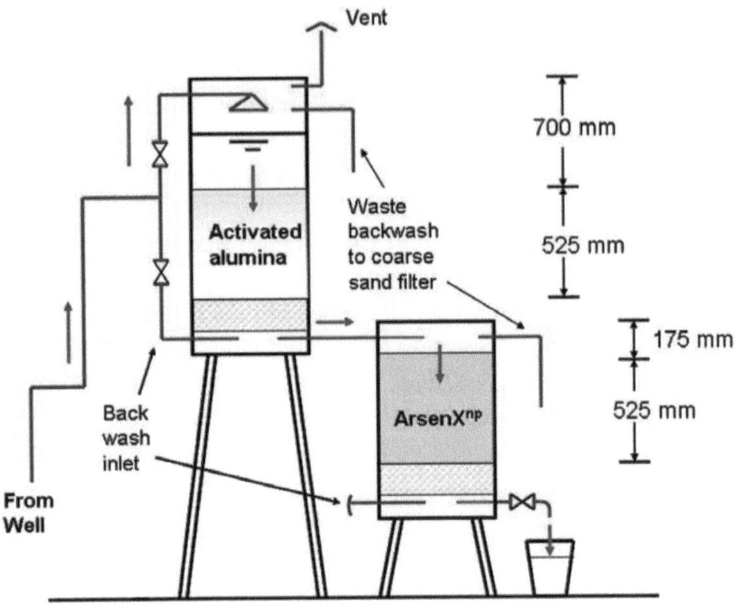


**Figure 2.3** Schematic detail of construction and operation of a split column unit used in the field. Reprinted from Ref. [34], Copyright 2007, with permission from Elsevier.

However, ArsenX$_{np}$ is a bead nanocomposite and it actually suffers from limitations. Although an advantage of using such materials in static treatment systems is that the flow rates may be modified by changing the size and packing density of the beads, water must flow through channels between beads, which generates a high back-flow resistance. Furthermore, these products tend to contain a significant mass of wasted reactive material within their volume, which has not reacted with the water as it is physically entrapped within the structure of the bead and is, therefore, not exposed to the polluted water.

#### 2.2.1.3 Porous 3D structures: The way forward?

The third and final solution is the incorporation of NPs into 3D porous, continuous, bulk structures. In theory, the continuous structure results in limited unreactive volumes, and the large structure can be applied to previously established infrastructure such as column filters and permeable reactive barriers. Furthermore, higher flow rates should also be facilitated as the composite maintains the mechanical properties of the parent support. Despite these logical advantages, however, there is relatively little literature covering this topic, especially when compared with the vast array for membrane and bead nanocomposites. Among the few examples available, support materials include polymers [37], graphene [38], carbon glass [39], and chitosan [23].

In 2011, Savina et al. [40] developed a macroporous polymer containing iron oxide NPs ($\alpha$-$Fe_2O_3$ and $Fe_3O_4$), prepared via a simple cryopolymerization process (Fig. 2.4). The flow rate (measured at constant hydrostatic pressure of approximately 0.1 bar) was 2.29 ± 0.34 and 2.78 ± 0.33 ($\times 10^{-3}$) m/s for the gels containing $\alpha$-$Fe_2O_3$ and $Fe_3O_4$, respectively, when packed in a glass column, flow rates close to those of gels containing no NPs. This is significantly better than the recommended flow rate for the commercially available $ArsenX_{np}$ beads (described above) [41]. However, the gels did not perform as well as "free" NPs for $As^{3+}$ removal tests, attributable to the NPs being embedded in the wall of the gel and, therefore, being less accessible for the contaminant solution. This result suggests that the reactivity would be improved if the NPs were only on the surface of the parent support and not embedded within it.

Investigating the up-scaling of the synthesis process, Savina et al. (2016) [42] created hydrogels of 100–500 mL (compared with 1–10 mL). When using the original method (described above) for larger volumes, it is difficult to control the ice nucleation, freezing kinetics, and crystal formation, resulting in poor control of the morphology and porosity of the gels. Furthermore, the inclusion of the high-density NPs within a large volume of low-density precursor solution results in particle separation, causing uneven distribution or failure of the composite structure. To solve this problem, a pre-freezing step was introduced. The resulting composite not only contained an even distribution of NPs and a smaller range of pore sizes, but also demonstrated dual porosity and, therefore, increased surface area. As well as the large macropores (20–200 μm) seen in the original method, there were smaller pores within the polymer walls caused by phase separation of the polymer formed in the non-frozen liquid stage. However, the material still contains entrapped NPs that are unable to react with contaminant solutions. A provisional $As^{3+}$ removal test, however, shows performance improvement compared to the original method [42], perhaps due to the added dual porosity.

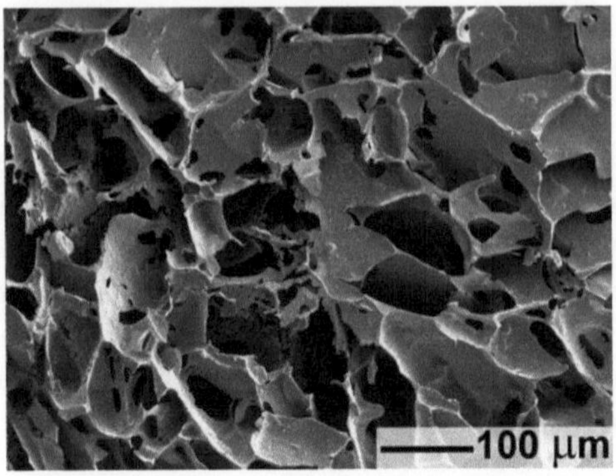


**Figure 2.4** SEM of macroporous polymer containing iron oxide NPs. Adapted from Ref. [40], copyright 2016 Macmillan Publishers Limited, part of Springer Nature.

Another excellent continuous example is by Sankar et al. (2013) [43], who developed a filter system where NPs are embedded within a nanocrystalline metal oxyhydroxide–chitosan structure, demonstrating how the composite can be easily tailored to the target contaminant by simply altering the composition. For example, silver NPs within an AlOOH–chitosan nanostructure removed bacteria and viruses, while composites incorporating magnesium oxide NPs within the AlOOH–chitosan nanostructure targeted heavy metals and an FeOOH–chitosan structure was demonstrated for arsenic removal. Furthermore, Sankar et al. developed a point-of-use filter system to incorporate their nanocomposite, which was a great demonstration of how such continuous porous structures can be easily incorporated into simple systems. When containing the antibacterial composite, the filter demonstrated outstanding results, cleansing 1500 L of water with a bacterial load of 105 CFU mL before needing to be replaced or reactivated. Assuming consumption of 10 L per day for a family, 120 g of composite was estimated to provide safe drinking water for a year at a cost of only US$ 2. Whether a similar performance is seen for other nanomaterial compositions (such as nZVI) is yet to be seen.

The added benefit of such a filtration system is that other nanocomposites could be placed within the multilayer axial block to set up a treatment train. This is an ideal example of a setup for performance comparison tests between different continuous composite materials.

More recent work, performed by a group of the Bristol University [39, 44, 45], produced a nanocomposite via the electrodeposition of nZVI onto carbon bulk substrates. The work demonstrated that by having an electrolyte containing both iron chloride ($FeCl_3$) and nZVI, a nanostructured layer containing both quasi-spherical and cubic features could be produced on the carbon surface, as shown in Fig. 2.5. Between these features, an ultrathin layer of homogeneous iron was identified. The cubic features were shown to be metallic iron crystals that had grown where single nZVI acted as nucleation sites for heterogeneous crystal growth.

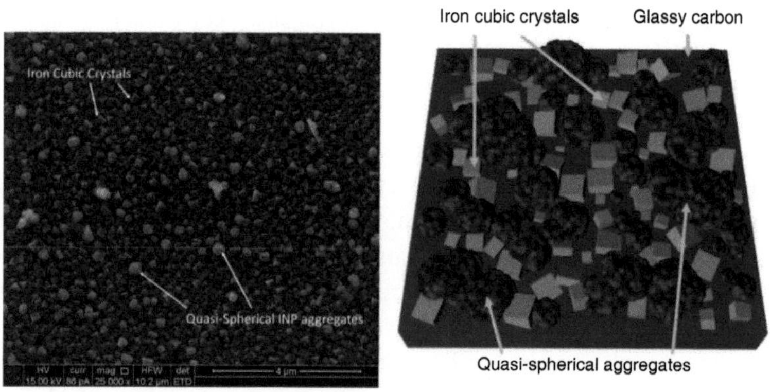


**Figure 2.5** SEM and schematic diagram highlighting the cubic and spherical features deposited on the RVCF surface.

The quasi-spherical features were either aggregates of nZVI or the result of disordered crystal growth around a polycrystalline NP core that has lost its oxide shell. It was also found that the size, shape, and number density of the nanostructured features significantly altered with deposition time, repetition, and sonication. The resulting nanocomposite has been shown to be effective for the treatment of BTEX (at rates competitive with more traditional adsorptive agents) [39, 44] and uranium [39]. This method is particularly useful as any adsorbed contaminants post-treatment could be reclaimed via reverse electroplating.

Unlike membranes and beads, continuous porous nanocomposites can be applied to a wider range of remediation applications. For example, some have been developed to remove oil from water. In order to clean up oil spills in large volumes of water, ideally a material with superhydrophobicity and superoleophilicity is required. Calcagnile et al. (2012) [37] developed a "white graphene" foam that could selectively adsorb oil from the surface of water, using commercial polyurethane foams that were functionalized with submicrometer polytetrafluroethylene (PTFE) particles and superparamagnetic iron oxide NPs via triboelectric charging and deposition, respectively. Resultantly, the overall synthesis process was both simple and cheap. The resulting composite effectively removed the oil and, because of the magnetic properties of the NPs, could then be removed using a magnetic field.

Graphene is another interesting structure material. It is an engineered material comprising $sp^2$-hybridized carbon atoms and formed with only one atom thickness. As large-scale production of graphene is still challenging [46], most hybrid composites begin with graphene oxide (GO), which is readily available from natural graphite. GO is ideal for iron compounds that have been shown to crosslink with oxygen groups on activated carbon surfaces. This is because on the surface of GO, there are multiple oxygen-containing groups, such as alcohols, ketones, and carboxyl groups. Using graphene, Cong et al. (2012) [38] produced a graphene/iron oxide NP hydrogel (Fig. 2.6). GO sheets were reduced by ferrous ions, inducing the self-assembly of a graphene hydrogel containing either α-FeOOH nanorods or magnetic $Fe_3O_4$ NPs. The resulting structure contained interconnecting networks and displayed significant adsorption of oils and heavy metal ions.

**Figure 2.6** A photo displaying the visual self-assembly of a graphene/iron oxide hydrogel. Reprinted with permission from Ref. [38]. Copyright 2012 American Chemical Society.

Although theoretically superior to other nanocomposite structures, few consumer products have been developed using this material; this, combined with a lack of scientific literature in this area, indicates that this technology is far from being commercially ready. This may be due to prohibitive manufacturing costs, but there are practical limitations that any water filter will need to overcome, such as balancing reactivity, flow rate, and structural integrity [40].

#### 2.2.1.4 What is holding back static nanocomposites?

As very well known, there is a high variety of water remediation technologies, including adsorbents, flocculants and coagulants, ion exchange resins, and size-exclusion filters. As the treated water is often used for drinking, there are multiple stringent regulatory requirements that these technologies must fulfill to achieve commercial maturity.

In the US, the quality of drinking water is protected under the Safe Drinking Water Act of 1974 [47] and is regulated by the United States Environmental Protection Agency, the Office of Ground Water and Drinking Water, and other organizations at the local level. Also, NSF International, a third-party organization, often plays an important role by certifying all products that come into contact with drinking water, including water filters, chemical treatments, and plumbing [48]. In order to gain certification, the claims made about the performance of a filter must be understood and proved to be realistic. Furthermore, the NSF Joint Committee on Drinking Water Treatment Units has developed various key standards for evaluation and certification, which must be passed depending on the claims [49].

Meanwhile, within the UK, these regulations are outlined and maintained by a combination of authorities: the Drinking Water Inspectorate (the drinking quality regulator for England and Wales) and the Drinking Water Quality Regulator for Scotland, the Health Protection Agency (Department of Health), the Department for Regional Development, the Department for Environment, Food and Rural Affairs (DEFRA), as well as local authorities and water suppliers [50]. The remediation technologies, alongside with the materials used for storage or transport of water, are judged for suitability on the basis of the demonstrable impact they have on the water quality from the point of view of consumption, i.e., if the resulting water suitable for human consumption, etc. The UK Drinking Water Inspectorate releases an annual report detailing which products have fulfilled these conditions and are approved for use in public water supply [51]. This list is purely based on the safety of the drinking water and does not assess the technologies or materials for fitness of purpose.

In all cases, the cost of regulatory testing for filter products to be used for drinking water is costly (ranging from US$ 20,000 to US$ 50,000 and above in the US) and likely to be prohibitive for individual academics or even universities to take forward. Correspondingly, the pathway from a promising prototype nanocomposite to achieve up-scaling and authorization as a market product is likely to be protracted and challenging, requiring one or more rounds of investment.

Furthermore, nanocomposites pose an additional problem. Mobile nanocomposites are still under scrutiny regarding long-term toxicity in the environment and on human health. They are, therefore, severely limited in commercial applications currently. Meanwhile, the more promising bulk nanocomposites would need further testing procedures to confirm that no potentially harmful NPs are being released into the drinking water. This may further increase the cost of regulatory testing, but this is in line with recommendations outlined in the Royal Society report "Nanoscience and nanotechnologies: Opportunities and Uncertainties" [18].

Furthermore, to be commercially viable, research will need to be pursued to ensure the arising technology is as sustainable as possible. Ideally, to maximize green credentials, the nanocomposites should be recyclable, with relatively simple methods available to remove adsorbed contaminants and exhausted NPs and then reuse the substrates. From the resulting NP-contaminant mix, desirable metals could then be harvested for further use, therefore providing an economic return, and the residual waste material would need to be suitably disposed of. Currently, many organizations using NPs follow the traditional chemical safety procedures for "hazardous materials" throughout the NP life cycle [52]. Although this is a good basis for safely handling and disposing of nanomaterials, it would be unsurprising if more specific regulations are drawn up in the future to account for new findings that arise regarding their toxicology. In doing so, it is ensured that nanocomposites are environmentally friendly throughout their life cycle.

Finally, a key problem, highlighted within this chapter, as the new materials and methods are being researched by academic groups, is that there are neither standard testing procedures, nor set of standardized test pollutants for laboratory experiments, nor size requirements for the overall volume of reactive material. Laboratory tests also often overestimate the performance of the product by testing simplistic water systems [5]. This leaves the comparison of efficacy very difficult and is currently unsatisfactory for bodies investing in further R&D in this area because it is unclear which is the best product.

## 2.3 Conclusions and Perspectives

The above review has examined the current state of play for research into nanocomposites for water treatment, and it is evident that this field of research is rapidly developing. Possible applications vary from small-scale uses, such as domestic point-of-use treatment systems, to environmental in situ methods akin to permeable reactive barriers, and finally to much larger industrial applications, including facility-scale batch or flow-through systems. Based on the arguments presented here, reactive filter composites that achieve the greatest future commercial success will likely display the following properties:

- Continuous bulk structure to avoid the disadvantages of "free" particles and to ensure that surface area and reactivity are maximized.

- Strong NP adhesion to ensure that no NPs are released into the wider environment.
- Sound mechanical properties of the parent structure to allow for optimal flow rates.
- High internal surface area to maximize reactivity.
- Ability to remove a large range of contaminants.
- Rechargeable/recyclable for sustainability with recovery of contaminants (metals, etc.) to convert waste to a potentially valuable commodity for economic gain.
- Low production cost to ensure widespread application is realistic.

To date, no "magic bullet" nanocomposite for water treatment has emerged commercially. However, given the present speed of technological development and market uptake, the future looks extremely promising for nanocomposites, not only for water treatment, but for many other important global industries.

## References


1. Cantrell, K., Kaplan, D., and Wietsma, T. (1995). Zero-valent iron for the in situ remediation of selected metals in groundwater, *J. Hazard. Mater.*, **42**, 201–212.
2. Warren, K., Arnold, R., Bishop, T., Lindholm, L., and Betterton, E. (1995). Kinetics and mechanism of reductive dehalogenation of carbon tetrachloride using zero-valence metals, *J. Hazard. Mater.*, **41**, 217–227.
3. MetalPrices.com. Reporting 600+ metal prices and news from around the globe. (Accessed: 17.01.17).
4. Schwertmann, U. and Cornell, R. (2008). *Iron Oxides in the Laboratory: Preparation and Characterization* (John Wiley & Sons).
5. Crane, R. and Scott, T. (2012). Nanoscale zero-valent iron: Future prospects for an emerging water treatment technology, *J. Hazard. Mater.*, **211–212**, 112–125.
6. Masciangioli, T. and Zhang, W. (2003). Peer reviewed: Environmental technologies at the nanoscale, *Environ. Sci. Technol.*, **37**, 102A–108A.
7. Zhang, W. (2003). Nanoscale iron particles for environmental remediation: An overview, *J. Nanoparticle Res.*, **5**, 323–332.

8. Phenrat, T., Saleh, N., Sirk, K., Tilton, R., and Lowry, G. (2006). Aggregation and sedimentation of aqueous nanoscale zerovalent iron dispersions, *Environ. Sci. Technol.*, **41**, 284–290.

9. Tratnyek, P. and Johnson, R. (2006). Nanotechnologies for environmental cleanup, *Nano Today*, **1**, 44–48.

10. Crane, R., Pullin, H., Macfarlane, J., Silion, M., Popescu, I., Andersen, M., Calen, V., and Scott, T. (2015). Field application of iron and iron-nickel nanoparticles for the ex situ remediation of a uranium-bearing mine water effluent, *J. Environ. Eng.*, **141**, 1–12.

11. Kanel, S., Nepal, D., Manning, B., and Choi, H. (2015). Transport of surface-modified iron nanoparticle in porous media and application to arsenic(III) remediation, *J. Nanoparticle Res.*, **9**, 725–735.

12. Biswal, J., Ramnani, S., Shirolikar, S., and Sabharwal, S. (2009). Synthesis of guar-gum-stabilized nanosized silver clusters with gamma radiation, *J. Appl. Polym. Sci.*, **114**, 2348–2355.

13. Chandra, V., Park, J., Chun, Y., Lee, J., Hwang, I., and Kim, K. (2010). Water-dispersible magnetite-reduced graphene oxide composites for arsenic removal, *ACS Nano*, **4**, 3979–3986.

14. He, D., He, X., Wang, K., Zhao, Y., and Zou, Z. (2013). Regenerable multifunctional mesoporous silica nanocomposites for simultaneous detection and removal of mercury(II), *Langmuir*, **29**, 5896–5904.

15. Frost, R., Xi, Y., and He, H. (2010). Synthesis of palygorskite supported zero-valent iron, *J. Colloid Interface Sci.*, **341**, 153–161.

16. Bardos, P., Bone, B., Elliott, D., Hartog, N., Henstock, J., and Nathanail, P. (2011). A risk/benefit approach to the application of iron nanoparticles for the remediation of contaminated sites in the environment. (DEFRA).

17. Phenrat, T., Saleh, N., Sirk, K., Kim, H., Lowry, G., and Lowry, G. (2008). Stabilization of aqueous nanoscale zerovalent iron dispersions by anionic polyelectrolytes, *J. Nanoparticle Res.*, **10**, 795–814.

18. The Royal Society and The Royal Academy of Engineering. (2004). Nanoscience and nanotechnologies: Opportunities and uncertainties. (The Royal Society and The Royal Academy).

19. Xu, J. and Bhattacharyya, D. (2008). Modeling of Fe/Pd nanoparticle-based functionalized membrane reactor for PCB dechlorination at room temperature, *J. Phys. Chem. C.*, **112**, 9133–9144.

20. Xiao, S., Wu, S., Shen, M., Guo, R., Huang, Q., Wang, S., and Shi, S. (2009). Polyelectrolyte multilayer-assisted immobilization of zero-valent iron

nanoparticles onto polymer nanofibers, *ACS Appl. Mater. Interfaces,* **1**, 2848–2855.

21. Xiao, S., Shen, M., Guo, R., Wang, S., and Shi, X. (2009). Immobilization of zerovalent iron nanoparticles into electrospun polymer nanofibers, *J. Phys. Chem. C.,* **113**, 18062–18068.

22. Daraei, P., Madaeni, S., Ghaemi, N., Salehi, E., Khadivi, M., Moradian, R., and Astinchap, B. (2012). Novel polyethersulfone nanocomposite membrane prepared by PANI/$Fe_3O_4$ nanoparticles, *J. Memb. Sci.*, **415–416**, 250–259.

23. Horzum, N., Demir, M., Nairat, M., and Shahwan, T. (2013). Chitosan fiber-supported zero-valent iron nanoparticles as a novel sorbent for sequestration of inorganic arsenic, *RSC Adv.,* **3**, 7828–7837.

24. Zendehnam, A., Arabzadegan, M., Hosseini, S., Robatmili, N., and Madaeni, S. (2013). Fabrication and modification of polyvinylchloride based heterogeneous cation exchange membranes, *Korean J. Chem. Eng.,* **30**, 1265–1271.

25. Ma, H., Huang, Y., Shen, M., Guo, R., Cao, X., and Shi, X. (2012). Enhanced dechlorination of trichloroethylene using electrospun polymer nanofibrous mats immobilized with iron/palladium bimetallic nanoparticles, *J. Hazard. Mater.,* **211–212**, 349–356.

26. Parshetti, G. and Doong, R. (2009). Dechlorination of trichloroethylene by Ni/Fe nanoparticles immobilized in PEG/PVDF and PEG/nylon 66 membranes, *Water Res.,* **43**, 3086–3094.

27. Noubactep, C., Care, S., Togue-Kamga, F., Schöner, A., and Woafo, P. (2010). Extending service life of household water filters by mixing metallic iron with sand, *CLEAN Soil, Air, Water,* **38**, 951–959.

28. Shi, L., Zhang, X., and Chen, Z. (2011). Removal of chromium (VI) from wastewater using bentonite-supported nanoscale zero-valent iron, *Water Res.,* **45**, 886–892.

29. DeMarco, M., SenGupta, A., and Greenleaf, J. (2003). Arsenic removal using a polymeric/inorganic hybrid sorbent, *Water Res.,* **37**, 164–176.

30. Agrawal, P. and Bajpai, A. (2011). Biosorption of chromium(VI) ions from aqueous solutions by iron oxide-impregnated alginate nanocomposites: Batch and column studies, *Toxicol. Environ. Chem.,* **93**, 1277–1297.

31. Bezbaruah, A., Krajangpan, S., Chisholm, B., Khan, E., and Bermudez, J. (2009). Entrapment of iron nanoparticles in calcium alginate beads for groundwater remediation applications, *J. Hazard. Mater.,* **166**, 1339–1343.

32. Bezbaruah, A., Shanbhogue, S., Simsek, S., and Khan, E. (2011). Encapsulation of iron nanoparticles in alginate biopolymer for trichloroethylene remediation, *J. Nanoparticle Res.,* **13**, 6673–6681.

33. Bezbaruah, A., Kalita, H., Almeelbi, T., Capecchi, C., Jacob, D., Ugrinov, A., and Payne, A. (2013). Ca-alginate-entrapped nanoscale iron: Arsenic treatability and mechanism studies, *J. Nanoparticle Res.,* **16**, 1–10.

34. Sarkar, S., Blaney, L., Gupta, A., Ghosh, D., and SenGupta, A. (2007). Use of ArsenX$^{np}$, a hybrid anion exchanger, for arsenic removal in remote villages in the Indian subcontinent, *React. Funct. Polym.,* **67**, 1599–1611.

35. Kim, H., Hong, H., Jung, J., Kim, S., and Yang, J. (2010). Degradation of trichloroethylene (TCE) by nanoscale zero-valent iron (nZVI) immobilized in alginate bead, *J. Hazard. Mater.,* **176**, 1038–1043.

36. Cumbal, L. and SenGupta, A. (2005). Arsenic removal using polymer-supported hydrated iron(III) oxide nanoparticles: Role of donnan membrane effect, *Environ. Sci. Technol.,* **39**, 6508–6515.

37. Calcagnile, P., Fragouli, D., Bayer, I., Anyfantis, G., Martiradonna, L., Cozzoli, P., Cingolani, R., and Athanassiou, A. (2012). Magnetically driven floating foams for the removal of oil contaminants from water, *ACS Nano.,* **6**, 5413–5419.

38. Cong, H., Ren, X., Wang, P., and Yu, S. (2012). Macroscopic multifunctional graphene-based hydrogels and aerogels by a metal ion induced self-assembly process, *ACS Nano.,* **6**, 2693–2703.

39. Tesh, S., Macfarlane, J., Hallam, K., and Scott, T. (2012). Nano-iron filters for reactive barrier and in-line filtration applications, *8$^{th}$ International Conference on Remediation of Chlorinated and Recalcitrant Compounds,* 21–24 May 2012, Monterey, California, USA.

40. Savina, I., Ingavle, G., Cundy, A., and Mikhalovsky, S. (2016). A simple method for the production of large volume 3D macroporous hydrogels, *Sci. Rep.,* **6**, 1–9.

41. Sylvester, P., Westerhoff, P., Möller, T., Badruzzaman, M., and Boyd, O. (2007). A hybrid sorbent utilizing nanoparticles of hydrous iron oxide for arsenic removal from drinking water, *Environ. Eng. Sci.,* **24**, 104–112.

42. Savina, I., English, C., Whitby, R., Zheng, Y., Leistner, A., Mikhalovsky, S., and Cundy, A. (2011). High efficiency removal of dissolved As(III) using iron nanoparticle-embedded macroporous polymer composites, *J. Hazard. Mater.,* **192**, 1002–1008.

43. Sankar, M., Aigal, S., Maliyekkal, S., Chaudhary, A., Kumar, A., Chaudhari, K., and Pradeep, T. (2013). Biopolymer-reinforced synthetic granular

nanocomposites for affordable point-of-use water purification, *Proc. Natl. Acad. Sci.*, **110**, 8459–8464.

44. Macfarlane, J., Tesh, S., Crane, R., Hallam, K., and Scott, T. (2014). Synthesis of nano-composite surfaces via the co-deposition of metallic salts and nanoparticles, *Mater. Sci. Eng. B Solid-State Mater. Adv. Technol.*, **182**, 59–68.
45. Tesh, S. (2017). Development of iron-based nano-composites for advanced water treatment (University of Bristol).
46. Zhang, K., Dwivedi, V., Chi, C., and Wu, J. (2010). Graphene oxide/ferric hydroxide composites for efficient arsenate removal from drinking water, *J. Hazard. Mater.*, **182**, 162–168.
47. United States Environmental Protection Agency. (2013). Drinking water contaminants: Standards and regulations. Available at: http://water.epa.gov/ (Accessed: 16.01.17).
48. NSF-International. (2013). About NSF. Available at: http://nsf.org/business/ (Accessed: 17.01.17).
49. NSF-International. (2013). Drinking water treatment units. Available at: http://nsf.org/ (Accessed: 17.01.17).
50. The Drinking Water Inspectorate. (2009). Drinking Water Safety: Guidance to health and water professionals.
51. The Drinking Water Inspectorate. (2013). List of approved products for use in public water supply in the United Kingdom.
52. Amoabediny, G., Naderi, A., Malakootikhah, J., Koohi, M., Mortazavi, S., Naderi, M., and Rashedi, H. (2009). Guidelines for safe handling, use and disposal of nanoparticles, *J. Phys. Conf. Ser.*, **170**, 1–12.

# Chapter 3

# Adsorption of Groundwater Pollutants by Iron Nanomaterials

**Dimitris Dermatas,[a] Thanasis Mpouras,[a] Nymphodora Papassiopi,[b] Christiana Mystrioti,[b] Aikaterini Toli,[b] and Iraklis Panagiotakis[a]**

[a]*School of Civil Engineering, National Technical University of Athens, Zografou Campus, 9 Iroon Polytexneiou Str.*

[b]*School of Mining and Metallurgical Engineering, National Technical University of Athens, Zografou Campus, 9 Iroon Polytexneiou Str. Zografou, Athens, 15780, Greece*

dermatas@gmail.com

## 3.1 Introduction

As an advent of population growth and climate change, more than half of the extracted groundwater is used for covering domestic needs, whereas globally 25–40% is used as drinking water, indicating that the demand for clean groundwater has already grown dramatically and it is bound to grow even further. Groundwater can be contaminated either by geogenic or by anthropogenic sources, rendering it unusable unless it is effectively remediated. Geogenic groundwater contamination is exclusively related to the

*Iron Nanomaterials for Water and Soil Treatment*
Edited by Marta I. Litter, Natalia Quici, and Martín Meichtry

ISBN 978-981-4774-67-3 (Hardcover), 978-981-4669-49-8 (eBook)
www.panstanford.com

geochemical background that feeds groundwater with heavy metals at concentrations, sometimes capable of posing significant adverse effects on humans and ecosystems. Anthropogenic contamination includes the presence of heavy metals (arsenic, chromium, cadmium, cobalt, manganese, zinc, copper, mercury, etc.) and/or organic compounds (volatile organic compounds, petroleum hydrocarbons, dyes and pigments, insecticides and herbicides, pharmaceutical compounds, organohalides, etc.) in groundwater as a result of waste generation and neglectful disposal [1]. The first indication of the origin of contamination is the level of contamination since in the case of anthropogenic pollution, the concentrations detected can be several orders of magnitude higher than those of geogenic origin [2]. Groundwater and soil contamination are strictly related, and as a consequence, methods employed for soil decontamination would indirectly affect groundwater and vice versa [3].

Soil and groundwater remediation technologies are typically classified into (a) in situ technologies or (b) ex situ technologies, based on whether contaminated soil or groundwater is treated in place or after removal, respectively. The most common in situ method is the installation of permeable reactive barriers (PRBs). PRBs are treatment zones consisting of the selected material for the decontamination, installed vertically in the flow path of a contaminated plume [4]. Ex situ remediation mainly includes different variations of the "pump and treat" method, where groundwater is extracted from the aquifer to the surface and is subsequently treated using conventional removal and/or filtration decontamination methods. Conventional removal methods are based on either photocatalytic processes, especially for degradation of organic compounds, or on chemical precipitation (hydrolysis) and coagulation processes, for removal of heavy metals. However, these processes generate large amounts of sludge that requires further treatment and are thus considered mostly ineffective, especially for the treatment of large quantities of dilute aqueous streams such as groundwater. Filtration techniques are processes that gained significant attention in recent years, but they are plagued by significant disadvantages. The very low sorptive capacity, the fact that some processes usually include precipitation as an intermediate step, and the very high cost are some characteristics that render them as non-desirable technologies in removing toxic trace elements from groundwater.

The combination of targeted adsorption-based processes with the previously mentioned filtration techniques (e.g., by coating the filter sand with nanoscale iron oxides or hydroxides) has thus emerged as a viable and promising alternative for enhancing the efficiency of "pump and treat" methods, thus making them both technologically and economically effective [1].

Adsorption is a well-known and effective separation process where a substance (adsorbate) that exists in the liquid phase attaches (adsorbs) on the surface or in the internal pores of another substance (adsorbent) that exists in the solid phase, via physical and/or chemical interactions. Adsorption processes are being applied globally in the sector of environmental protection and restoration due to their ability to concentrate several contaminants such as heavy metals and organics on the surface of the solid phase, thus providing with clean, decontaminated water. In the field of groundwater decontamination, sorption techniques include adsorption and ion exchange processes [5]. Specifically, the adsorption-based process is considered a superior technique in terms of cost, flexibility, and simplicity of design/operation and insensitivity to toxic pollutants, producing a high-quality effluent without the additional formation of any harmful substances [6]. However, conventional low-cost materials used as adsorbents are characterized by limited efficiency, especially for removing trace quantities of pollutants that, in combination with the increasingly stringent groundwater quality standards, may fail to provide the ultimate solution to the water treatment problem.

In an attempt to overcome the aforementioned limitations, the use of nanomaterials in adsorption-based environmental remediation has recently emerged as a viable alternative and is closely being examined with respect to both the associated benefits and resulting impacts. Consequently, funding for nanotechnology research has expanded rapidly around the world over the last decade, with global public investment reaching $8.4 billion and a private financing of $8.6 billion in 2008 [7]. Apart from the efficiency for removal of contaminants, the main target of adsorption-based nanotechnology is the development of low-cost and environmentally friendly adsorbent nanomaterials able to be regenerated and reused. A high proportion of efficient and cost-effective groundwater remediation methods now rely on using nanotechnology adsorbent products with a predominant use of iron nanoparticles [8]. Overall, research has shown that nanoscale iron oxides and nanozerovalent iron (nZVI) are very promising materials for the effective remediation of a wide range of groundwater pollutants. Moreover, iron oxide nanoparticles occur naturally in the earth's crust by various environmental sources such as volcanoes and fires, while nZVI is more rarely found in the geoenvironment, since it only occurs under specific geological conditions. The contaminant removal efficiency of these iron forms and their natural occurrence, combined with their easy synthesis in large scale, are important factors that led to their preferred application in contaminated groundwater sites with almost negligible risks of secondary contamination [9, 10]. This chapter aims to elucidate the adsorption efficiency and mechanisms of some of the most common nanoscale iron oxides such as magnetite ($Fe_3O_4$), maghemite ($\gamma$-$Fe_2O_3$), hematite ($\alpha$-$Fe_2O_3$), goethite ($\alpha$-FeOOH), various spinel ferrites ($M^{2+}Fe_2{}^{3+}O_4$), and of nZVI toward some common groundwater pollutants. In parallel, the relative significance of the natural occurrence of iron oxides in aquifers will also be revealed.

## 3.2 Adsorption of Pollutants on Nanoscale Iron Oxides

The application of iron oxide nanomaterials as adsorbents for several contaminants has received much attention due to their unique properties, compared to their bulk form, such as (i) the infinitely small size, (ii) the high surface-area-to-volume ratio, (iii) the surface modifiability, (iv) the excellent magnetic properties, since particles with diameter size between 2 and 20 nm and high surface area are uniquely super-paramagnetic, (v) the great biocompatibility, (vi) the ease of separation using an external magnetic field, (vii) reusability after removal of the adsorbed toxic contaminants, and (viii) low production cost compared with other nanomaterials [11]. These properties make iron nanoparticles excellent adsorbents for various pollutants such as heavy metals and organic compounds. The adsorption of pollutants onto iron oxide surfaces can be mainly attributed to two different mechanisms:

- van der Waals interactions with the oxide surface (physisorption) or chemical reactions with the functional groups (chemisorption)
- ion exchange of pollutant ions in aqueous solution with iron ions in the iron oxide lattice structure

### 3.2.1 Magnetite ($Fe_3O_4$)

Bulk $Fe_3O_4$ is an ubiquitous magnetic iron oxide that occurs in the lithosphere, pedosphere, and biosphere and usually contains both ferrous and ferric (Fe(II) and Fe(III)) iron species; its general type is $Fe(II)Fe(III)_2O_4$ [12]. The crystal structure of $Fe_3O_4$ follows an inverse spinel structure where octahedral and tetrahedral/octahedral layers are being alternated. $Fe_3O_4$ nanoparticles exhibit amphoteric surface activity, high dispersibility, and high metal adsorption capacity, and they can be easily separated by using an external magnetic field, making their regeneration and subsequent reuse much easier [13].

$Fe_3O_4$ has been extensively studied as a heavy metal adsorbent for both oxyanionic and cationic forms. Cr and As are common metals widely found in groundwater due to geogenic contamination, resulting from the oxidation of naturally occurring Cr(III) in certain geologic formations such as ultramafic rocks [14], and due to its release through rock weathering and volcanism [15], respectively. In the geoenvironment, Cr occurs either in its highly soluble hexavalent (Cr(VI)) or in the insoluble trivalent (Cr(III)) form. X-ray photoelectron spectroscopy (XPS) and X-ray absorption near edge structure (XANES) analysis have shown that Cr(VI) adsorption is a strongly pH-dependent process with adsorption capacity being maximized at pH values lower than the point of zero charge (PZC) of $Fe_3O_4$ ($pH_{PZC}$ = 6.5). Cr(VI) adsorption is based on electrostatic attractions between the Cr(VI) anions (chromates) and the positively charged surface. However, at alkaline pH values, where electrostatic repulsions between chromates and surface occur, ion exchange is the predominant mechanism since chromates replace the hydroxyl groups due to their higher affinity with the iron surface [16]. With regard to As, it occurs in the geoenvironment in two oxidation states: arsenite (As(III)) and arsenate (As(V)). Several studies have shown that As(V) adsorption on $Fe_3O_4$ is strongly affected by the solution pH, and the mechanism is based on inner-sphere surface complexation

(Lewis acid–base interactions), where the iron of $Fe_3O_4$ acts as the electron-pair acceptor and As(V) as the donor of oxygen [17]. In addition, the results suggested that after the complexation on the $Fe_3O_4$ surface, As(V) is progressively incorporated in the structure of the nanoparticles as the As(V) concentration increases [18]. Similar to As(V), As(III) binding is attributed to the formation of an inner-sphere complex, exhibiting a tridentate bonding geometry [19]. Overall, Cr(VI) and As adsorption is favored at pH lower than 7. For higher groundwater pH values, $Fe_3O_4$ may not be the preferred adsorbent.

Several studies have investigated the adsorption of divalent metal cations such as lead (Pb(II)) [13, 20, 21], copper (Cu(II)) [13, 20, 21], zinc (Zn(II)) [13, 21], cadmium (Cd(II)) [20, 21], mercury (Hg(II)) [20], nickel (Ni(II)) [21], and manganese (Mn(II)) [13] on $Fe_3O_4$. All these studies reported that the adsorption of these metal cations is mainly a function of pH, with the maximum adsorption capacity observed usually at pH about 7–8. FTIR, XPS, and zeta potential analyses showed that the main adsorption mechanism is the occurrence of electrostatic interactions, since with decreasing pH, and for pH values lower than the $pH_{PZC}$, adsorption efficiency decreases. This phenomenon can be attributed to: (a) the increase in proton concentration with decreasing pH, inhibiting the sorption of the metal cations, and (b) the decrease in the negative surface charge of $Fe_3O_4$ at acidic conditions. In addition, it must be mentioned that for pH values higher than 6, the precipitation of the metals as hydroxides plays an important role in their removal [13, 20, 21]. Moreover, electronegativity is an important factor affecting the adsorption efficiency, since different ways of attraction take place between the cations and the adsorption sites, depending on the metal hydrated ionic radii (e.g., Pb(II) > Cu(II) > Zn(II) > Mn(II)) [13]. Overall, the $Fe_3O_4$-based removal of cationic heavy metals can be satisfactory for the full pH range of groundwater (pH 6–8.5).

Several organic compounds that are usually detected in groundwater have been tested for their adsorption affinity with $Fe_3O_4$ surfaces, for example, *cis*-dichloro-diamine-platinum [22] and pentachlorophenol (PCP) [23]. Especially for PCP, the surface of $Fe_3O_4$ is used as a heterogeneous catalyst, at neutral pH, for undergoing a Fenton-like oxidation. The process was controlled by surface reactions, and the species compete each other for adsorption on the surface active sites [23]. Polycyclic aromatic hydrocarbons (PAHs), another category of hydrophobic and persistent organic compounds, are composed of at least two condensed aromatic rings; the simplest PAHs are naphthalene, anthracene, and phenanthrene. In these cases, bare $Fe_3O_4$ has been proved as an insufficient adsorbent (5% adsorption) for such compounds [24]. Dyes and pigments are another category of organic compounds used, mostly in the textiles, paper, plastics, leather, food, and cosmetic industry to color products, and most of them are considered mutagenic and carcinogenic. Several studies have reported the adsorption efficiency of $Fe_3O_4$ toward organic dyes such as the reactive red 120, 4-(2-pyridylazo) resorcinol and a neutral red dye proving simultaneously that the adsorption mechanism is based on ionic adsorption and electrostatic attraction between the positively charged dye molecules and the hydroxyl groups of the $Fe_3O_4$ surface at neutral pH [25–28]. A different mechanism, this of intraparticle diffusion, is proposed for dyes that contain hydroxyl groups in their structure (eriochrome black-T, bromophenol blue, bromocresol green, and fluorescein), without excluding the contribution of surface adsorption [28]. Finally, the most abundant organic factor in aquatic systems on earth, natural or dissolved organic matter (NOM or DOM), is thought to contribute toward water contamination, but also to possess a crucial role for the migration of several organic pollutants [29]. NOM constitutes a diverse group of hydrophobic and hydrophilic organic compounds, and in groundwater, it is present at concentrations ranging from 0.5 up to 10 mg/L as organic carbon. The interactions between metal oxides and aquatic/soil organic matter are important in environmental monitoring and remediation since, to some extent, soil minerals are coated with organic matter. Depending on their chemistry, these coatings can modify entirely or partially the mineral (hydr)oxide surface affecting its properties and reactivity and enhancing the stability of the nanomaterials by preventing oxidation [27, 30]. The adsorption of organic matter is found to be a pH-dependent process since, with decreasing pH, the adsorption efficiency was significantly increased. FTIR analysis showed that the adsorption mechanism is based on electrostatic attractions between the positively charged surface and the negatively charged humic acid (HA) molecules. At alkaline pH values, the ligand-exchange reaction between the functional groups of HA and the active surface sites is considered to be the dominant mechanism [31]. The efficiency of HA-$Fe_3O_4$ particles on Rhodamine B adsorption was attributed to the attractions of the positively charged dye and the negative charge of HA [27].

### 3.2.2 Maghemite ($\gamma$-$Fe_2O_3$)

$\gamma$-$Fe_2O_3$ is a weathering product of $Fe_3O_4$ since, in ambient atmosphere, $Fe_3O_4$ gradually oxidizes to $\gamma$-$Fe_2O_3$. It is the ferrimagnetic cubic form of Fe(III) oxide and differs from the $Fe_3O_4$ inverse spinel structure by vacancies on the cation sublattice. However, it exhibits a crystalline structure similar to $Fe_3O_4$, and the same chemical composition as $\alpha$-$Fe_2O_3$, being a metastable phase between $Fe_3O_4$ and $\alpha$-$Fe_2O_3$ [9, 32, 33].

Cr(VI) adsorption on $\gamma$-$Fe_2O_3$ has been extensively studied and is generally accepted as a strongly pH-dependent process with adsorption capacity being maximized at pH values about 3–4 [16, 34, 35]. For pH values lower than the PZC of $\gamma$-$Fe_2O_3$ ($pH_{PZC} \sim 6.3$), the surface of $\gamma$-$Fe_2O_3$ is positively charged due to the presence of protons in the solution, and Cr(VI) exists predominantly as chromates and hydrogen chromates. At these acidic conditions, attractive electrostatic interactions between the Cr(VI) anionic species and the positively charged surface control the adsorption process. Increasing the pH above the PZC, the surface becomes negatively charged, and Cr(VI) exists mainly as fully dissociated chromate anion. Under these conditions, electrostatic repulsions occur and adsorption decreases with increasing pH. However, even at alkaline pH values (higher than 8), the adsorption phenomena are not eliminated due to the electrostatic repulsions and, as confirmed by Raman spectroscopy, the ion exchange between chromates and the hydroxyl groups of the surface takes place. Since Cr(VI) adsorption on $\gamma$-$Fe_2O_3$ is of electrostatic nature, the low $pH_{PZC}$ compared with the groundwater pH (about 7–8) is a restrictive factor for Cr(VI) adsorption. In addition, the adsorption efficiency of $\gamma$-$Fe_2O_3$ toward cationic metals such as Pb(II) [36], Cu(II) [34–36], Zn(II) [35], Cd(II) [36], Ni(II) [34], and Hg(II) [37] has been reported, with the process described as fast, highly selective, and pH dependent. Adsorption efficiency was maximized at pH values in the range 6.5–8.5, indicating that $\gamma$-$Fe_2O_3$ could act as an effective adsorbent for these cations. XPS and TEM analyses showed that the adsorption mechanism is not the same for all cations. Cu(II) adsorption is due to electrostatic attraction and ion exchange, while, in the case of Ni(II), adsorption is based only on electrostatic attraction. Another parameter affecting the adsorption efficiency of divalent cations is electronegativity, with the most electronegative metals tending to form stronger covalent bonds with oxygen atoms of the solid surface, for example, Pb(II) > Cu(II) > Cd(II) [36].

The adsorption efficiency of bare $\gamma$-$Fe_2O_3$ nanoparticles toward organic compounds has been scarcely reported. Organic dyes such as the Congo red and the anionic titan yellow have been tested for their adsorption affinity for $\gamma$-$Fe_2O_3$, with the results showing that this iron oxide can be used as an effective adsorbent of these dyes at acidic or slightly acidic conditions [6, 38]. As the pH of the dye solution increased, a proportional decrease in adsorption took place probably due to the successive deprotonation of hydroxyl groups on the adsorbent and the occurrence of electrostatic repulsion between the negatively charged sites of $\gamma$-$Fe_2O_3$ and the dye anions. In addition, competition between the hydroxyls and the dye anions for the positively charged adsorption sites possibly exists. The application of diffuse reflectance Fourier transform spectroscopy showed that trimethyl phosphate (TMP) bonds mainly to Lewis acid Fe sites through the O phosphoryl atom on $\gamma$-$Fe_2O_3$. The hydrogen bonding is not excluded as a bonding mechanism. After the initial adsorption, a surface oxidation pathway dominates, which results in the formation of surface methoxy and formate [33].

### 3.2.3 Hematite ($\alpha$-$Fe_2O_3$)

$\alpha$-$Fe_2O_3$ is a red iron oxide that is widespread in soils and rocks. $\alpha$-$Fe_2O_3$ is often the end product in the transformation of other iron oxides due to its thermodynamic stability [33]. Under environmental conditions, $\gamma$-$Fe_2O_3$ turns into $\alpha$-$Fe_2O_3$ crystalline structure. In contrast to $Fe_3O_4$ and $\gamma$-$Fe_2O_3$, $\alpha$-$Fe_2O_3$ exhibits a rhombohedral structure where oxygen ions are hexagonally close packed, and iron is situated only in octahedral sites. The unique properties of nano $\alpha$-$Fe_2O_3$ are mostly associated with the oxidation state and its coordination geometry [32].

Since $\alpha$-$Fe_2O_3$ is considered a raw source of iron, it has been extensively studied as a material for water treatment. $\alpha$-$Fe_2O_3$

nanoparticles have been efficiently used as adsorbents for Cr(VI). Despite the adsorption capacity being maximized at acidic conditions, the adsorption capacity of α-$Fe_2O_3$ remains significant even at alkaline conditions, most likely due to its high PZC point ($pH_{PZC}$ ~ 7.5–9). The adsorption mechanism is predominantly the formation of inner-sphere complexes with the solid surface since desorption experiments showed that Cr(VI) is not easily desorbed in the aqueous phase; without excluding the presence of outer-sphere complexes since experimental data showed that adsorption is affected by the ionic strength of the solution [39]. Thus, it can be concluded that α-$Fe_2O_3$ can be used as an efficient adsorbent for metal anions even at alkaline pH values. The significant adsorption capacity of α-$Fe_2O_3$ toward other metals such as As(III), Cd(II), Co(II), Cu(II), and Ni(II), even at conditions that occur in groundwater, has also been reported. In all cases, the adsorption mechanism is based on electrostatic interactions between the α-$Fe_2O_3$ surface and metal ions [40, 41].

As in the case of γ-$Fe_2O_3$, only a few studies have reported the adsorption of organic compounds on α-$Fe_2O_3$. The adsorption mechanism of TMP and asphaltene on α-$Fe_2O_3$ nanoparticles is similar to that described in the case of using γ-$Fe_2O_3$ as the adsorbent (Section 3.2.2) [33, 42]. Similar results were observed for the pharmaceutical compound salicylic acid, the metabolite of aspirin, since its adsorption affinity was optimized under very acidic conditions and was significantly decreased at alkaline pH values [43]. On the contrary, the application of α-$Fe_2O_3$ as adsorbent for Congo red showed a good efficiency even at neutral pH values [6]. Finally, the adsorption of HA, fulvic acid (FA), and other naturally occurring organic substances on α-$Fe_2O_3$ can create a natural coating able to modify its surface properties and, in combination with the high PZC of α-$Fe_2O_3$, to enhance the adsorption of other pollutants. Adsorption of such substances is proportionally affected by their molecular weight, their hydrophobicity, and the aromatic carbon content, and inversely proportional to the oxygen/carbon ratio of the organic compound. The adsorption mechanism is suggested to be based on electrostatic interactions between the negatively charged organic matter and the positively charged surface of α-$Fe_2O_3$ at acidic and slight acidic conditions but, under neutral and slightly alkaline conditions, a ligand-exchange mechanism between the carboxylic acid groups on the substances and the hydroxylated surface sites on the solids occurs. These observations suggest that, in a shallow aquifer, HA from either terrestrial or aquatic sources is likely to form organic coatings on mineral surfaces, thus creating an important sorbing phase for hydrophobic organic compounds in low-carbon substrates, which are typically found in aquifers [30, 44, 45].

### 3.2.4 Goethite (α-FeOOH)

α-FeOOH is a yellow-brownish anti-ferromagnetic material. Its orthorhombic structure has been confirmed as a hexagonally close-packed array of $O^{2-}$ and $OH^-$ anions with Fe(III) in the center of the octahedral. It is a widespread soil mineral and a major component of many ores, sediments, and soils, since it is one of the most thermodynamically stable iron oxide resulting from rock weathering. α-FeOOH formed as a result of rock weathering is often poorly crystalline and rich in defects and impurities, characteristics that enhance its surface reactivity and thus its adsorption capacity. Moreover, α-FeOOH displays nanometer-sized particles in width and several microns in length, either if it is naturally formed or laboratory synthesized [46].

α-FeOOH adsorption capacity for anions such as As(V) and Cr(VI) decreases with increasing pH and is usually maximized at slightly acidic conditions. However, the high $pH_{PZC}$ (7.5–9) renders α-FeOOH a highly potential adsorbent for metal ions. XAFS analysis showed that the adsorption mechanism of As(V) on α-FeOOH is based on the formation of a bidentate inner-sphere surface complex, following the typical adsorption behavior of oxyanions. Similar results were obtained regarding the adsorption of Cr(VI) [47]. Moreover, the authors mentioned that the initial concentration of the anion affects its binding on the surface, and thus the adsorption mechanism may differ in the case of geogenic or anthropogenic pollution. A surface complexation reaction was found to be the main adsorption mechanism in the case that metal cations are the adsorbates. The adsorption of divalent metal cations on α-FeOOH is directly affected by the release of a water molecule from the hydration sphere of the divalent metal cation [10]. This leads to the formation of an inner-sphere surface complex, which means that the cation is directly, specifically, and strongly bound on the solid surface [47]. Interestingly, a significant adsorption capacity is determined for neutral pH values that usually exist in groundwater (pH 6–8.5), thus indicating the major role of α-FeOOH in aquifers [48].

Regarding the adsorption of organic compounds, despite the fact that several organic compounds have been tested for their affinity with α-FeOOH surfaces, only some of them have shown significant adsorption affinity. Antibiotics and generally pharmaceutical compounds are some of the most persistent organic compounds in aquatic systems. The adsorption mechanism of tylosin and sulfamethazine is rather complicated and may consist of multiple consecutive steps like diffusion on external surfaces, then intraparticle diffusion or pore diffusion, and finally, sorption on the surface interior, indicating a low mobility in iron-rich aquifers [49]. On the contrary, adsorption of the xenobiotic compounds bisphenol A, 17α-ethynylestradiol, and estrone on α-FeOOH remained at low percentages in the pH range 3–11, indicating that the surface charge does not play a significant role in the adsorption and, thus, that iron oxides cannot be considered a restrictive factor for their distribution and transport in groundwater [50]. PAHs adsorption is thought to be affected by the structure/shape of the molecules, with linear PAHs (anthracene), exhibiting the strongest sorption. The main mechanism is probably the dispersion/polarization interactions of the $\pi$-electrons of PAHs by surface OH groups and the formation of weak hydrogen bonds where the $\pi$-system acts as a proton acceptor [51]. However, at neutral pH values, anthracene adsorption did not exceed 4% [52]. Finally, as already mentioned, the adsorption of naturally occurring organic compounds on iron oxides can contribute to the alteration of their surface properties, enhancing their adsorption affinity toward specific compounds. The significant role of HA adsorption on α-FeOOH, as an intermediate step for adsorption of pesticides (2-methyl-4-chlorophenoxyacetic acid, MCPA) and herbicides (paraquat, $PQ^{2+}$), was reported by Brigante et al. HA adsorption led to the formation of ionic pairs or outer-sphere complexes between the negatively charged groups of HA and $PQ^{2+}$, revealing that the natural occurrence of α-FeOOH–HA in the geoenvironment may act as an adsorbent for cationic herbicides, causing their deactivation and reducing their transport in groundwater [53].

### 3.2.5 Spinel Ferrites ($M^{2+}Fe_2^{3+}O_4$)

Spinel ferrites (SFs) are naturally occurring composites of metal oxides containing ferric ions and having the general formula $M^{2+}Fe_2^{3+}O_4$, whereas $M^{2+}$ can be $Mg^{2+}$, $Co^{2+}$, $Ni^{2+}$, $Zn^{2+}$, $Fe^{2+}$, $Cu^{2+}$, $Mn^{2+}$, etc. SFs exhibit unique physicochemical properties and have been extensively studied because of their stronger magnetic features, higher chemical resistance to oxidation, and larger surface area compared with iron oxide nanoparticles. SFs are classified as normal ($ZnFe_2O_4$, $CdFe_2O_4$), inverse ($MgFe_2O_4$, $NiFe_2O_4$, $CoFe_2O_4$, $CuFe_2O_4$), and mixed ($MnFe_2O_4$), depending on the distribution of cations in tetrahedral and octahedral sites [5].

$MnFe_2O_4$ and $CoFe_2O_4$ have been extensively studied because of their high physical and chemical stability and their excellent magnetization, properties that render them good adsorbents for various metals such as Cr(VI), As(III), As(V), Pb(II), Cu(II), Se(IV), and Se(VI) [54–56]. As(V) or As(III) adsorption on the solid surface of both SFs is based on the formation of inner-sphere complexes. In addition, the results suggested a partial oxidation of As(III) into As(V) onto $MnFe_2O_4$ after its adsorption [56]. Regarding adsorption of divalent cations such as Pb(II) and Cu(II), FTIR analysis revealed that the adsorption mechanism is based on the binding of hydroxyl groups to the metal surface [54]. The possible difference on adsorption capacity of a specific adsorbate on the surface of different ferrites was investigated using the highly mobile and toxic metal selenium as adsorbate and $MnFe_2O_4$, $CuFe_2O_4$, and $CoFe_2O_4$ as adsorbents. Each SF contains two types of hydroxyl binding sites: the M-OH and Fe-OH sites. The M-OH sites, which differ at each SF, are able to act by providing adsorption sites directly or indirectly by affecting the adsorption of Fe-OH. The adsorption capacity of the tested SFs for selenium was, in decreasing order: $CuFe_2O_4 > CoFe_2O_4 >> MnFe_2O_4$. Zeta potential, FTIR, and XPS analyses confirmed the formation of inner and outer-sphere complexes as the adsorption mechanism for Se(IV) and Se(VI) ions, respectively. The affinity of Se(IV) and Se(VI) ions for the metal surface strongly depends on the amount of hydroxyl groups and the surface charge of each metal surface, respectively [55]. Apart from Se, the inverse $CuFe_2O_4$ SF was extensively studied as an adsorbent for several metals such as As(V), As(III), Cd(II), and Mo [55, 57, 58]. As(V) adsorption on $CuFe_2O_4$ is a result of electrostatic attractions, and at pH values occurring in groundwater, the electrostatic repulsions and the limited binding sites on the adsorbent surface almost eliminate the adsorption efficiency. On the other hand, the adsorption efficiency for As(III) was significantly higher in the pH range 7–8 compared with that of As(V), while simultaneous oxidation of As(III) to the less toxic As(V) occurred on the $CuFe_2O_4$ surface [57]. The electrostatic interactions were also expected to be the main mechanism for the adsorption of Mo(VI) on $CuFe_2O_4$, indicating that, in natural environments, the presence of $CuFe_2O_4$ cannot act efficiently for Mo adsorption, but it can be applied in an ex situ groundwater treatment where the pH can be adjusted to the optimized values [58]. Finally, several studies have applied different SFs such as $NiFe_2O_4$ [59], $ZnFe_2O_4$ [60], and $MgFe_2O_4$ [61], in order to investigate their adsorption efficiency for Cr(VI). The results suggest that $NiFe_2O_4$ is a promising favorable adsorbent for Cr(VI) in groundwater, since it exhibited significant adsorption capacity even at pH value equal to 7 [59]. Generally, the relatively high $pH_{PZC}$ point of SFs, which ranges between 7 and 8.5, is another characteristic that renders ferrites as potential adsorbents for metal anions in groundwater.

As already mentioned, organic dyes and their intermediates are among the most common organic pollutants found in surface waters or groundwater due to inadequate treatment practices of industrial effluents. Several SFs have been used as adsorbents for organic dyes with satisfactory results in some cases. $CaFe_2O_4$ exhibits very high affinity for anionic dyes such as Congo red mainly via hydrogen-binding interactions between the $-NH_2$ functional groups of the dye and the surface hydroxyl groups. The absence of $-NH_2$ groups in cationic dyes such as crystal violet, methylene blue, and Rhodamine B is probably the reason for the low adsorption affinity with the surface, indicating the high adsorption selectivity of $CaFe_2O_4$ [62]. Similar observations were obtained in the case of $ZnFe_2O_4$ [63], $NiFe_2O_4$ [59], $CoFe_2O_4$ [64], or $MnFe_2O_4$ [65], used as adsorbents for the removal of dyes, demonstrating that the main adsorption mechanism is the electrostatic attraction between the dye ions and the SFs surface. However, different adsorption capacities even for the same dye have been observed at each SF, due to different cation distribution in the structure [66].

## 3.3 Nanozerovalent Iron

Various metals in the zero oxidation state, such as $Fe^0$, $Zn^0$, $Sn^0$, $Ni^0$, $Mg^0$, and $Al^0$, can be effective for the remediation of polluted groundwater [67]. Elemental iron ($Fe^0$) is the most commonly used because it combines many advantages in comparison to other alternatives, such as high efficiency, low cost, widespread availability, and environmental compatibility. Zerovalent iron nanoparticles (nZVI) are typically less than 100 nm in diameter, with a spherical shape, and they form chain-type agglomerates. The surface activity, the catalytic ability, and the mechanical properties can be increased by 10 to 100 times in comparison with microscale metallic iron [68]. nZVI particles exhibit a core–shell structure, which is due to the typical corrosion reactions taking place when metallic iron is immersed in an aquatic environment. Elemental iron is a strong reductant with standard reduction potential equal to −0.44 V for the pair $Fe^{2+}/Fe^0$:

$$Fe^0_{(s)} \rightarrow Fe^{2+}_{(aq)} + 2\ e^- \tag{3.1}$$

$$E^0 = -0.44\ V \tag{3.2}$$

In an aqueous environment, metallic iron is oxidized and electrons are released, the electron acceptor being (a) dissolved $O_2$ in aerobic conditions, (b) water under anaerobic conditions, or (c) hydrogen cations at very acidic pH [69, 70]:

(a) $Fe^0$ oxidation in aerobic aqueous conditions:

$$2\ Fe^0_{(s)} + O_{2(aq)} + 2\ H_2O \rightarrow 2\ Fe^{2+}_{(aq)} + 4\ OH^- \tag{3.3}$$

(b) $Fe^0$ oxidation in anaerobic aqueous conditions:

$$Fe^0_{(s)} + 2\ H_2O \rightarrow 2\ Fe^{2+}_{(aq)} + H_{2(aq)} + 2\ OH^- \tag{3.4}$$

(c) $Fe^0$ oxidation in acidic pH:

$$Fe^0_{(s)} + 2\ H^+ \rightarrow Fe^{2+}_{(aq)} + H_{2(aq)} \tag{3.5}$$

The primary product of elemental iron oxidation is ferrous iron, as presented in Eqs. (3.3–3.5). In the presence of oxygen, the ferrous iron is oxidized to ferric ions, which then precipitate and form insoluble oxides and hydroxides. In oxygenated natural waters, the most stable iron phase, as predicted thermodynamically, is $\alpha$-$Fe_2O_3$. However, several other metastable ferric hydroxides and oxides are usually formed, such as the amorphous ferrihydrite ($Fe(OH)_3$) or crystalline α-FeOOH. Under anaerobic conditions, when the redox potential becomes negative, the most stable iron phase is $Fe_3O_4$. Due to the above corrosion reactions, iron nanoparticles develop rapidly the core–shell structure. The presence of these nanocomponents in the core–shell structure gives combined properties for the removal of contaminants. The metallic iron core acts as an electron donor source, rendering it a very efficient reducing agent, and the outer layer of iron oxides and hydroxides constitutes a sorptive medium, maintaining various contaminants on the surface of the shell by surface complexation. Electron transfer through the metal core takes place (a) directly by surface defects such as pits or holes, (b) indirectly through the conduction band (oxide conduction band), or (c) by absorption by the structural or adsorbed $Fe^{2+}$, thus preserving the reducing ability of the nanoparticles [71].

### 3.3.1 nZVI Reactivity Toward various Categories of Contaminants: Removal Mechanisms

Numerous studies have shown that nZVI can be applied for the removal of a wide spectrum of contaminants, both organic and inorganic, combining various removal mechanisms, such as reduction, adsorption, and even oxidation through the Fenton mechanism. nZVI reactivity is primarily related with the strong reducing capacity of metallic iron. Data for selected redox reactions involving characteristic inorganic and organic contaminants are given in Table 3.1. Two columns with values of standard reduction potentials are included in the table. The first column corresponds to the standard potentials ($Eh^0$), with proton activity = 1 M. $Eh^0$ values in the second column have been calculated assuming pH 7. This series of values is considered to be more representative of the conditions prevailing in groundwater. Data are presented in order of decreasing $Eh^0$ (pH 7). The reducing strength of the species to the right side of the reactions tends to increase toward the table bottom. As seen, $Fe^0$ is situated very low in this order, and theoretically, it can reduce all the oxidized species situated above, i.e., with higher $Eh^0$ (pH 7) values.

**Table 3.1** Standard reduction potential for characteristic pollutants

| Standard reduction potential | | |
|---|---|---|
| **Half reactions** | **$E_h^0$ (V)** | **$E_h^0$ (pH 7)(V)** |
| **Metals and other inorganic contaminants** | | |
| $Hg^{2+} + 2e^- \rightarrow Hg$ | 0.86 | 0.86 |
| $Ag^+ + e^- \rightarrow Ag$ | 0.80 | 0.80 |
| $UO_2^{2+} + 2e^- \rightarrow UO_{2\,(s)} + 2H_2O$ | 0.42 | 0.42 |
| $CrO_4^{2-} + 8H^+ + 3e^- \rightarrow Cr^{3+} + 4H_2O$ | 1.51 | 0.41 |
| $Cr_2O_7^{2-} + 14H^+ + 6e^- \rightarrow 2Cr^{3+} + 7H_2O$ | 1.36 | 0.39 |
| $NO_3^- + 10H^+ + 8e^- \rightarrow NH_4^+ + 3H_2O$ | 0.88 | 0.36 |
| $Cu^{2+} + 2e^- \rightarrow Cu$ | 0.34 | 0.34 |
| $Cu^{2+} + e^- \rightarrow Cu^+$ | 0.16 | 0.16 |
| $Pb^{2+} + 2e^- \rightarrow Pb$ | −0.13 | −0.13 |
| $H_3AsO_3 + 3H^+ + 3e^- \rightarrow As + 3H_2O$ | 0.24 | −0.17 |
| $H_2AsO_4^- + 3H^+ + 2e^- \rightarrow H_3AsO_3 + 3H_2O$ | 0.42 | −0.20 |
| $Ni^{2+} + 2e^- \rightarrow Ni$ | −0.25 | −0.25 |
| $Cd^{2+} + 2e^- \rightarrow Cd$ | −0.40 | −0.40 |
| $Fe^{2+} + 2e^- \rightarrow Fe$ | −0.44 | −0.44 |
| $Zn^{2+} + 2e^- \rightarrow Zn$ | −0.76 | −0.76 |
| $Ba^{2+} + 2e^- \rightarrow Ba$ | −2.92 | −2.92 |
| **Organic contaminants** | | |
| $ClH_2C\text{-}CH_2Cl + 2e^- \rightarrow H_2C{=}CH_2 + 2Cl^-$ | 0.74 | 0.74 |
| $CCl_4 + H^+ + 2e^- \rightarrow CHCl_3 + Cl^-$ | 0.67 | 0.46 |
| $Cl_2C{=}CCl_2 + H^+ + 2e^- \rightarrow Cl_2C{=}CHCl + Cl^-$ | 0.57 | 0.36 |
| $Cl_2C{=}CHCl + H^+ + 2e^- \rightarrow Cl_2C{=}CH_2 + Cl^-$ | 0.53 | 0.32 |
| $ClHC{=}CH_2 + H^+ + 2e^- \rightarrow H_2C{=}CH_2 + Cl^-$ | 0.45 | 0.24 |
| $Cl_2C{=}CH_2 + H^+ + 2e^- \rightarrow ClHC{=}CH_2 + Cl^-$ | 0.42 | 0.21 |

The reductive capacity of nZVI is the preponderant removal mechanism for metal cations with positive $Eh^0$ well above the standard potential of iron. A characteristic case is mercury, $Hg^{2+}$, with $Eh^0 = 0.86$ V [72]. It is also the crucial mechanism for elements that are soluble and mobile in their high valence state, while forming stable solid phases in their reduced state. The most characteristic example are the oxyanions of Cr(VI), which can be removed from the aqueous phase when reduced to Cr(III), due to the formation of $Cr(OH)_3$ or mixed Fe(III)-Fe(II)-Cr(III) hydroxides [69, 73–75] (see Chapters 8 and 9 for a detailed mechanism of Cr(VI) removal with nZVI). The reductive action of nZVI is also the primary mechanism for remediation of halogenated organic compounds. As seen in Table 3.1, the reduction potentials of characteristic chlorinated hydrocarbons range between 0.21 and 0.74 V, well above that of iron [76, 77].

When the metals have a redox potential more negative or close to that of iron, for example, Zn, Cd, the preponderant removal mechanism is adsorption and precipitation or coprecipitation with Fe(II) and Fe(III). This is also the case of oxyanions such as As(III) and As(V). As previously discussed, the metallic core of nZVI is surrounded by a layer of iron oxides and hydroxides providing an excellent substrate for adsorption, both for cations such as $Cd^{2+}$ and $Zn^{2+}$ and anions such as $H_2AsO_4^-$ and $H_2AsO_3^-$. Careful examination of Eqs. (3.3–3.5) reveals that iron corrosion produces hydroxyls or consumes protons. In all cases, pH is increased, and this helps in the precipitation of various compounds such as metal hydroxides and carbonates. In the case of As, the pH increase favors the precipitation of ferric and ferrous As(V) phases like scorodite ($FeAsO_4 \cdot 2H_2O$) and symplesite [$Fe_3(AsO_4)_2 \cdot 8H_2O$]. Cd and As are two characteristic contaminants, whose removal from the aqueous phase using nZVI occurs primarily through adsorption and coprecipitation mechanisms. The main published studies evaluating As [78–82] and Cd [83–85] removal were carried out using commercial nZVI products or composite nZVI materials. The composite (hybrid) materials consisted of nZVI incorporated on a porous matrix (e.g., mesoporous carbon) or supported on another carrier material (e.g., montmorillonite). The removal capacity ranged from 18.2 to 114.9 mg/g for As(III) and from 12 to 91.42 mg/g for As(V). In the case of Cd, the reported removal varied between 67 and 770 mg/g. In almost all the studies, the removal of As species and Cd was attributed to physical adsorption and chemisorption processes (see Chapter 8 for a detailed mechanism of As(III) and As(V) removal with nZVI).

## 3.4 Conclusion

Adsorption of heavy metals on iron oxides and nZVI surfaces is mainly based on van der Waals interactions between the ions and the surface, or on chemical reactions with the functional groups. The adsorption capacity is strongly affected by the pH of groundwater, the initial concentration, and the electronegativity of the metal pollutant. Iron oxides exhibit significant adsorption capacity at close to neutral pH values like those occurring in groundwater. Their adsorption capacity is strongly affected by their structural properties such as surface area, crystallinity, and the oxide PZC. Among the reported iron oxides, hematite, goethite, and ferrites are considered potentially better adsorbents than other iron oxides for heavy metals due to their usually higher $pH_{PZC}$. However, the crystallinity and thus the surface area of the oxide are other factors that affect sorption capacity. Since the crystallinity is inversely proportional to the surface area, and taking into account that goethite presents a usually higher degree of amorphicity compared with hematite, it is expected to present a higher adsorption capacity. Spinel ferrites, due to their high specific surface area, high chemical stability, and superior magnetic characteristics compared with iron oxides, can be also considered good adsorbents for heavy metals. Regarding organic pollutants, bare iron oxides exhibit a limited efficiency, and surface modifications are necessary. The presence of naturally occurring organic matter can act as an organic coating on the surface of iron oxides, able to modify its properties toward several organic pollutants, thus enhancing the adsorption process of organics. Aquifers rich in iron oxides gain a natural advantage toward groundwater decontamination, since iron oxides can act as natural adsorbents for pollutants. In addition, there is much recent interest in the use of engineered nanoscale iron oxides and nZVI as nanoadsorbents, not only for ex situ but also for in situ groundwater treatment. Their natural occurrence can offset secondary contamination effects caused by their addition and possible subsequent release in the aquifer, since their addition simply constitutes an "enrichment" of the aquifer with iron oxides. However, uncertainties over the health impacts of iron nanomaterials and their environmental fate must be addressed before their widespread application. Regarding ex situ treatments, these nanomaterials exhibit significant advantages as adsorbents, compared with other engineered nanoparticles, due to their high adsorption capacity, low-cost synthesis, reusability, and easy separation/regeneration due to their magnetic properties.

## Acknowledgments

Christiana Mystrioti received financial support by IKY fellowships of excellence for postgraduate students under the Greece Siemens program.

## References

1. Dermatas, D. and Meng, X. (2004). Removal of As, Cr and Cd by adsorptive filtration, *Global Nest Int. J.*, **6**, pp. 3–80.
2. Dermatas, D., Panagiotakis, I., Mpouras, T., and Tettas, K. (2016). The origin of hexavalent chromium as a critical parameter for remediation of contaminated aquifers, *Bull. Environ. Contam. Toxicol.*, **98**, pp. 331–337.
3. Dermatas, D. and Panagiotakis, I. (2012). Remediation of contaminated soil. In *Encyclopedia of Sustainability Science and Technology*, Meyers, R. A. (ed.) (Springer), pp. 2430–2454.
4. Rajan, C. S. (2011). Nanotechnology in groundwater remediation, *IJESD*, **2**, pp. 1–6.
5. Reddy, H. K. and Yun, Y. S. (2016). Spinel ferrite magnetic adsorbents: Alternative future materials for water purification? *Coord. Chem. Rev.*, **315**, pp. 90–111.
6. Dhal, J. P., Mishra, B. G., and Hota, G. (2015). Ferrous oxalate, maghemite and hematite nanorods as efficient adsorbents for decontamination of Congo red dye from aqueous system, *IJEST*, **12**, pp. 1845–1856.
7. Shapira, P. and Wang, J. (2010). Follow the money, *Nature*, **468**, pp. 627–628.
8. USEPA. 2008c. (2009). Office of Solid Waste and Emergency Response. Nanotechnology for Site Remediation Fact Sheet. Report number: EPA 542-F-08-009. Available at: http://www.epa.gov/tio/download/remed/542-f-08-009.pdf.
9. Neyaz, N., Siddiqui, W. A., and Nair, K. K. (2013). Application of surface functionalized iron oxide nanomaterials as nanosorbents in extraction of toxic heavy metals from ground water: A review, *Int. J. Environ. Sci.*, **4**, pp. 472–483.

10. Hua, M., Zhang, S., Pan, B., Zhang, W., Lv, L., and Zhang, Q. (2012). Heavy metal removal from water/wastewater by nanosized metal oxides: A review, *J. Hazard. Mater.*, **211–212**, pp. 317–331.

11. Wang, X., Guo, Y., Yang, L., Han, M., Zhao, J., and Cheng, X. (2012). Nanomaterials as sorbents to remove heavy metal ions in wastewater treatment, *JEAT*, **2**, pp. 2–7.

12. Cornell, R. M. and Schwertmann, U. (2003). *The Iron Oxides: Structure, Properties, Reactions, Occurrences, and Uses*, 2nd Ed. (Wiley-VCH, Germany).

13. Giraldo, L., Erto, A., and Moreno-Pirajan, J. C. (2013). $Fe_3O_4$ nanoparticles for removal of heavy metals from aqueous solutions: Synthesis and characterization, *Adsorption*, **19**, pp. 465–474.

14. Chrysochoou, M., Theologou, E., Bompoti, N., Dermatas, D., and Panagiotakis, I. (2016). Occurrence, origin and transformation processes of geogenic chromium in soils and sediments, *Curr. Pollut. Rep.*, **2**, pp. 224–235.

15. Dermatas, D., Moon, D. H., Menounou, N., Meng, X., and Hires, R. (2004). An evaluation of arsenic release from monolithic solids using a modified semi-dynamic leaching test, *J. Hazard. Mater.*, **B116**, pp. 25–38.

16. Weilong, W. and Xiaobo, F. (2013). Efficient removal of Cr(VI) with Fe/Mn mixed metal oxide nanocomposites synthesized by a grinding method, *J. Nanomater.*, http://dx.doi.org/10.1155/2013/514917.

17. An, B., Liang, Q., and Zhao, D. (2011). Removal of arsenic(V) from spent ion exchange brine using a new class of starch-bridged $Fe_3O_4$ nanoparticles, *Water Res.*, **45**, pp. 1961–1972.

18. Wang, Y., Morin, G., Ona-Nguema, G., Juillot, F., Calas, G., and Brown, G. E. Jr. (2011). Distinctive arsenic(V) trapping modes by $Fe_3O_4$ nanoparticles induced by different sorption processes, *Environ. Sci. Technol.*, **45**, pp. 7258–7266.

19. Morin, G., Wang, Y., Ona-Nguema, G., Juillot, F., Calas, G., Menguy, N., Aubry, E., Bargar, J. R., and Brown, G. E. Jr. (2009). EXAFS and HRTEM evidence for As(III)-containing surface precipitates on nanocrystalline $Fe_3O_4$: Implications for As sequestration, *Langmuir*, **25**, pp. 9119–9128.

20. Liu, J. F., Zhao, Z. S., and Jiang, G. B. (2008). Coating $Fe_3O_4$ magnetic nanoparticles with humic acid for high efficient removal of heavy metals in water, *Environ. Sci. Technol.*, **42**, pp. 6949–6954.

21. Karami, H. (2013). Heavy metal removal from water by $Fe_3O_4$ nanorods, *Chem. Eng. J.*, **219**, pp. 209–216.

22. Petranovska, A. L., Abramov, N. V., Turanska, S. P., Gorbyk, P. P., Kaminskiy, A. N., and Kusyak, N. V. (2015). Adsorption of cis-dichlorodiammineplatinum by nanostructures based on single-domain $Fe_3O_4$, *J. Nanostructure Chem.*, **5**, pp. 275–285.

23. Xue, X., Hanna, K., Abdelmoula, M., and Deng, N. (2009). Adsorption and oxidation of PCP on the surface of $Fe_3O_4$: Kinetic experiments and spectroscopic investigations, *Appl. Catal. B: Environ.*, **89**, pp. 432–440.

24. Zhang, S., Niu, H., Hu, Z., Cai, Y., and Shi, Y. (2010). Preparation of carbon coated $Fe_3O_4$ nanoparticles and their application for solid-phase extraction of polycyclic aromatic hydrocarbons from environmental water samples, *J. Chromatogr. A*, **1217**, pp. 4757–4764.

25. Absalan, G., Asadi, M., Kamran, S., Sheikhian, L., and Goltz, D. M. (2011). Removal of reactive red-120 and 4-(2-pyridylazo) resorcinol from aqueous samples by $Fe_3O_4$ magnetic nanoparticles using ionic liquid as modifier, *J. Hazard.Mater.*, **192**, pp. 476–484.

26. Iram, M., Guo, C., Guan, Y., Ishfaq, A., and Liu, H. (2010). Adsorption and magnetic removal of neutral red dye from aqueous solution using $Fe_3O_4$ hollow nanospheres, *J. Hazard. Mater.*, **181**, pp. 1039–1050.

27. Peng, L., Qin, P., Lei, M., Zeng, Q., Song, H., Yang, J., Shao, J., Liao, B., and Gu, J. (2012). Modifying $Fe_3O_4$ nanoparticles with humic acid for removal of Rhodamine B in water, *J. Hazard. Mater.*, **209–210**, pp. 193–198.

28. Saha, B., Das, S., Saikia, J., and Das, G. (2011). Preferential and enhanced adsorption of different dyes on iron oxide nanoparticles: A comparative study, *J. Phys. Chem. C*, **115**, pp. 8024–8033.

29. Amin, M. T., Alazba, A. A., and Manzoor, U. (2014). A review of removal of pollutants from water/wastewater using different types of nanomaterials, *Adv. Mater. Sci.*, 2014, pp. 1–24.

30. Ko, I., Kim, J. Y., and Kim, K. W. (2005). Adsorption properties of soil humic and fulvic acids by hematite, *Chem. Speciation Bioavailability*, **17**, pp. 41–48.

31. Rahman, M. S., Whalen, M., and Gagnon, G. A. (2013). Adsorption of dissolved organic matter (DOM) onto the synthetic iron pipe corrosion scales (goethite and $Fe_3O_4$): Effect of pH, *Chem. Eng. J.*, **234**, pp. 149–157

32. Chirita, M. and Grozescu, I. (2009). $Fe_2O_3$–nanoparticles, physical properties and their photochemical and photoelectrochemical applications, *Chem. Bull. "POLITEHNICA" Univ. (Timisoara)*, **54**, pp. 1–8.

33. Makie, P., Westin, G., Persson, P., and Osterlund, L. (2011). Adsorption of trimethyl phosphate on maghemite, hematite, and goethite nanoparticles, *J. Phys. Chem. A*, **115**, pp. 8948–8959.

34. Hu, J., Chen, G., and Lo, I. (2006). Selective removal of heavy metals from industrial wastewater using maghemite nanoparticle: Performance and mechanisms, *J. Environ. Eng.*, **132**, pp. 709–715.

35. Predescu, A. and Nicolae, A. (2012). Adsorption of Zn, Cu and Cd from waste waters by means of maghemite nanoparticles, *U.P.B. Sci. Bull. B*, **74**, pp. 255–264.

36. Fialova, D., Kremplova, M., Melichar, L., Kopel, P., Hynek, D., Adam, V., and Kizek, R. (2014). Interaction of heavy metal ions with carbon and iron based particles, *Materials*, **7**, pp. 2242–2256.

37. Vélez, E., Campillo, G. E., Morales, G., Hincapié, C., Osorio, J., Arnache, O., Uribe, J. I., and Jaramillo, F. (2016). Mercury removal in wastewater by iron oxide nanoparticles, *J. Phys.: Conf. Series*, **687**.

38. Akrami, A. and Niazi, A. (2016). Synthesis of maghemite nanoparticles and its application for removal of titan yellow from aqueous solutions using full factorial design, *Desalin. Water Treat.*, **57**, pp. 22618–22631.

39. Adegoke, H. I., Adekola, F. A., Fatoki, O. S., and Ximba, B. J. (2014). Adsorption of Cr(VI) on synthetic hematite ($\alpha$-$Fe_2O_3$) nanoparticles of different morphologies, *Korean J. Chem. Eng.*, **31**, pp. 142–154.

40. Al-Saad, K. A., Amr, M. A., Hadi, D. T., Arar, R. S., AL-Sulaiti, M. M., Abdulmalik, T. A., Alsahamary, N. M., and Kwak, J. C. (2012). Iron oxide nanoparticles: Applicability for heavy metal removal from contaminated water, *Arab J. Nuclear Sci. Appl.*, **45**, pp. 335–346.

41. Hafez, H. (2012). A study on the use of nano/micro structured goethite and hematite as adsorbents for the removal of Cr(III), Co(II), Cu(II), Ni(II), and Zn(II) metal ions from aqueous solutions, *IJEST*, **4**, pp. 3018–3028.

42. Shayan, N. N. and Mirzayi, B. (2015). Adsorption and removal of asphaltene using synthesized maghemite and hematite nanoparticles, *Energy Fuels*, **29**, pp. 1397–1406.

43. Kallay, N., Preocanin, T., Markovic, J., and Kovacevi, D. (2007). Adsorption of organic acids on metal oxides: Application of the surface potential measurements, *Colloids Surf. A: Physicochem. Eng. Aspects*, **306**, pp. 40–48.

44. Murphy, E. M., Zachara, J. M., Smith, S. C., and Phillips, J. L. (1992). The sorption of humic acids to mineral surfaces and their role in contaminant binding, *Sci. Total Environ.*, **117–118**, pp. 413–423.

45. Qin, X., Liu, F., Wang, G., and Huang, G. (2015). Adsorption of humic acid from aqueous solution by hematite: Effects of pH and ionic strength, *Environ. Earth Sci.*, **73**, pp. 4011–4017.

46. Liu, H., Chen, T., and Frost, R. L. (2014). An overview of the role of goethite surfaces in the environment, *Chemosphere*, **103**, pp. 1–11.

47. Grossl, P. R. and Sparks, D. L. (1995). Evaluation of contaminant ion adsorption/desorption on goethite using pressure-jump relaxation kinetics, *Geoderma*, **67**, pp. 87–101.

48. Mohapatra, M., Mohapatra, L., Singh, P., An, S., and Mishra, B. K. (2010). A comparative study on Pb(II), Cd(II), Cu(II), Co(II) adsorption from single and binary aqueous solutions on additive assisted nano-structured goethite, *Int. J. Eng. Sci. Technol.*, **2**, pp. 89–103.

49. Shareef, A., Angove, M. J., Wells, J. D., and Johnson, B. B. (2006). Sorption of bisphenol A, 17α-ethynylestradiol and estrone to mineral surfaces, *J. Colloid Interface Sci.*, **297**, pp. 62–69.

50. Guo, X., Yang, C., Dang, Z., Zhang, Q., Li, Y., and Meng, Q. (2013). Sorption thermodynamics and kinetics properties of tylosin and sulfamethazine on goethite, *Chem. Eng. J.*, **223**, pp. 59–67.

51. Tunega, D., Gerzabek, M. H., Haberhauer, G., Totsche, K. U., and Lischk, H. (2009). Model study on sorption of polycyclic aromatic hydrocarbons to goethite, *J. Colloid Interface Sci.*, **330**, pp. 244–249.

52. Angove, M. J., Fernandes, M. B., and Ikhsan, J. (2002). The sorption of anthracene onto goethite and kaolinite in the presence of some benzene carboxylic acids, *J. Colloid Interface Sci.*, **247**, pp. 282–289.

53. Brigante, M., Zanini, G., and Avena, M. (2010). Effect of humic acids on the adsorption of paraquat by goethite, *J. Hazard. Mater.*, **184**, pp. 241–247.

54. Ren, Y., Li, N., Feng, J., Luan, T., Wen, Q., Li, Z., and Zhang, M. (2012). Adsorption of Pb(II) and Cu(II) from aqueous solution on magnetic porous ferrospinel $MnFe_2O_4$, *J. Colloid Interface Sci.*, **367**, pp. 415–421.

55. Sun, W., Pan, W., Wang, F., and Xu, N. (2015). Removal of Se(IV) and Se(VI) by $MFe_2O_4$ nanoparticles from aqueous solution, *Chem. Eng. J.*, **273**, pp. 353–362.

56. Zhang, S., Niu, H., Cai, Y., Zhao, X., and Shi, Y. (2010). Arsenite and arsenate adsorption on coprecipitated bimetal oxide magnetic nanomaterials: $MnFe_2O_4$ and $CoFe_2O_4$, *Chem. Eng. J.*, **158**, pp. 599–607.

57. Tu, Y. J, You, C. F., Chang, C. K., Wang, S. L., and Chan, T. S. (2013). Adsorption behavior of As(III) onto a copper ferrite generated from printed circuit board industry, *Chem. Eng. J.*, **225**, pp. 33–439.

58. Tu, Y. J., You, C. F., Chang, C. K., Chan, T. S., and Li, S. H. (2014). XANES evidence of molybdenum adsorption onto novel fabricated nano-magnetic $CuFe_2O_4$, *Chem. Eng. J.*, **244**, pp. 343–349.

59. Hou, X., Feng, J., Liu, X., Ren, Y., Fan, Z., Wei, T., Meng, J., and Zhang, M. (2011). Synthesis of 3D porous ferromagnetic $NiFe_2O_4$ and using as novel adsorbent to treat wastewater, *J. Colloid Interface Sci.*, **362**, pp. 477–485.

60. Jia, Z., Qin, Q., Liu, J., Shi, H., Zhang, X., Hu, R., Li, S., and Zhu, R. (2015). The synthesis of hierarchical $ZnFe_2O_4$ architecture and their application for Cr(VI) adsorption removal from aqueous solution, *Superlattices Microstruct.*, **82**, pp. 174–187.

61. Kaur, M., Kaur, N., Jeet, K., and Kaur, P. (2015). $MgFe_2O_4$ nanoparticles loaded on activated charcoal for effective removal of Cr(VI)–A novel approach, *Ceram. Int.*, **41**, pp. 13739–13750.

62. Liu, X., An, S., Wang, Y., Yang, Q., and Zhang, L. (2015). Rapid selective separation and recovery of a specific target dye from mixture consisted of different dyes by magnetic Ca-ferrites nanoparticles, *Chem. Eng. J.*, **262**, pp. 517–526.

63. Samoila, P., Cojocaru, C., Cretescu, I., Stan, C. D., Nica, V., Sacarescu, L., and Harabagiu, V. (2015). Nanosized spinel ferrites synthesized by sol-gel autocombustion for optimized removal of azo dye from aqueous solution, *J. Nanomater.*, pp. 1–13.

64. Ghaemi, M., Absalan, G., and Sheikhian, L. (2014). Adsorption characteristics of Titan yellow and Congo red on $CoFe_2O_4$ magnetic nanoparticles, *JICS*, DOI 10.1007/s13738-014-0448-0.

65. Wu, R. and Qu, J. (2005). Removal of water-soluble azo dye by the magnetic material $MnFe_2O_4$, *J. Chem. Technol. Biotechnol.*, **80**, pp. 20–27.

66. Wang, L., Li, J., Wang, Y., Zhao, L., and Jiang, Q. (2012). Adsorption capability for Congo red on nanocrystalline $MFe_2O_4$ (M= Mn, Fe, Co, Ni) spinel ferrites, *Chem. Eng. J.*, **181–182**, pp. 72–79.

67. Powell, R. M., Puls, R. W., Blowes, D. W., Vogan, J. L., Gillham, R. W., Powell, P. D., Schultz, D., Landis, R., and Sivavic, T. (1998). Permeable reactive barrier technologies for contaminated remediation, *U.S. Environmental Protection Agency*, EPA/600/R-98/125, Ada, OK.

68. O'Carroll, D., Sleep, B., Krol, M., Boparai, H., and Kocur, C. (2013). Nanoscale zerovalent iron and bimetallic particles for contaminated site remediation, *Adv. Water Resour.*, **51**, pp. 104–122.

69. Gheju, M. (2011). Hexavalent chromium reduction with zero-valent iron (ZVI) in aquatic systems, *Water Air Soil Poll.*, **222**, pp. 103–148.

70. Noubactep, C. (2010). The fundamental mechanism of aqueous contaminant removal by metallic iron, *Water SA*, **36**, pp. 663–670.

71. Li, L., Fan, M., Brown, R. C., Van Leeuwen, H. J., Wang, J., Wang, W., Song, Y., and Zhang, P. (2006). Synthesis, properties, and environmental applications of nanoscale iron-based materials: A review, *Crit. Rev. Env. Sci. Technol.*, **36**, pp. 405–431.

72. Liu, T., Wang, Z. L., Yan, X., and Zhang, B. (2014). Removal of mercury(II) and chromium(VI) from wastewater using a new and effective composite: Pumice-supported nanoscale zero-valent iron, *Chem. Eng. J.*, **245**, pp. 34–40.

73. Mystrioti, C., Sparis, D., Papasiopi, N., Xenidis, A., Dermatas, D., and Chrysochoou, M. (2015b). Assessment of polyphenol coated nano zerovalent iron for hexavalent chromium removal from contaminated waters, *Bull. Environ. Contam. Toxicol.*, **94**, pp. 302–307.

74. Mystrioti, C., Xanthopoulou, T. D., Tsakiridis, P., Papassiopi, N., and Xenidis, A. (2016). Comparative evaluation of five plant extracts and juices for nanoiron synthesis and application for hexavalent chromium reduction, *Sci. Total Environ.*, **539**, pp. 105–113.

75. Toli, A., Chalastara, K., Mystrioti, C., Xenidis, A., and Papassiopi, N. (2016). Incorporation of zerovalent iron nanoparticles in the matrix of cationic resin beads for the remediation of Cr(VI) contaminated waters, *Environ. Pollut.*, **214**, pp. 419–429.

76. Chang, M. C., Shu, H. Y., Hsieh, W. P., and Wang, M. C. (2005). Using nanoscale zero-valent iron for the remediation of polycyclic aromatic hydrocarbons contaminated soil, *J. Air Waste Manag. Assoc.*, **55**, pp. 1200–1207.

77. Liu, Y. and Lowry, G. V. (2006). Effect of particle age ($Fe^0$ content) and solution pH on nZVI reactivity: $H_2$ evolution and TCE dechlorination, *Environ. Sci. Technol.*, **40**, pp. 6085–6090.

78. Morgada, M. E., Levy, I. K., Salomone, V., Farıas, S. S., Lopez, G., and Litter, M. I. (2009). Arsenic (V) removal with nanoparticulate zerovalent iron: Effect of UV light and humic acids, *Catal. Today*, **143**, pp. 261–268.

79. Zhu, H., Ji, Y., Wu, X., and Wang, H. (2009). Removal of arsenic from water by supported nano zero-valent iron on activated carbon, *J. Hazard. Mater.*, **172**, pp. 1591–1596.

80. Kim, K. R., Lee, B. T. T., and Kim, K. W. (2012). Arsenic stabilization in mine tailings using nano-sized $Fe_3O_4$ and zerovalent iron with the enhancement of mobility by surface coating, *J. Geochem. Explor.*, **113**, pp. 124–129.

81. Bhowmick, S., Chakraborty, S., Mondal, P., Van Renterghem, W., Van den Berghe, S., Roman-Ross, G., Chatterjee, D., and Iglesias, M. (2014). Montmorillonite-supported nanoscale zero-valent iron for removal of arsenic from aqueous solution: Kinetics and mechanism, *Chem. Eng. J.*, **243**, pp. 14–23.

82. Sikder, M. T., Tanaka, S., Saito, T., and Kurasaki, M. (2014). Application of zerovalent iron impregnated chitosan-caboxymethyl-bcyclodextrin composite beads as arsenic sorbent, *JECE*, **2**, pp. 370–376.

83. Wang, C., Luo, H., Zhang, Z., Wu, Y., Zhang, J., and Chen, S. (2014). Removal of As(III) and As(V) from aqueous solutions using nanoscale zerovalent iron-reduced graphite oxide modified composites, *J. Hazard. Mater.*, **268**, pp. 124–131.

84. Baikousi, M., Georgiou, Y., Daikopoulos, C., Bourlinos, A. B., Filip, J., Zboril, R., Deligiannakis, Y., and Karakassides, M. A. (2015). Synthesis and characterization of robust zerovalent iron/mesoporous carbon composites and their applications in arsenic removal, *Carbon*, **93**, pp. 636–647.

85. Nakseedee, P., Tanboonchuy, V., Pimpha, N., Khemthong, P., Liao, C.-H., and Grisdanurak, N. (2015). Arsenic removal by nanoiron coupled with gas bubbling system, *J. Taiwan Inst. Chem. Eng.*, **47**, pp. 182–189.

# Chapter 4

# Application of Nanozerovalent Iron for Water Treatment and Soil Remediation: Emerging Nanohybrid Approach and Environmental Implications

**Nirupam Aich,[a] Chunming Su,[b] Ijung Kim,[c] and Arvid Masud[a]**
[a]*Department of Civil, Structural and Environmental Engineering, University at Buffalo, The State University of New York, Buffalo, NY 14260, USA*
[b]*United States Environmental Protection Agency, National Risk Management Research Laboratory, Ada, OK 74820, USA*
[c]*Department of Civil and Environmental Engineering, Western New England University, Springfield, MA 01119, USA*
nirupama@buffalo.edu

## 4.1 Introduction

Nanoscale zerovalent iron (nZVI, $Fe^0$) with 10–100 nm particle size has an extraordinary reductive capability of degrading the broadest range of organic contaminants and immobilizing inorganic/heavy metal ions [1, 2]. Thus, nZVI has become the most commonly used

*Iron Nanomaterials for Water and Soil Treatment*
Edited by Marta I. Litter, Natalia Quici, and Martín Meichtry

ISBN 978-981-4774-67-3 (Hardcover), 978-981-4669-49-8 (eBook)
www.panstanford.com

engineered nanomaterial (ENM) for in situ soil and groundwater remediation technologies, operating in total 77 pilot and field-scale polluted sites worldwide (> 70% in the United States) [1, 3]. The projected spending for site cleanup and restoration is \$250 billion by 2040 in the United States [1] and €38 billion (~\$41.7 billion) a year in Europe [4, 5]. Moreover, nZVI use in industrial wastewater treatment plants is being demonstrated [6, 7]. Reasonably, significant research efforts are currently in progress to improve nZVI contaminant degradation efficacy by reducing its high aggregation propensity [i.e., due to van der Waals (vdW) and magnetic attraction forces], increasing the surface reactivity, and protecting from surface oxidation in aqueous environment [2, 5, 8, 9]. These include surface modification, metal addition, and solid support attachment.

The advancement of ENM synthesis and modification has led to the development of nanohybrids (NHs)—an emerging class of ENMs where a single nanoscale entity is composed of two or more ENMs providing enhanced or multifunctional activities [10]. Various metal (Pd, Pt, Ni, etc.) ENM deposition on the surface of nZVI has led to the development of bimetallic nZVI nanohybrids (BM-nZVI NHs) [2]. The attached secondary metal ENM acts as a catalyst to improve the oxidation–reduction reactions via the formation of excess hydrogen on the BM-nZVI surface and improve degradation of contaminants in groundwater remediation and wastewater treatment. Nanoscale hybridization approach has enabled solid support attachment for nZVI performance enhancement via conjugation with various nanocarbons (nCs), including one-dimensional (1D) carbon nanotubes (CNTs), two-dimensional (2D) reduced graphene oxides (rGO), etc. [11]. It should be noted that solid supports can be of macro- and mesoscales too, including clay materials, biochars, activated carbons; however, this chapter will focus solely on nanoscale hybridization. These hybridizations allow for the nZVI pollutant degradation capability to be improved via multifaceted achievements—reduction in aggregation propensity, high reactive surface area, enhanced adsorption of contaminants on the CNT or rGO surfaces, and more electron transport efficiency due to excellent conductivity of nCs. Thus, these emerging nC-nZVI hybrids show promises with better organic and inorganic contaminant removal and degradation in sub-surface and wastewater. However, with their altered physicochemical properties due to hybridization, concerns regarding unprecedented environmental risks are also posed.

The increasing usage of nZVI and their hybridized forms (e.g., BM-nZVI or nC-nZVI) leads to the concern about their environmental health and safety. Typically, NHs present with uniquely altered and emergent properties different from those of their component ENMs. These altered/emergent properties of NHs may lead to uncertain environmental consequences in terms of altered fate, transport, and toxicity behavior. For example, nZVI while combined with the solid nanosupports CNTs or graphenes can have altered particle size, dimensionality (0D nZVI versus 1D CNT-nZVI or 2D rGO-nZVI), and surface chemistry. These properties may change their aggregation behavior, transport through porous media, and even interactions with biological entities found in the environment leading to uncertain environmental toxicity. Similarly, the presence of Pd, Pt, Ni, and other metallic ENMs on the surfaces of BM-nZVI alters their biochemical interactions significantly and can alter their interactions with microorganisms and animal cells.

Nanozerovalent iron (nZVI), due to its high activity toward persistent organic contaminant degradation and heavy metal removal, has attracted applications in industrial wastewater treatment as well as in situ soil remediation techniques. Enhancement of the catalytic activity of nZVI is pursued by employing different surface modification strategies and most recently conjugating or hybridizing with other carbonaceous or metallic nanomaterials. NH formation has multifaceted advantages toward its redox activity increases, electronic property alterations, providing with higher surface area, and better adsorption capabilities. Achievement of multifunctionality through the hybridization approach, on the contrary, can bring about uncertainty in understanding the environmental fate, transport, and toxicity of conjugated nZVI due to the alterations of physicochemical properties.

This chapter will discuss the emergence of nanoscale hybridization of nZVI-based materials, their multifunctional capabilities, and their importance in environmental remediation and water treatment. Moreover, the chapter will provide a brief review on the environmental fate and toxicity concerns from nZVI-based materials. Then, the chapter will also discuss the complexity in assessing environmental risk of nZVI-based NHs from their altered and emergent properties. Finally, the chapter will put forth some strategies for proactive measures to determine the environmental risks associated with novel hierarchical nZVI-based NHs.

## 4.2 Pollutant Removal Mechanisms of nZVI and Its Hybrids

The major mechanisms of pollutant degradation by nZVI can be classified into chemical reduction and adsorption [11]. nZVI becomes oxidized and reduces organic or inorganic pollutants into less toxic compounds. For example, a typical reductive degradation process of chlorinated organic contaminants by nZVI can be characterized as follows:

$$Fe^0 + RCl + H^+ \rightarrow Fe^{2+} + RH + Cl^- \quad (4.1)$$

Because of the high surface-area-to-volume ratio, nZVI has been expected to significantly enhance the contaminant removal efficiency than their microscale counterparts. However, its surface can be rapidly oxidized by contact with water, forming the iron oxide shell ($Fe(OH)_2$), which limits the reduction process of organic contaminants (as shown in the following equation):

$$Fe^0 + 2H_2O \rightarrow Fe(OH)_2 + H_2 \uparrow \quad (4.2)$$

However, the iron oxide shell provides more oxygen groups for chemical bonding with other pollutants. This specifically renders not only the adsorption of metalloids such as arsenic [12], but also the simultaneous adsorption–reduction of heavy metals (e.g., lead [13], chromium [2], etc., as shown below) along with the adsorption of organic contaminants [14].

$$\equiv FeOH + Cr_2O_7^{2-} \rightarrow \equiv Fe - Cr_2O_7^- + OH^- \quad (4.3)$$

To overcome the inefficiency due to deterioration of chemical reduction capability, deposition of other metal ENMs on nZVI surfaces has been pursued to form bimetallic nZVI or BM-nZVI. Examples of them include Fe/Pd, Fe/Pt, Fe/Ag, Fe/Ni, etc. The primary metal nZVI ($Fe^0$) plays the role of electron donor, while the secondary metal serves as a catalyst that not only promotes the reduction via hydrogenation, but also prevents the oxide film formation [15]. Since the primary and the secondary metals play the roles of an anode and a cathode of a galvanic cell, electron transfer between two metals continues inhibiting the oxide film formation [16]. For example, the following pathway shows the dehalogenation via hydrogenation with Fe/Pd bimetallic nanoparticles [17].

$$Fe^0 \rightarrow Fe^{2+} + 2e^- \quad (4.4)$$

$$2H_2O + 2e^- \rightarrow H_{2(g)} + 2OH^- \quad (4.5)$$

$$Pd^{2+} + 2e^- \rightarrow Pd^0 \quad (4.6)$$

$$H_{2(g)} \xrightarrow{Pd} PdH_2 \quad (4.7)$$

$$PdH_2 + RCl \rightarrow RH + PdClH \quad (4.8)$$

However, both nZVI and BM-nZVI suffer from high agglomeration tendency due to their high vdW attraction and magnetic attraction forces, resulting in the loss of reactive surface area and contaminant degradation ability. Thus, surface modification with polymer or surfactant coating has been pursued extensively with the recent development of attachment of nZVI or BM-nZVI to solid supports (e.g., activated carbons, zeolites, clay materials, etc.). The most recent trend has been the use of carbon-based ENMs, i.e., CNT or graphene as solid supports for nZVI or BM-nZVI. nZVI deposition on 1D CNT or 2D rGO via reduction of iron salt precursors has shown to restrict nZVI aggregate size within 10–80 nm [8, 18–26]. The specific surface area (SSA) of these hybrids is always found to be significantly greater than the SSA of nZVI only, and closer to that of CNT or rGO [8, 18–22]. Moreover, the hybrids tend to have overall reduction in aggregation propensity, though that has not been systematically studied. The domination of nCs dimensionality further reduces the oxidation susceptibility of nZVI and causes alterations of crystal structures [27]. Both CNT and graphene (or rGO) provide large surfaces with $\pi$-electron clouds, rendering a high transfer rate for electron hopping between well-dispersed nZVI, subsequently resulting in a lower redox potential in the aqueous media [18, 27]. Recent advances in producing hierarchical 3D CNT/rGO/metal nanostructures have proven to further reduce the sizes for iron-based metal nanoparticles and further reduce the aggregation of nC-nZVI hybrids by the intercalating effects between CNT and rGO [28–30]. Moreover, the porous nature of the heterogeneous structures provides protection from outside environment. The CNT/rGO network remarkably improves the electrical conductivity by conjugation through the defects of CNT and edges of rGO [28–31]. As a result, 3D CNT/rGO structures have enhanced redox activities for iron-based particles compared with 1D CNT and 2D rGO. These solid supports also provide functional groups for enhanced adsorption of contaminants.

CNTs and their oxidized forms have been studied as a promising adsorbent with their high adsorption capacity in water and wastewater treatment due to their hydrophobic nature and presence of functional groups. When coupled with nZVI, hybrid CNT-nZVIs have shown their effectiveness in the treatment of Cr(VI) [32, 33], Se(IV) [22], and Sb(III) [27]. For example, the removal of Cr(VI) by CNT-nZVI occurs not only by the direct adsorption of Cr(VI) onto CNTs via electrostatic interaction (Eq. 4.9) [32], but also by the reduction of Cr(VI) to Cr(III) with the contact of nZVI (Eq. 4.10) followed by the electrostatic adsorption (Eqs. 4.11 and 4.12), which is facilitated at higher pH, producing negatively charged adsorbent and positively charged Cr(III) [33–35]. Also, CNTs are capable of organic adsorption via $\pi$–$\pi$ interaction and hydrogen bonding [36]. Therefore, the decoration with functional groups as well as environmental conditions is determinant in the CNTs adsorption capacity.

$$\equiv OH_2^+ + HCrO_4^- \rightarrow \equiv OH_2CrO_4^- + H^+ \quad \text{(direct adsorption)} \quad (4.9)$$

$$2HCrO_4^- + 3Fe^0 + 14H^+ \rightarrow 3Fe^{2+} + 2Cr^{3+} + 8H_2O \quad (4.10)$$

$$Cr^{3+} + 2OH^- \rightarrow Cr(OH)_2^+ \quad (4.11)$$

$$\equiv COO^- + Cr(OH)_2^+ \rightarrow \equiv COOCr(OH)_2 \quad \text{(adsorption through reduction)} \quad (4.12)$$

Simultaneous reduction and adsorption are also available with rGO-nZVI NHs [37]. In fact, rGO-nZVIs with oxygen-containing functional groups such as epoxy (C-O-C), hydroxyl (OH), and carboxyl (COOH) groups have resulted in the significant adsorption of Pb(II) [38], As(III) and As(V) [39], U(VI) [40], Cr(VI) [41], Re(VII) [42], disinfection by-products [43], colorizing agents [8, 44], phosphorus [45], and chlorinated organic pollutants [46]. As illustrated in Fig. 4.1, the nZVI removes U(VI) through reductive adsorption. This results in the formation of magnetite ($Fe_3O_4$) and hematite ($Fe_2O_3$) layer on nZVI core, which adsorbs the reduction product ($U_3O_7$) of U(VI). In addition, the adsorption of U(VI) (hydr)oxide also occurs, mainly caused by the dissolution of the core nZVI ($Fe^0$) into $Fe^{2+}$ ion. When graphene support matrix is added, the reduction rate of U(VI) is enhanced and the support matrix provides additional active sites for the sorption of U(VI) (hydr)oxide, which results in the higher removal rate of U(VI) [40].

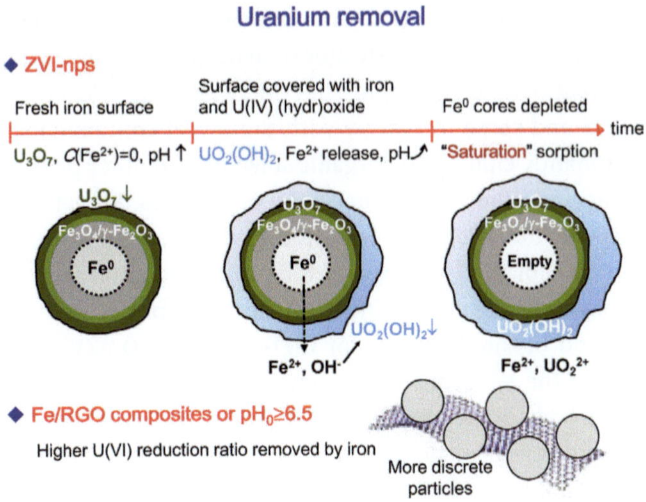


**Figure 4.1** Mechanism of removal of U(VI) by nZVI and reduced graphene supported nZVI. Reprinted from Ref. [40], Copyright 2015, with permission from Elsevier.

Since graphene is also known to adsorb organics via $\pi$–$\pi$ interaction [47], the selective removal for the pollutants with C=C double bonds or benzene rings such as polycyclic aromatic hydrocarbons [39] is also possible. Further modifications for the simultaneous reduction and adsorption can take place using BM-nZVI on carbonaceous-ENMs to treat chlorinated organic pollutants [48]. Though the chemical reduction might be considered a primary mechanism over the adsorption in such cases as antibiotic removal [49], more quantitative data need to be accumulated to generalize such a comparison. Also, minor contributions from other mechanisms such as aggregation, ion exchange, hydroxylation, and precipitation need to be considered as well [11].

## 4.3 Applications of nZVI-Based Nanohybrids for Site Remediation

The contaminant treatment technology using nZVI was first developed for groundwater remediation at various organic contaminant polluted sites. In situ delivery or injection of nZVI into the groundwater and nZVI incorporation in the underground permeable reactive barriers (PRBs) have been the major modes of applications of this technology [1]. As mentioned earlier, the total number of contaminant soil and groundwater remediation sites involving nZVI-based treatment technologies has increased significantly, mostly in the US and some in Europe with the target contaminants being predominantly chlorinated organic solvents and other polyaromatic hydrocarbons. The recent emergence of NHs has not yet fully been appreciated; however, their inclusions are also increasing. Currently, around 30% of nZVIs developed for site remediation in the US are in the bimetallic form or as BM-nZVI; however, no such reports have been found in the literature for CNT-nZVI or rGO-nZVI since they are still in the lab-scale experimentation stage. Some of the pilot- and field-scale demonstrations of BM-nZVI NH-based contaminant treatments have been described below. Initially, during 2001, a field-scale feasibility test was performed by injecting ~1.7 kg Fe/Pd BM-nZVI in 1340 L slurry into a trichloroethylene (TCE)-contaminated industrial site in New Jersey, US, where within a 4-week period, 96.5% TCE degradation was achieved. The schematic diagram of the injection process of the BM-nZVI into the groundwater is illustrated in Fig. 4.2. The injection well (DGC–15) was used to inject BM-nZVI, and the piezometers were used to collect sample downgradient of the injection well [50].

In 2002, a former waste disposal area in Research Triangle Park in North Carolina, US, was treated by 11.2 kg of Fe/Pd nanoparticles in 6056 L slurry, and 99% of TCE was reduced in a few days [51]. Since 1994, the Naval Air Station in Jacksonville, Florida, US (approximately 135 $m^2$ area with a thickness of 5.5 m) was reported to be contaminated by waste solvent storage tanks. In 2003, 136 kg of bimetallic Fe/Pd nanoparticles were injected through 10 injection points, and 99% TCE reduction was achieved.

Similarly, the Naval Air Engineering Station in Lakehurst, New Jersey, was highly contaminated by chlorinated organic pollutants deep into 70 ft below groundwater table. A total of 136 kg of bimetallic Fe/Pd nanoparticles was injected through 15 locations over the contaminated sites, treating 79% of TCE and 83% of DCE, respectively [52]. The Zhao research group at Auburn University in the US developed carboxymethyl cellulose (CMC) stabilized Fe/Pd BM-nZVI for site remediation and applied them to the fields in California [53], Alabama [54], and Utah [55], respectively. In 2009, 20 kg of Fe/Si nanoparticles were injected to an industrial site in West Lafayette, Ohio, resulting in up to 90% removal of TCE [56]. Injection of CMC-stabilized nZVI at a pilot site located in the South of the USA resulted in up to 87% removal of chlorinated solvents (TCE, PCE, and PCB) over 596 days after the injection [54]. These results are attributed to the initial rapid abiotic degradation (within the first 29 days) of the chlorinated solvents by the CMC-stabilized nZVI and later by a slow process of biotransformation by the microbial community enhanced by the nZVI. However, the duration of initial abiotic degradation could be enhanced by improving nZVI reactivity or removal capacity for an extended period as well as for an expanded area.

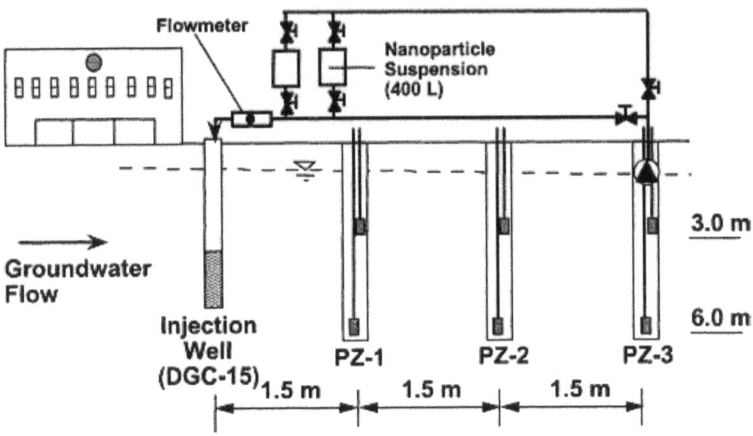


**Figure 4.2** Schematic diagram of in situ injection of Fe/Pd BM-nZVI into TCE-contaminated groundwater site in New Jersey, US. Reprinted with permission from Ref. [50]. Copyright 2001 American Chemical Society.

## 4.4 Applications of nZVI-Based Nanohybrids for Wastewater Treatment

The usage of nZVIs and their NHs in wastewater treatment processes has not yet been fully realized. There have been some pilot studies for the removal of heavy metals from metal-finishing industrial wastewater [6, 57, 58] and power plant wastewater [59] with some development of fundamental lab- and bench-scale studies. The target contaminants are mainly heavy metals such as Cr, Pb, etc., and metalloids such as As [60]; in some cases, nitrates [59, 61] and dyes [62] are also removed by nZVI. Their removal efficiency was significantly improved by the simple mixing of the nZVI or the NH nZVI with the wastewater in either sequential batch reactors or continuous flow stirred tank reactors. The key removal mechanisms are adsorption and chemical reduction. Moreover, nZVI-based materials offer the unique opportunities for magnetic separation for adsorbed contaminants and the recycling of the nanoadsorbent. Figure 4.3(a–d) shows a setup for pilot-scale plant for treating Cu(II)-contaminated wastewater with nZVI. The pilot plant consisted of three tanks, a reactor for nZVI, a clarifier, and a coagulation–sedimentation tank. Greater than 96% Cu(II) removal efficiency was achieved with nZVI dosage of 0.20 g/L in the pilot plant [59].

Hybridized nZVI, for example BM-nZVI composed of Ni/Fe, was introduced to the waste metal working fluids, one of the recalcitrant wastewater, to facilitate the advanced chemical oxidation by providing reactive oxygen species and to further advance the biological treatment [57]. On the other hand, Ni/Fe BM-nZVIs have shown to improve hexavalent chromium reduction due to synergistic iron passivation and electron transfer by the Ni [63]. Though CNT- or rGO-nZVIs have not been tested in pilot- or field-scale industrial wastewater treatment facilities, bench-scale studies [64, 65] have shown significant promises for heavy metal removal. With the combination of carbon nanomaterials, the nZVI surface is able to remain as the active sites for the heavy metal removal due to the formation of small electrodes (cathode: carbon, anode: iron) preventing the oxide filming on the nZVI surface. Furthermore, the aggregation of nZVI can be prevented by the immobilization of the material onto another surface in the case of hybrid nZVI, which further enhances its capacity as a reactive agent especially for heavy metal removal.

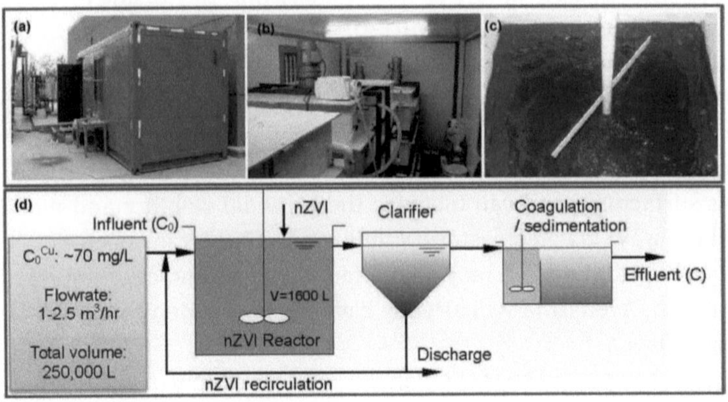


**Figure 4.3** Pilot plant for Cu(II)-contaminated wastewater treatment (a–c). The reactors were placed inside a shipping container (a); nZVI reactor (b); nZVI suspension in the reactor (c); scheme of the treatment steps of the pilot plant (d). Reproduced from Ref. [58] with permission of The Royal Society of Chemistry.

## 4.5 Environmental Fate and Transport of nZVI and Its Nanohybrids

The efficacy of practically applying nZVIs and their NHs in wastewater treatment and groundwater remediation depends on their aggregation behavior in aquatic media and transport properties through porous media. Due to their high vdW and magnetic attraction forces, nZVI tends to aggregate fast and, therefore, also can get filtered out losing the delivery efficiency for groundwater injection wells. In the case of most applications, different surface coating (hydrophilic biopolymers [66], polyelectrolytes [67], amphyphyles [68], emulsions [69], etc.) are used with nZVI, which also results in reduced tendency of attachment and higher mobility through porous soil grains when exposed to groundwater environment [70]. Various environmental parameters such as dissolved oxygen, ionic strength, organic matter, pH, and hydraulic conductivity also play a significant role in controlling fate and transport of nZVI [9, 71]. For example, higher dissolved oxygen in water can cause higher rate of chemical oxidation of nZVI; the iron oxide layer will become hydrophilic, and with hydroxylation of the surface, this can result in reduced agglomeration and deposition [70]. BM-nZVIs such as Fe/Pd or Fe/Ni also show similar results in terms of aggregation and porous media transport and, thus, are required to undergo surface modification to attain better delivery to the target pollution location. Conversely, natural organic matter present in the aquatic environment can act as surfactants and can increase the colloidal stability and mobility of the nZVI via electrosteric stabilization [72]. Transformation such as aging of the nZVI may also affect the environmental interaction of nZVI. Aged nZVI will already have higher portion of the $Fe^{2+}/Fe^{3+}$ oxide layer.

It is expected that the nanoscale solid supports, including CNT and rGO, will offer unique physicochemical properties to alter nZVI aggregation and transport behavior. In the case of nC-nZVI hybrids, the surface interactions would be quite complex; the colloidal interactions might be predominantly influenced by the larger surface presence of CNT or rGO rather than the nZVI. The intercalating effects might help to reduce the influences of the magnetic forces and reduce aggregation of these NHs and thereby influence their porous media transport. However, so far, no systematic studies have been performed to investigate such behavior in detail and the potential fate and transport of nC-nZVI hybrids are thus difficult to predict based on the available literature. It is important to note that nZVI does not share the similar mechanisms for aggregation and deposition as CNTs or rGOs. While the extended Derjaguin, Landau, Verwey, and Overbeek (DLVO) theory can successfully explain the interfacial interactions of quasi-spherical nZVI by including electrostatic/steric, magnetic, and vdW forces; the type of behavior of tubular CNTs and planar rGOs cannot be explained by DLVO due to the existence of non-DLVO forces (e.g., hydrophobic) in their interfacial interactions. Thus, the hybrid aggregation and transport behavior might as well be rather more complex than usual. Systematic studies need to be performed to determine how, when, and which altered or emerging physicochemical properties will influence the behaviors and what fundamental parameters need to be measured for understanding their interfacial interactions.

## 4.6 Environmental Toxicity of nZVI and Its Nanohybrids

Surface-modified and hybridized nZVIs can cause significant ecotoxic response. Although there is very limited toxicological data available for hybridized nZVI, toxicological studies done on bare nZVI have reported toxic response on biological species by affecting cellular activity, viability, and integrity [70]. A large share of the toxicological studies has focused on the toxic effects of nZVI on different microbial species (e.g., bacteria [73, 74], virus [75], fungi [76, 77], etc.). However, certain studies have also revealed adverse effect of nZVI on algae [78], soil and aquatic invertebrates [79, 80], and fish [81]. Structural and chemical transformation of nZVI in the ecological and biological media (via Fenton reactions) leads to the production of dissolved iron species ($Fe^{2+}$) or ROS (reactive oxygen species), presenting a concern due to their harmful interactions with environmental microorganisms and other organism cells [82]. Recent extensive studies with aquatic microorganisms identified $Fe^{2+}$ and ROS generation as the major contributors to nZVI cytotoxic effects [9, 71]. Cell membrane integrity disruption by the highly reductive nZVI direct interaction with membrane molecules leads to the release of $Fe^{2+}$ ions inside the cell, which then reacts with mitochondrial $H_2O_2$ to produce more unstable ROS components. Besides, direct interactions or internalization of nZVI in the cell can lead to ROS generation. In both cases, produced ROS can react with cell organelles and cause oxidative stress, leading to cell death. A limited number of nZVI cytotoxicity studies with bronchial epithelial cells also indicated production of $Fe^{2+}$ and ROS and their potential cytotoxic effects; however, interaction mechanisms at cellular and tissue level have not been reported [83, 84]. Toxicity mechanisms imply that direct contact between nZVI and microorganisms is a prerequisite for cellular redox reactions and subsequent generation of ROS. Surface coatings on nZVI inhibit direct interaction with microorganisms, increasing mobility, which suppresses the toxic effect of nZVI [85]. A recent study reported that with an increasing dose of polyelectrolyte carboxymethyl cellulose (CMC) coating on nZVI, the cytotoxic response of nZVI to *Escherichia coli* is significantly reduced. Above a certain dose of coating (1.6 wt% of nZVI), no cytotoxic effect was observed [86]. Similarly, the presence of natural organic matter (NOM) in water also can result in reduced antimicrobial effect of nZVI due to hindrance against the direct contact caused by electrosteric repulsion [87].

There is a dearth of knowledge about the unique potential of ecotoxicity that nZVI can exert when conjugated with other nanomaterials. Limited studies done on nZVI-based bimetallic NHs (BM-nZVI) have shown variable antibacterial effects toward *E. coli* based on the other conjugating metal [88]. For example, Fe/Cu BM NH caused almost the double cell death of *E. coli* compared to Fe/Pd, with the same dosage [88]. The possible mechanism was postulated to be the direct nanoparticle–cell interaction followed by the cell membrane damage and organismal dysfunction. However, this is a preliminary study without any considerations for influence of various environmental and material parameters, thus requiring further investigations. On the other hand, there is an absence of understanding regarding the toxicological effects of nC-nZVI hybrids, i.e., CNT-nZVI or rGO-nZVI or in general any carbon-metallic NHs. As discussed earlier, nZVI conjugation to CNT or rGO significantly changes the physicochemical properties of nZVI, including their particle size, surface area, aggregation propensity, ROS generation capability, dissolution behavior, and redox potential, which might affect their cellular interactions. Moreover, the large surface presence of tubular CNT or sheet-like rGO may dominate the physical interactions with cells. High aspect ratio, needle-like, and stiff CNTs are known to physically cross the cell membranes via diffusion, membrane damage, or direct pore transport and also delineate asbestos-like toxicity behavior [89]. Two-dimensional graphene, though more flexible than CNT, also physically perturbs the cell membrane [90, 91]. Furthermore, both CNT and graphene have exhibited ROS generation ability due to their excellent electronic properties and low-to-zero bandgap. Their capacity to conduct electrons at high speed can help mediate electron transfer from intracellular components of a cell to the extracellular environments resulting in the generation of ROS. These ROS can cause lipid peroxidation on the cell membrane or oxidative damages to intracellular organelles, including nucleic acids. The interactions of CNT and graphene with cell membrane and subsequent internalization, distribution, and toxicity are also modulated by various material properties, including length and diameter (CNT), lateral sizes and number of layers (graphene or rGO), mechanical stiffness, hydrophobicity, surface chemistry or coatings, presence and dissolution of metal impurities (e.g., iron), aggregation state, etc. Upon deposition of nZVI on CNT or graphene, many of these properties are likely to be altered (as mentioned above) and thus will likely have significantly altered toxicological implications and necessitate immediate attention.

## 4.7 Conclusion

While novel nanoscale hybridization approaches to nZVI-based treatment technologies promise more efficient pollutant degradation, their unprecedented environmental implications need to be investigated and underlying mechanisms need to be deciphered. Since, there is a high likelihood of nZVI-based NHs to be applied in large scale in the environmental remediation and thereby they will directly be exposed to the ecosystem, their potential environmental implications need to be determined along with mechanistic understanding. The research community that works on environmental health and safety of engineered nanomaterials needs to be proactive to determine both the application and implication potentials of these novel nanostructures (i.e., nZVI-based NHs) to ensure their sustainable design and usage.

## References


1. Karn, B., Kuiken, T., and Otto, M. (2009). Nanotechnology and in situ remediation: A review of the benefits and potential risks, *Environ. Health Perspect.*, **117**, 12, pp. 1823–1831.
2. O'Carroll, D., Sleep, B., Krol, M., Boparai, H., and Kocur, C. (2013). Nanoscale zero valent iron and bimetallic particles for contaminated site remediation, *Adv. Water Resour.*, **51**, pp. 104–122.
3. Bardos, P., Bone, B., Černík, M., Elliott, D. W., Jones, S., and Merly, C. (2015). Nanoremediation and international environmental restoration markets, *Remediation J.*, **25**, 2, pp. 83–94.
4. EEB, D. A., "Report: Soil: worth standing your ground for. Arguments for the Soil Framework Directive.," Ed., DNR (German League for Nature and Environment) and EEB (European Environmental Bureau), 2011.

5. Bardos, P., Bone, B., Daly, P., Elliott, D., Jones, S., Lowry, G., and Merly, C., "Report: A risk/benefit appraisal for the application of nano-scale zero valent iron (nZVI) for the remediation of contaminated sites," in *Taking Nanotechnological Remediation Processes from Lab Scale to End User Applications for the Restoration of a Clean Environment*, Ed., Nanotechnology for Contaminated Land Remediation (NanoREM), 2014.
6. Li, S., Wang, W., Liu, Y., and Zhang, W.-X. (2014). Zero-valent iron nanoparticles (nZVI) for the treatment of smelting wastewater: A pilot-scale demonstration, *Chem. Eng. J.*, **254**, pp. 115–123.
7. Li, S., Wang, W., Yan, W., and Zhang, W.-X. (2014). Nanoscale zero-valent iron (nZVI) for the treatment of concentrated Cu(II) wastewater: A field demonstration, *Env. Sci. Process. Impacts*, **16**, 3, pp. 524–533.
8. Wang, W., Cheng, Y., Kong, T., and Cheng, G. (2015). Iron nanoparticles decoration onto three-dimensional graphene for rapid and efficient degradation of azo dye, *J. Hazard. Mater.*, **299**, pp. 50–58.
9. Yirsaw, B. D., Megharaj, M., Chen, Z., and Naidu, R. (2016). Environmental application and ecological significance of nano-zero valent iron, *J. Environ. Sci.*, **44**, pp. 88–98.
10. Aich, N., Plazas-Tuttle, J., Lead, J. R., and Saleh, N. B. (2014). A critical review of nanohybrids: Synthesis, applications and environmental implications, *Environ. Chem.*, **11**, 6, pp. 609–623.
11. Zou, Y., Wang, X., Khan, A., Wang, P., Liu, Y., Alsaedi, A., Hayat, T., and Wang, X. (2016). Environmental remediation and application of nanoscale zero-valent iron and its composites for the removal of heavy metal ions: A review, *Environ. Sci. Technol.*, **50**, 14, pp. 7290–7304.
12. Carabante, I., Grahn, M., Holmgren, A., Kumpiene, J., and Hedlund, J. (2009). Adsorption of As(V) on iron oxide nanoparticle films studied by in situ ATR-FTIR spectroscopy, *Colloids Surf., A*, **346**, 1, pp. 106–113.
13. Liu, Q., Bei, Y., and Zhou, F. (2009). Removal of lead(II) from aqueous solution with amino-functionalized nanoscale zero-valent iron, *Cent. Eur. J. Chem.*, **7**, 1, pp. 79–82.
14. Xu, P., Zeng, G. M., Huang, D. L., Feng, C. L., Hu, S., Zhao, M. H., Lai, C., Wei, Z., Huang, C., and Xie, G. X. (2012). Use of iron oxide nanomaterials in wastewater treatment: A review, *Sci. Total. Environ.*, **424**, pp. 1–10.
15. Gunawardana, B., Singhal, N., and Swedlund, P. (2011). Degradation of chlorinated phenols by zero valent iron and bimetals of iron: A review, *Environ. Eng. Res.*, **16**, 4, pp. 187–203.
16. Schrick, B., Blough, J. L., Jones, A. D., and Mallouk, T. E. (2002). Hydrodechlorination of trichloroethylene to hydrocarbons using

bimetallic nickel-iron nanoparticles, *Chem. Mater.*, **14**, 12, pp. 5140–5147.

17. Choi, H., Al-Abed, S. R., Agarwal, S., and Dionysiou, D. D. (2008). Synthesis of reactive nano-Fe/Pd bimetallic system-impregnated activated carbon for the simultaneous adsorption and dechlorination of PCBs, *Chem. Mater.*, **20**, 11, pp. 3649–3655.
18. Wang, C., Luo, H., Zhang, Z., Wu, Y., Zhang, J., and Chen, S. (2014). Removal of As(III) and As(V) from aqueous solutions using nanoscale zero valent iron-reduced graphite oxide modified composites, *J. Hazard. Mater.*, **268**, pp. 124–131.
19. Fan, M., Li, T., Hu, J., Cao, R., Wu, Q., Wei, X., Li, L., Shi, X., and Ruan, W. (2016). Synthesis and characterization of reduced graphene oxide-supported nanoscale zero-valent iron (nZVI/rGO) composites used for Pb(II) removal, *Materials*, **9**, 8, pp. 687–707.
20. Cheng, J., "Thesis: Carbon-nZVI nanocomposites for dechlorination of halogenated hydrocarbons," Ed., The University of Western Ontario, 2013.
21. Jiao, W., Feng, Z., Liu, Y., and Jiang, H. (2016). Degradation of nitrobenzene-containing wastewater by carbon nanotubes immobilized nanoscale zerovalent iron, *J. Nanopart. Res.*, **18**, 7, pp. 198–206.
22. Sheng, G., Alsaedi, A., Shammakh, W., Monaquel, S., Sheng, J., Wang, X., Li, H., and Huang, Y. (2016). Enhanced sequestration of selenite in water by nanoscale zero valent iron immobilization on carbon nanotubes by a combined batch, XPS and XAFS investigation, *Carbon*, **99**, pp. 123–130.
23. Reza Sohrabi, M., Mansouriieh, N., Khosravi, M., and Zolghadr, M. (2015). Removal of diazo dye Direct Red 23 from aqueous solution using zero-valent iron nanoparticles immobilized on multi-walled carbon nanotubes, *Wat. Sci. Tech.*, **71**, 9, pp. 1367–1374.
24. Guo, J., Wang, R., Tjiu, W. W., Pan, J., and Liu, T. (2012). Synthesis of Fe nanoparticles@graphene composites for environmental applications, *J. Hazard. Mater.*, **225–226**, pp. 63–73.
25. Jabeen, H., Kemp, K. C., and Chandra, V. (2013). Synthesis of nano zerovalent iron nanoparticles–graphene composite for the treatment of lead contaminated water, *J. Environ. Management*, **130**, pp. 429–435.
26. Sun, Y., Ding, C., Cheng, W., and Wang, X. (2014). Simultaneous adsorption and reduction of U(VI) on reduced graphene oxide-supported nanoscale zerovalent iron, *J. Hazard. Mater.*, **280**, pp. 399–408.

27. Mishra, S., Dwivedi, J., Kumar, A., and Sankararamakrishnan, N. (2016). Removal of antimonite (Sb(III)) and antimonate (Sb(V)) using zerovalent iron decorated functionalized carbon nanotubes, *RSC Adv.*, **6**, 98, pp. 95865–95878.

28. Shen, L., Zhang, X., Li, H., Yuan, C., and Cao, G. (2011). Design and tailoring of a three-dimensional $TiO_2$–graphene–carbon nanotube nanocomposite for fast lithium storage, *J. Phys. Chem. Lett.*, **2**, 24, pp. 3096–3101.

29. Wang, C., Cao, M., Wang, P., Ao, Y., Hou, J., and Qian, J. (2014). Preparation of graphene–carbon nanotube–$TiO_2$ composites with enhanced photocatalytic activity for the removal of dye and Cr (VI), *Appl. Catal., A*, **473**, pp. 83–89.

30. Vadahanambi, S., Lee, S.-H., Kim, W.-J., and Oh, I.-K. (2013). Arsenic removal from contaminated water using three-dimensional graphene-carbon nanotube-iron oxide nanostructures, *Environ. Sci. Technol.*, **47**, 18, pp. 10510–10517.

31. Lee, S.-H., Sridhar, V., Jung, J.-H., Karthikeyan, K., Lee, Y.-S., Mukherjee, R., Koratkar, N., and Oh, I.-K. (2013). Graphene–nanotube–iron hierarchical nanostructure as lithium ion battery anode, *ACS Nano*, **7**, 5, pp. 4242–4251.

32. Jung, C., Heo, J., Han, J., Her, N., Lee, S.-J., Oh, J., Ryu, J., and Yoon, Y. (2013). Hexavalent chromium removal by various adsorbents: Powdered activated carbon, chitosan, and single/multi-walled carbon nanotubes, *Sep. Purif. Technol.*, **106**, pp. 63–71.

33. Lv, X., Xu, J., Jiang, G., and Xu, X. (2011). Removal of chromium (VI) from wastewater by nanoscale zero-valent iron particles supported on multiwalled carbon nanotubes, *Chemosphere*, **85**, 7, pp. 1204–1209.

34. Kumar, A. S. K., Jiang, S.-J., and Tseng, W.-L. (2015). Effective adsorption of chromium(VI)/Cr(III) from aqueous solution using ionic liquid functionalized multiwalled carbon nanotubes as a super sorbent, *J. Mater. Chem. A*, **3**, 13, pp. 7044–7057.

35. Santhana Krishna Kumar, A. and Rajesh, N. (2013). Exploring the interesting interaction between graphene oxide, Aliquat-336 (a room temperature ionic liquid) and chromium(VI) for wastewater treatment, *RSC Adv.*, **3**, pp. 2697–2709.

36. Pan, B. and Xing, B. (2008). Adsorption mechanisms of organic chemicals on carbon nanotubes, *Environ. Sci. Technol.*, **42**, 24, pp. 9005–9013.

37. Sun, Y., Ding, C., Cheng, W., and Wang, X. (2014). Simultaneous adsorption and reduction of U(VI) on reduced graphene oxide-

supported nanoscale zerovalent iron, *J. Hazard. Mater.*, **280**, pp. 399–408.

38. Yang, X., Chen, C., Li, J., Zhao, G., Ren, X., and Wang, X. (2012). Graphene oxide-iron oxide and reduced graphene oxide-iron oxide hybrid materials for the removal of organic and inorganic pollutants, *RSC Advances*, **2**, 23, pp. 8821–8826.

39. Wang, C., Luo, H., Zhang, Z., Wu, Y., Zhang, J., and Chen, S. (2014). Removal of As(III) and As(V) from aqueous solutions using nanoscale zero valent iron-reduced graphite oxide modified composites, *J. Hazard. Mater.*, **268**, pp. 124–131.

40. Li, Z.-J., Wang, L., Yuan, L.-Y., Xiao, C.-L., Mei, L., Zheng, L.-R., Zhang, J., Yang, J.-H., Zhao, Y.-L., and Zhu, Z.-T. (2015). Efficient removal of uranium from aqueous solution by zero-valent iron nanoparticle and its graphene composite, *J. Hazard. Mater.*, **290**, pp. 26–33.

41. Jabeen, H., Chandra, V., Jung, S., Lee, J. W., Kim, K. S., and Kim, S. B. (2011). Enhanced Cr(VI) removal using iron nanoparticle decorated graphene, *Nanoscale*, **3**, 9, pp. 3583–3585.

42. Li, J., Chen, C., Zhang, R., and Wang, X. (2016). Reductive immobilization of Re(VII) by graphene modified nanoscale zero-valent iron particles using a plasma technique, *Sci. China Chem.*, **59**, 1, pp. 150–158.

43. Chen, H., Cao, Y., Wei, E., Gong, T., and Xian, Q. (2016). Facile synthesis of graphene nano zero-valent iron composites and their efficient removal of trichloronitromethane from drinking water, *Chemosphere*, **146**, pp. 32–39.

44. Guo, J., Wang, R., Tjiu, W. W., Pan, J., and Liu, T. (2012). Synthesis of Fe nanoparticles@graphene composites for environmental applications, *J. Hazard. Mater.*, **225**, pp. 63–73.

45. Liu, F., Yang, J., Zuo, J., Ma, D., Gan, L., Xie, B., Wang, P., and Yang, B. (2014). Graphene-supported nanoscale zero-valent iron: Removal of phosphorus from aqueous solution and mechanistic study, *J. Environ. Sci.*, **26**, 8, pp. 1751–1762.

46. Farooq, U., Danish, M., Lu, S., Naqvi, M., Gu, X., Fu, X., Zhang, X., and Nasir, M. (2016). Synthesis of nZVI@ reduced graphene oxide: An efficient catalyst for degradation of 1,1,1-trichloroethane (TCA) in percarbonate system, *Res. Chem. Intermediat.*, **23**, pp. 1–18.

47. Wu, T., Cai, X., Tan, S., Li, H., Liu, J., and Yang, W. (2011). Adsorption characteristics of acrylonitrile, p-toluenesulfonic acid, 1-naphthalenesulfonic acid and methyl blue on graphene in aqueous solutions, *Chem. Eng. J.*, **173**, 1, pp. 144–149.

48. Xu, J., Lv, X., Li, J., Li, Y., Shen, L., Zhou, H., and Xu, X. (2012). Simultaneous adsorption and dechlorination of 2,4-dichlorophenol by Pd/Fe nanoparticles with multi-walled carbon nanotube support, *J. Hazard. Mater.*, **225**, pp. 36–45.

49. Chen, J., Qiu, X., Fang, Z., Yang, M., Pokeung, T., Gu, F., Cheng, W., and Lan, B. (2012). Removal mechanism of antibiotic metronidazole from aquatic solutions by using nanoscale zero-valent iron particles, *Chem. Eng. J.*, **181**, pp. 113–119.

50. Elliott, D. W. and Zhang, W.-X. (2001). Field assessment of nanoscale bimetallic particles for groundwater treatment, *Environ. Sci. Technol.*, **35**, 24, pp. 4922–4926.

51. Zhang, W.-X. (2003). Nanoscale iron particles for environmental remediation: An overview, *J. Nanopart. Res.*, **5**, 3–4, pp. 323–332.

52. Gavaskar, A., Tatar, L., and Condit, W., "Report: Cost and performance report nanoscale zero-valent iron technologies for source remediation," Ed., DTIC Document, 2005.

53. Bennett, P., He, F., Zhao, D., Aiken, B., and Feldman, L. (2010). In situ testing of metallic iron nanoparticle mobility and reactivity in a shallow granular aquifer, *J. Contam. Hydrol.*, **116**, 1, pp. 35–46.

54. He, F., Zhao, D., and Paul, C. (2010). Field assessment of carboxymethyl cellulose stabilized iron nanoparticles for in situ destruction of chlorinated solvents in source zones, *Water Res.*, **44**, 7, pp. 2360–2370.

55. Zhang, M., "Thesis: In situ degradation of trichloroethylene in soil and groundwater with stabilized zero valent iron nanoparticles and catalytic hydrodechlorination with supported palladium nanoparticles," Ed., Auburn University, 2012.

56. Comba, S., Di Molfetta, A., and Sethi, R. (2011). A comparison between field applications of nano-, micro-, and millimetric zero-valent iron for the remediation of contaminated aquifers, *Water, Air, & Soil Pollution*, **215**, 1–4, pp. 595–607.

57. Jagadevan, S., Jayamurthy, M., Dobson, P., and Thompson, I. P. (2012). A novel hybrid nano zerovalent iron initiated oxidation–biological degradation approach for remediation of recalcitrant waste metalworking fluids, *Water Res.*, **46**, 7, pp. 2395–2404.

58. Li, S., Wang, W., Yan, W., and Zhang, W.-X. (2014). Nanoscale zero-valent iron (nZVI) for the treatment of concentrated Cu (II) wastewater: A field demonstration, *Environ. Sci. Process Impacts.*, **16**, 3, pp. 524–533.

59. Huang, Y. H., Peddi, P. K., Tang, C., Zeng, H., and Teng, X. (2013). Hybrid zero-valent iron process for removing heavy metals and nitrate from

flue-gas-desulfurization wastewater, *Sep. Purif. Technol.*, **118**, pp. 690–698.

60. Li, S., Wang, W., Liang, F., and Zhang, W.-X. (2016). Heavy metal removal using nanoscale zero-valent iron (nZVI): Theory and application, *J. Harzard. Mater.*, **322**, pp. 163–171.

61. Kassaee, M., Motamedi, E., Mikhak, A., and Rahnemaie, R. (2011). Nitrate removal from water using iron nanoparticles produced by arc discharge vs. reduction, *Chem. Eng. J.*, **166**, 2, pp. 490–495.

62. Fu, F., Dionysiou, D. D., and Liu, H. (2014). The use of zero-valent iron for groundwater remediation and wastewater treatment: A review, *J. Hazard. Mater.*, **267**, pp. 194–205.

63. Zhou, S., Li, Y., Chen, J., Liu, Z., Wang, Z., and Na, P. (2014). Enhanced Cr(VI) removal from aqueous solutions using Ni/Fe bimetallic nanoparticles: Characterization, kinetics and mechanism, *RSC Adv.*, **4**, 92, pp. 50699–50707.

64. Bystrzejewski, M., Pyrzyńska, K., Huczko, A., and Lange, H. (2009). Carbon-encapsulated magnetic nanoparticles as separable and mobile sorbents of heavy metal ions from aqueous solutions, *Carbon*, **47**, 4, pp. 1201–1204.

65. Zhang, W., Shi, X., Zhang, Y., Gu, W., Li, B., and Xian, Y. (2013). Synthesis of water-soluble magnetic graphene nanocomposites for recyclable removal of heavy metal ions, *J. Mater. Chem. A*, **1**, 5, pp. 1745–1753.

66. He, F., Zhao, D., Liu, J., and Roberts, C. B. (2007). Stabilization of Fe-Pd nanoparticles with sodium carboxymethyl cellulose for enhanced transport and dechlorination of trichloroethylene in soil and groundwater, *Ind. Eng. Chem. Res.*, **46**, 1, pp. 29–34.

67. Kanel, S., Goswami, R., Clement, T., Barnett, M., and Zhao, D. (2007). Two-dimensional transport characteristics of surface stabilized zero-valent iron nanoparticles in porous media, *Environ. Sci. Technol.*, **42**, 3, pp. 896–900.

68. Hydutsky, B. W., Mack, E. J., Beckerman, B. B., Skluzacek, J. M., and Mallouk, T. E. (2007). Optimization of nano- and microiron transport through sand columns using polyelectrolyte mixtures, *Environ. Technol.*, **41**, 18, pp. 6418–6424.

69. Quinn, J., Geiger, C., Clausen, C., Brooks, K., Coon, C., O'Hara, S., Krug, T., Major, D., Yoon, W.-S., and Gavaskar, A. (2005). Field demonstration of DNAPL dehalogenation using emulsified zero-valent iron, *Environ. Sci. Technol.*, **39**, 5, pp. 1309–1318.

70. Jang, M.-H., Lim, M., and Hwang, Y. S. (2014). Potential environmental implications of nanoscale zero-valent iron particles for environmental remediation, *Environ. Health Toxicol.*, **29**, pp. e2014022.
71. Lefevre, E., Bossa, N., Wiesner, M. R., and Gunsch, C. K. (2016). A review of the environmental implications of in situ remediation by nanoscale zero valent iron (nZVI): Behavior, transport and impacts on microbial communities, *Sci. Total Environ.*, **565**, pp. 889–901.
72. Chen, J. W., Xiu, Z. M., Lowry, G. V., and Alvarez, P. J. J. (2011). Effect of natural organic matter on toxicity and reactivity of nano-scale zero-valent iron, *Water Res.*, **45**, 5, pp. 1995–2001.
73. Chaithawiwat, K., Vangnai, A., McEvoy, J. M., Pruess, B., Krajangpan, S., and Khan, E. (2016). Impact of nanoscale zero valent iron on bacteria is growth phase dependent, *Chemosphere*, **144**, pp. 352–359.
74. Kim, J. Y., Park, H.-J., Lee, C., Nelson, K. L., Sedlak, D. L., and Yoon, J. (2010). Inactivation of *Escherichia coli* by nanoparticulate zerovalent iron and ferrous ion, *Appl. Environ. Microbiol.*, **76**, 22, pp. 7668–7670.
75. You, Y., Han, J., Chiu, P. C., and Jin, Y. (2005). Removal and inactivation of waterborne viruses using zerovalent iron, *Environ. Sci. Technol.*, **39**, 23, pp. 9263–9269.
76. Otero-González, L., García-Saucedo, C., Field, J. A., and Sierra-Álvarez, R. (2013). Toxicity of $TiO_2$, $ZrO_2$, Fe0, $Fe_2O_3$, and $Mn_2O_3$ nanoparticles to the yeast, *Saccharomyces cerevisiae*, *Chemosphere*, **93**, 6, pp. 1201–1206.
77. Shah, V., Dobiášová, P., Baldrian, P., Nerud, F., Kumar, A., and Seal, S. (2010). Influence of iron and copper nanoparticle powder on the production of lignocellulose degrading enzymes in the fungus *Trametes versicolor*, *J. Hazard. Mater.*, **178**, 1–3, pp. 1141–1145.
78. Kadar, E., Rooks, P., Lakey, C., and White, D. A. (2012). The effect of engineered iron nanoparticles on growth and metabolic status of marine microalgae cultures, *Sci. Total Environ.*, **439**, pp. 8–17.
79. El-Temsah, Y. S. and Joner, E. J. (2012). Ecotoxicological effects on earthworms of fresh and aged nano-sized zero-valent iron (nZVI) in soil, *Chemosphere*, **89**, 1, pp. 76–82.
80. Sacca, M. L., Fajardo, C., Costa, G., Lobo, C., Nande, M., and Martin, M. (2014). Integrating classical and molecular approaches to evaluate the impact of nanosized zero-valent iron (nZVI) on soil organisms, *Chemosphere*, **104**, pp. 184–189.
81. Li, H., Zhou, Q., Wu, Y., Fu, J., Wang, T., and Jiang, G. (2009). Effects of waterborne nano-iron on medaka (*Oryzias latipes*): Antioxidant

enzymatic activity, lipid peroxidation and histopathology, *Ecotoxicol. Environ. Saf.*, **72**, 3, pp. 684–692.

82. Stefaniuk, M., Oleszczuk, P., and Ok, Y. S. (2016). Review on nano zerovalent iron (nZVI): From synthesis to environmental applications, *Chem. Eng. J.*, **287**, pp. 618–632.

83. Sun, Z., Yang, L., Chen, K.-F., Chen, G.-W., Peng, Y.-P., Chen, J.-K., Suo, G., Yu, J., Wang, W.-C., and Lin, C.-H. (2016). Nano zerovalent iron particles induce pulmonary and cardiovascular toxicity in an in vitro human co-culture model, *Nanotoxicology*, **10**, 7, pp. 881–890.

84. Keenan, C. R., Goth-Goldstein, R., Lucas, D., and Sedlak, D. L. (2009). Oxidative stress induced by zero-valent iron nanoparticles and Fe(II) in human bronchial epithelial cells, *Environ. Sci. Technol.*, **43**, 12, pp. 4555–4560.

85. Zhao, X., Liu, W., Cai, Z. Q., Han, B., Qian, T. W., and Zhao, D. Y. (2016). An overview of preparation and applications of stabilized zero-valent iron nanoparticles for soil and groundwater remediation, *Water Res.*, **100**, pp. 245–266.

86. Dong, H. R., Xie, Y. K., Zeng, G. M., Tang, L., Liang, J., He, Q., Zhao, F., Zeng, Y. L., and Wu, Y. A. (2016). The dual effects of carboxymethyl cellulose on the colloidal stability and toxicity of nanoscale zero-valent iron, *Chemosphere*, **144**, pp. 1682–1689.

87. Chen, J., Xiu, Z., Lowry, G. V., and Alvarez, P. J. J. (2011). Effect of natural organic matter on toxicity and reactivity of nano-scale zero-valent iron, *Water Res.*, **45**, 5, pp. 1995–2001.

88. Kim, E.-J., Le Thanh, T., and Chang, Y.-S. (2014). Comparative toxicity of bimetallic Fe nanoparticles toward *Escherichia coli*: Mechanism and environmental implications, *Environ. Sci. Nano*, **1**, 3, pp. 233–237.

89. Donaldson, K., Poland, C. A., Murphy, F. A., MacFarlane, M., Chernova, T., and Schinwald, A. (2013). Pulmonary toxicity of carbon nanotubes and asbestos: Similarities and differences, *Adv. Drug Deliv. Rev.*, **65**, 15, pp. 2078–2086.

90. Sanchez, V. C., Jachak, A., Hurt, R. H., and Kane, A. B. (2012). Biological interactions of graphene-family nanomaterials: An interdisciplinary review, *Chem. Res. Toxicol.*, **25**, 1, pp. 15–34.

91. Wang, Z., Zhu, W., Qiu, Y., Yi, X., von dem Bussche, A., Kane, A., Gao, H., Koski, K., and Hurt, R. (2016). Biological and environmental interactions of emerging two-dimensional nanomaterials, *Chem. Soc. Rev.*, **45**, 6, pp. 1750–1780.

Chapter 5

# An Integrated Experimental and Modeling Approach to Assess the Mobility of Iron-Based Nanoparticles in Groundwater Systems

**Tiziana Tosco, Carlo Bianco, and Rajandrea Sethi**
*DIATI, Department of Environment, Land and Infrastructure Engineering, Politecnico di Torino, C.so Duca degli Abruzzi 24, 10129 Torino, Italy*
rajandrea.sethi@polito.it

## 5.1 Introduction

Nanoremediation is a promising technology for the remediation of contaminated soils and aquifers. Engineered nanoparticles (NPs) are introduced into the subsurface in the form of a reactive suspension for the in situ degradation, transformation, or immobilization of pollutants [1]. The use of engineered NPs has been extensively studied in recent years as an alternative to more conventional approaches, such as pump and treat (P&T) and permeable reactive barriers (PRBs) [1, 2]. Despite the widespread use of P&T and PRBs, these remediation technologies are often costly and ineffective for

*Iron Nanomaterials for Water and Soil Treatment*
Edited by Marta I. Litter, Natalia Quici, and Martín Meichtry

ISBN 978-981-4774-67-3 (Hardcover), 978-981-4669-49-8 (eBook)
www.panstanford.com

the treatment of deep contaminations and/or in the presence of recalcitrant and weakly water-soluble pollutants. Moreover, P&T and PRBs can be used only for the treatment of the dissolved fraction of the pollutants (plume), while they are not suitable for the direct degradation of the contamination source. NP-based methods aim to overcome many of the drawbacks and limitations of P&T and PRBs. Several engineered NPs have been studied in the last years for groundwater remediation purposes. Even if the use of other materials has been explored, most of the particles that are currently being tested and show a good performance for groundwater remediation are iron-based NPs. Microparticles and NPs of zerovalent iron (respectively, MZVI and nZVI) have been extensively studied for the degradation of many recalcitrant contaminants, such as chlorinated hydrocarbons [2–5], metal ions [6, 7], PAH [8], PCBs, and pesticides [9]. nZVI has been investigated at the laboratory and field scale, both in the form of nZVI particles alone and as composite materials such as CARBO-IRON®, where nZVI is embedded in a carbon matrix to promote mobility and contaminant targeting [10], or as bimetallic particles [11, 12]. Moreover, a particular application of nZVI is represented by the emulsified zerovalent iron (EZVI) [13]. Nanosized iron oxides, such as goethite, have been largely studied because of their high sorption capacity toward heavy metals [14], such as chromium and arsenic. Moreover, ferrihydrite NPs have been employed as electron acceptors to stimulate the microbial assisted degradation of several organic contaminants, for example, BTEX [15].

NPs are usually dispersed in water-based, highly concentrated slurries, and injected directly into and/or in proximity of the core of the polluted zone to treat the plume or the source of contamination. Alternatively, NPs can be injected and immobilized close to the contamination to generate a downstream reactive zone for the treatment of the dissolved phase, similar to a PRB. NP suspensions are typically delivered into the target contaminated area through wells, trenches, or using appropriate tools such as direct push equipment. They can be injected at high pressure, thus producing fracturing injection (when the injection pressure overcomes the critical one of the porous medium) or at lower pressure, resulting in permeation injection [16–18]. If fracturing delivery is used, preferential migration pathways are generated, and the final distribution of

NPs in the subsurface is highly inhomogeneous. On the contrary, if permeation injection is adopted, the resulting distribution of NPs around the injection point can be fairy homogeneous and is controlled more easily. The capability of controlling the NP distribution by tuning the operating conditions during the injection (e.g., NP concentration in the injected slurry, injection rate, and duration) is a key issue in the design of an NP field application. As a consequence, the comprehension of the transport mechanisms and the development of a transport model are crucial to provide an estimate of the expected travel distance and iron distribution.

Great efforts have been devoted so far to the identification of the mechanisms controlling the transport of iron-based NPs: their mobility is determined by the combination of particle–particle and particle–porous medium physicochemical interactions, magnetic attraction among particles, filtration and straining of aggregates, sedimentation, rheological properties of the dispersant fluid, etc. These processes have been proven to be strongly influenced by both hydrodynamic (e.g., pore–water velocity, injection flow rate) [19] and hydrochemical parameters (e.g., pore–water ionic strength, pH) [20–22]. Several approaches have been proposed to model NP transport and retention in porous media [3, 4, 23–25]. However, in most cases, such models have been developed with the aim of reproducing and explaining NP transport at the laboratory scale, in well-controlled conditions, and are not readily applicable to the field scale. As a consequence, despite the existence of such advanced models, to the authors' knowledge, few simulation tools are available in the literature to simulate NP transport in three dimensions [26–28].

The design of a field-scale injection of engineered NP suspensions requires reliable procedures and approaches to effectively assess the expected NP mobility at the field scale, and for a reliable estimation of several operative parameters, such as particle distribution around the injection well, radius of influence for a target concentration, number of required injection wells, etc. This information can be typically obtained using an experimental approach, running a wide set of column transport tests under all different field-relevant conditions, and inferring the expected mobility at the field scale from the laboratory results. However, this approach may be time consuming and costly and does not guarantee a direct up-scalability to the field, if not supported by modeling. In this chapter, an integrated experimental and modeling approach is proposed for an effective and reliable design of field-scale applications of NPs to groundwater remediation. In the first part of the chapter, the mechanisms controlling NP transport in porous media are discussed. Two numerical tools are presented, which were developed at Politecnico di Torino in the framework of FP7 EU projects Nanorem (G.A. n. 309517) and Aquarehab (G.A. n. 226565), namely MNMs (Micro- and Nano-particle transport, filtration, and clogging Model Suite) and MNM3D (Micro and Nanoparticle transport Model in 3D geometries) [29], respectively, for the simulation of lab-scale and field-scale NP transport. It is then discussed how they can be integrated with laboratory tests and field monitoring for an effective design of an NP-based remediation, and an application of the proposed procedure is presented.

## 5.2 Mechanisms Controlling Particle Transport in Porous Media

When NPs are dispersed and transported within groundwater, they are subjected to processes such as filtration, straining, physical–chemical deposition, and aggregation. The dynamic processes of attachment and detachment are governed by physical laws that operate at the scale of grains and pores, but impact the transport behavior of NPs at the macroscale [30, 31]. Both physical and physicochemical mechanisms contribute to remove NPs from pore water. Deposition due to physical phenomena includes mechanical filtration and straining of single particles or of aggregates. The latter is also partly influenced by physicochemical interactions and can be a reversible process, while mechanical filtration is typically irreversible [2].

Different retention mechanisms may take place (Fig. 5.1) depending on particle–particle and particle–collector interactions [2]:

- **Linear reversible attachment**: When particle–particle and particle–collector interaction energies are similar, the deposition is not limited nor affected anyhow by the amount of deposed particles.

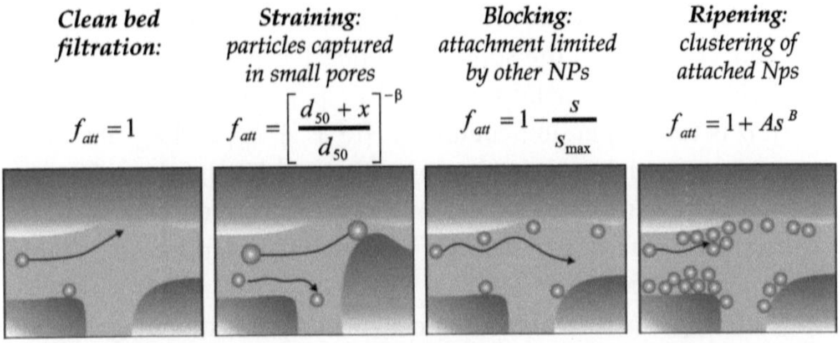


**Figure 5.1** Pore scale particle retention processes; $d_{50}$ is median grain size of the porous medium; $x$ is the distance from the inlet. Modified from Ref. [2], Copyright 2014, with permission from Elsevier.

- **Blocking**: When particle–particle interaction energies are repulsive, deposed particles exclude the immediate vicinity of the collector surface from further NP deposition and a maximum concentration of deposited particles $s_{\max}$ is reached [32].
- **Ripening**: When particle–particle interaction energies are attractive, deposed particles tend to attract the suspended ones, leading to a progressive increase in the attachment rate, until the porous medium is completely clogged. In the case of blocking, the presence of a layer of attached NPs does not affect the pore flow. On the contrary, in the case of ripening, the influence of attached and suspended particles on the fluid and porous medium properties (porosity, conductivity, fluid density, etc.) cannot be neglected, and clogging occurs, leading to a progressive reduction in porosity and permeability [25].

NP transport in porous media can be mathematically described by a modified advection–dispersion equation that takes into account the mass exchanges between liquid and solid phase due to physical and physicochemical interactions. For a one-dimensional problem, we have:

$$\frac{\partial}{\partial t}(\varepsilon c) + \sum_i \left( \rho_b \frac{\partial s_i}{\partial t} \right) + \frac{\partial}{\partial x}(qc) - \frac{\partial^2}{\partial x^2}(\varepsilon Dc) = 0 \tag{5.1}$$

$$\rho_b \frac{\partial s_i}{\partial t} = \varepsilon k_{a,i} f_{\mathrm{att},i} c - \rho_b k_{d,i} s_i \tag{5.2}$$

where $c$ is the NP concentration in the liquid phase [$L^{-3}$], $s$ is the NP concentration in the solid phase [$M^{-1}$], ε is porosity [-], $\rho_b$ is the bulk density of the solid matrix [M $L^{-3}$], $q$ is the Darcy velocity [L $T^{-1}$], $D$ is the dispersion coefficient [$L^2$ $T^{-1}$], $k_{a,1}$ and $k_{d,i}$ are the NP attachment and detachment rate coefficients [$T^{-1}$], $f_{att,i}$ are functions [-] depending on the process(es) being described (Fig. 5.1). The subscript $i$ refers to the $i$-th concurrent retention mechanism (usually, up to two).

The type of interaction mechanisms and the magnitude of attachment and detachment kinetics are strongly influenced by factors that depend on both operative and natural conditions. An NP transport simulation tool effective in assisting the design of a field-scale application of NPs has to take into account these effects in a quantitative, and coupled, way [17]. The following three parameters are of particular importance for iron-based NPs used in groundwater remediation:

- **Flow velocity**: During NP injection through a screened well, flow velocity decreases hyperbolically with increasing distance from the well. A formulation of $k_a$ and $k_d$ as a function of pore-water velocity $v = q/\varepsilon$ was proposed by Tosco et al. [17], based on the single collector removal efficiency $\eta_0$ [33]:

$$k_{a,i}(v)=\frac{3}{2}\frac{(1-\varepsilon)}{\varepsilon d_{50}}C_{a,i}\,\eta_0\,v \tag{5.3}$$

$$k_{d,i}(v)=C_{d,i}\,\mu v \tag{5.4}$$

  where $d_{50}$ is the average grain size of the porous medium, $C_{a,i}$ [-] and $C_{d,i}$ [T $M^{-1}$] are parameters to be determined from fitting of experimental data, and $\mu$ [M $L^{-1}$ $T^{-1}$] is the fluid viscosity. $\eta_0$ can be calculated using different correlations [19, 33–35].

- **Ionic strength**: Ionic strength (IS) significantly affects the particle–particle and particle–porous medium interactions. As a consequence, on a longer time frame, possible changes in the groundwater salt content may lead to immobilization or re-mobilization of the particles, thus affecting their long-term behavior. The semi-empirical constitutive equations proposed by Tosco et al. [22] are:

$$k_{a,i}(c_{\text{salt}}) = \frac{k_{a\infty,i}}{1+\left(\frac{CDC_i}{c_{\text{salt}}}\right)^{\beta_{a,i}}} \tag{5.5}$$

$$k_{d,i}(c_{\text{salt}}) = \frac{k_{d0,i}}{1+\left(\frac{c_{salt}}{CRC_i}\right)^{\beta_{d,i}}} \tag{5.6}$$

$$s_{\max,i}(c_{\text{salt}}) = \gamma_{s,i}\, c_{\text{salt}}^{\beta_{s,i}} \tag{5.7}$$

where the terms $k_{a\infty,i}$, $CDC_i$, $\beta_{a,i}$, $k_{d0,i}$, $CRC_i$, $\beta_{d,i}$, $\gamma_{S,i}$, and $\beta_{S,i}$ are empirical coefficients determined via fitting procedures; $c_{salt}$ is the salt concentration in pore water expressed in mM.

- **Fluid viscosity**: NPs suspended in water in most cases are to be stabilized by adding polymers. Polymers commonly used in field applications (e.g., guar gum, xanthan gum, CMC) have a non-Newtonian behavior (shear thinning, viscosity changes with changing flow velocity) [36, 37]. NP colloidal stability and, consequently, mobility are significantly affected by fluid viscosity, which also affects pore pressure in the aquifer during injection.

In addition, clogging, due to the accumulation of significant amounts of NPs in the pores, is simulated through the reduction in permeability and porosity induced by deposited particles. In the approach proposed by Tosco and Sethi [25], the pressure buildup arising from the particle deposition is calculated using a modified formulation of Darcy's law, taking into account the reduction in porosity, permeability, and the possible non-Newtonian nature of the fluid.

## 5.3 Numerical Tools for NP Transport Modeling

### 5.3.1 MNMs for One-Dimensional Modeling of NP Transport

MNMs is a complete tool for the simulation of particle transport in one-dimensional saturated porous media and for the interpretation of laboratory column transport tests. MNMs represents the evolution of MNM1D [22] and E-MNM1D [25], whose features are integrated here and extended in a user-friendly Matlab-based graphical interface. A comparison with other existing simulation tools for particle transport is reported in Table 5.1. MNMs can be downloaded from the web page of the Groundwater Engineering research group of Politecnico di Torino (http://areeweb.polito.it/ricerca/groundwater/software/MNMs.php). The main features of MNMs, summarized in Figs. 5.2 and 5.3, are:

- **Interaction energy profiles** calculation (Fig. 5.3a), which can be used to estimate particle–particle and particle–collector interaction energy profiles using the DLVO (Derjaguin and Landau, Verwey and Overbeek) and Extended-DLVO (accounting for Born repulsion, steric and magnetic interactions) theory. It is useful to forecast the micro- and nanoparticle behavior in terms of aggregation and mobility.

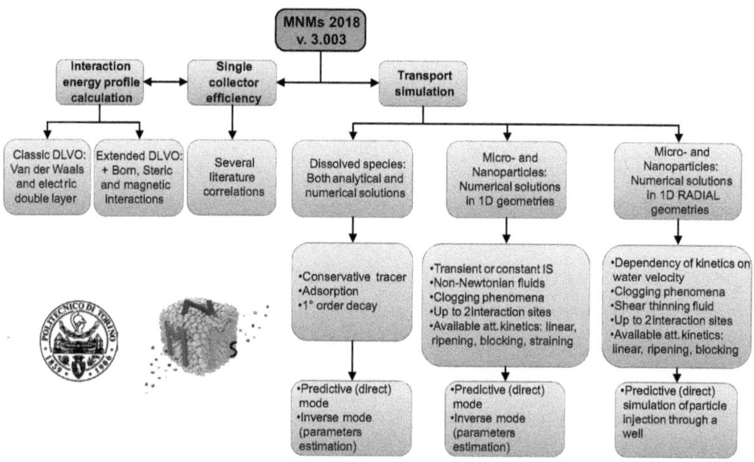


**Figure 5.2** Structure and main features of MNMs.

- Calculation of **single collector attachment efficiency $\eta_0$** (Fig. 5.3b) using up to seven different formulations: $\eta_0$ is one of the most important parameters for assessing the mobility of NPs in porous media.
- Simulation of the **transport of a dissolved species** accounting for equilibrium sorption and first-order degradation. MNMs

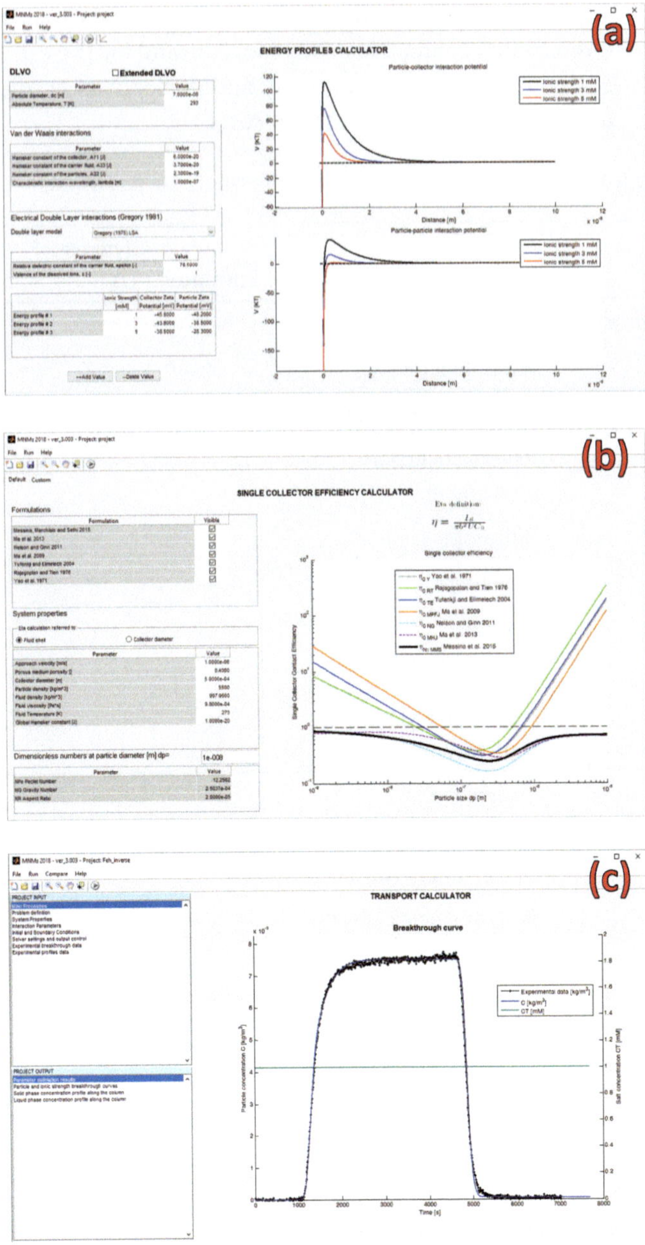


**Figure 5.3** The MNMs graphical interface: (a) DLVO interaction profile calculation; (b) single collector efficiency calculation; (c) particle transport simulation under transient ionic strength conditions.

implements both numerical and analytical solutions in one-dimensional geometries. This tool is useful for the estimation of the porous medium porosity and dispersivity through the interpretation of column tracer tests.

**Table 5.1** Software comparison: Stanmod, Hydrus-1D, and MNMs

| | | **Stanmod** | **Hydrus-1D** | **MNMs** |
|---|---|---|---|---|
| Interaction | Single-site | ✓ | ✓ | ✓ |
| | Multiple-sites | ✓ | ✓ | ✓ |
| Processes | Classical filtration theory | × | ✓ | ✓ |
| | Reversible linear attachment | ✓ | ✓ | ✓ |
| | Blocking | × | ✓ | ✓ |
| | Ripening | × | ✓ | ✓ |
| | Straining | × | ✓ | ✓ |
| | IS dependency | × | × | ✓ |
| | Pore–water velocity dependency | × | × | ✓ |
| Features | Inverse mode | ✓ | ✓ | ✓ |
| | Extended-DLVO interaction profiles | × | × | ✓ |
| | Tool for single collector efficiency | × | × | ✓ |
| | Flow-transport coupling (clogging) | × | × | ✓ |
| | Non-Newtonian carrier fluid | × | × | ✓ |
| | Radial simulation of particle injection | × | × | ✓ |

- Simulation of **particle transport under transient IS** conditions: MNMs solves NP transport equations taking into account the colloids deposition and release during transients of the carrier fluid (water) IS. Four different types of attachment kinetics can be simulated: linear, Langmuirian with blocking, ripening, and straining.

- Simulation of **porous medium clogging**, when the modification of the column porosity and permeability due to the deposition of colloidal particles strongly influences the flow field and cannot be neglected [32]. In this case, MNMs takes into account the variation of pressure drop along the column due to the medium clogging and solve the differential system coupling flow and transport equations.
- Simulation of particle transport in the presence of **non-Newtonian carrier fluids**. Shear-thinning fluids are characterized by high viscosity in static conditions, which hinder NP aggregation and sedimentation during the storage of the suspension, and viscosity close to water in dynamic conditions, thus reducing the injection pressures and promoting particle transport in the aquifer. MNMs can be used to interpret column tests where the influence of the suspension rheology cannot be neglected. A generalized Darcy's law is used to estimate the pressure buildup due to the non-Newtonian nature of the injected fluid, a critical issue for field injections.
- Simulation of **pilot-scale injection of micro- and nanoparticle slurries through a single well** (radial simulation tool). In field applications, fluids are typically injected into the subsurface via wells or direct push systems, generating a radial or radial-like flow, with decreasing velocity with increasing distance from the delivery point [17]. MNMs solves the transport of NPs injected in a homogeneous infinite porous medium assuming a one-dimensional geometry with radial symmetry (Fig. 5.4). The radial tool takes into account the influence of flow velocity (and then of injection rate) on the kinetics of deposition and release processes, the eventual clogging, and the non-Newtonian (shear thinning) rheological properties of the carrier fluid. The radial model is a useful support tool that can be used in the preliminary design of nanoparticle injection.

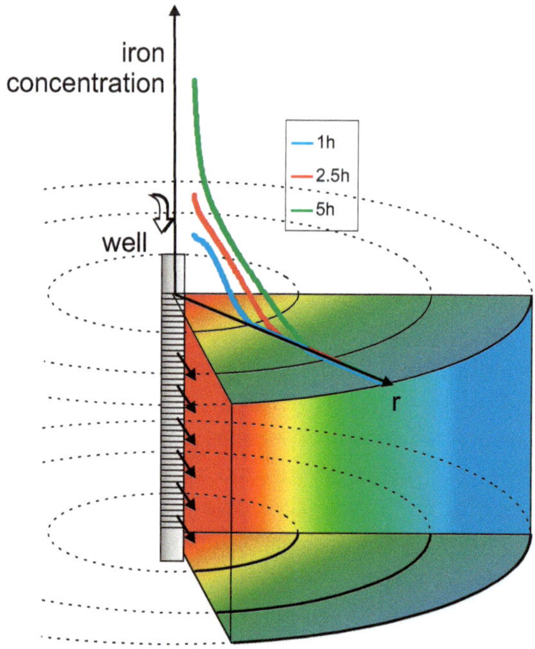


**Figure 5.4** Radial geometry: conceptual model and example of simulation results. Modified from Ref. [17], Copyright 2014, with permission from Elsevier.

Like many transport models, MNMs can be run in two modes:

- **Predictive or forward mode**: Assuming that transport-controlling mechanisms and transport parameters (attachment/detachment and clogging) are known, the model is run once to forecast NP transport.
- **Inverse mode**: Assuming transport mechanisms and determining the transport parameters (e.g., $k_a$ and $k_d$) by fitting experimental data (e.g., breakthrough curves and concentration profiles).

### 5.3.2 MNM3D for Three-Dimensional NP Transport Modeling

MNM3D is a FORTRAN code developed for the simulation of NP transport in real complex scenarios [29]. In MNM3D, the colloid transport equations and the dependencies of the attachment and detachment kinetic coefficients on transients in pore–water IS and velocity have been implemented. MNM3D combines MNMs with the well-known transport model RT3D [38], a numerical code that solves the reactive transport of multiple mobile and/or immobile dissolved species in three-dimensional saturated groundwater systems.

MNM3D implements a dual-site reversible attachment kinetics, where the deposition terms are expressed according to the general formulation proposed by Tosco and Sethi [25]. It solves the NP transport Eqs. 5.1 and 5.2, accounting for dependency of the attachment and detachment kinetics on the groundwater IS (Eqs. 5.5–5.7) and velocity (Eqs. 5.3 and 5.4). In addition, MNM3D implements a new formulation of the attachment and detachment coefficients to simulate the simultaneous effects of pore–water velocity and IS [29]. To this purpose, the dependency of the empirical parameters $C_{a,i}$ and $C_{d,i}$ on the IS has been explicitly expressed by coupling Eqs. (5.3 and 5.4) with Eqs. (5.5 and 5.6):

$$k_{a,i}(v, c_{salt}) = \frac{3}{2}\frac{(1-\varepsilon)}{\varepsilon d_s}\frac{C'_{a,i}}{1+\left(\frac{CDC_i}{c_{salt}}\right)^{\beta_{a,i}}}\eta_0 v \tag{5.8}$$

$$k_{d,i}(v, c_{salt}) = \frac{C'_{d,i}}{1+\left(\frac{c_{salt}}{CRC_i}\right)^{\beta_{d,i}}}\mu v \tag{5.9}$$

where $C'_{a,i}$ and $C'_{d,i}$ are fitting parameters, which account for all the other phenomena not explicitly considered here.

Heterogeneities in the hydrodynamic properties of the porous medium and of the particle–soil interactions can be taken into account by implementing space–variable hydrodynamic, transport, and kinetic parameters. MNM3D is composed of four different packages that can be alternatively selected according to the phenomena that must be introduced into the simulation: particle transport with reversible attachment, IS-dependent transport, velocity-dependent transport, coupled IS- and velocity-dependent transport. MNM3D can be easily implemented in many open source and commercial graphical interfaces, which already support RT3D, and will be included in a future release of the commercial software Visual Modflow Flex (Waterloo Hydrogeologic).

## 5.4 Using the Modeling Tools to Assist NP-Based Remediation

The implementation of an NP-based remediation technology at a contaminated site requires the support of quantitative modeling tools to predict the NP mobility at different stages of the technology application, both in the planning and design stages, and in a later stage to predict the long-term particle mobility after injection. Performing a nanoremediation, like any other technology application, is a stepwise process. Quantitative modeling tools and laboratory tests can support the design process by predicting the NP mobility at different stages of the technology application, both during the planning and implementation, and in a later stage, to predict the long-term particle mobility after injection.

In the first stages of the evaluation of the technology applicability, an NP transport model can be used in combination with column transport tests performed for a specific field application: by modeling the experimental results of selected tests, the key transport features of the NPs can be extrapolated and used to simulate NP transport under a wider set of conditions. In the second step, an NP transport model will be used to forecast placement of the NP with the injection, and to forecast NP long-term mobility, i.e., potential transport of particles out of the core area during and after injection. Regarding verification, model forecasts can guide how, where, and when to monitor, to prove that the expectations on NP placement are met.

The main advantages those modeling tools can offer during the design of a full-scale injection of nanoparticles in groundwater are:

- **Limiting the number of tests to be performed in the laboratory**: Column transport tests are usually performed to identify the driving forces playing a major role in the field-scale mobility of the particles. As an example, flow velocity is expected to be a key parameter affecting the result of a pilot injection through a screened well. Several column tests are usually necessary to fully characterize the influence of flow velocity on the particle mobility in all the field-relevant conditions, for example, near-well high velocity, decreasing velocity with increasing distance from the well, background flow, etc. When instead a numerical model is available, such as MNMs, only a few (typically 3 or 4) column tests injecting particles at different flow rates are sufficient. If the parameter of interest is the injected concentration, or the concentration of stabilizer, the same approach can be adopted, running a few column tests at different NP or stabilizer concentrations. MNMs is then applied to extract quantitative information on the key transport features associated with the investigated parameter (inverse mode), and also to simulate NP transport under a wider set of conditions (forward mode). Modeling may thus spare extensive costly and time-consuming experiments, while widening the range of conditions that can be investigated.

- **Exploring different implementation scenarios**: The information gathered from the modeling of the column tests, in terms of processes identified and associated rate parameters, has general validity. Simulations can, therefore, be run in the forward mode to forecast NP transport under various implementation scenarios. The model results can aid in optimizing NP and slurry properties and injections schemes (e.g., based on a desired radius of influence and NP concentration in the target area, which discharge rate should be applied, how long injection should last, which concentration of NPs and stabilizers is the most effective, etc.). On the other hand, models can also be used to foresee the injection performance when operating conditions are fixed and represent a limiting factor in the application of the remediation (e.g., maximum injectable discharge rate with the available pumps, longest possible injection time, etc.). Equally important, model results can point out what crucial but still missing information would contribute most to a better-informed decision and, therefore, suggest if and which additional laboratory transport tests should be performed to fill the knowledge gaps.
- **Guiding monitoring and testing assumptions**: A model-supported and reasoned monitoring plan can, therefore, help reducing the overall costs of the remediation technology, in terms of installation of new monitoring points, schedule of the sampling campaign, and number and position of core sampling.

Moreover, any discrepancy between actual observation and previous modeled prediction should rise questions on the reliability of the conceptual site model, providing information to revise it and adjust any follow-up activities.

### 5.4.1 Overall Strategy

Figure 5.5 summarizes a strategy for the model-assisted design of a pilot injection of NPs through a screened well in aquifer systems. The approach combines experiments and modeling through these steps:

- Porous medium (core sampling) and groundwater samples are collected at the contaminated site.
- A detailed physicochemical characterization of the NP slurry, groundwater, and porous medium is performed.
- Based on NP and porous medium properties, a first guess of the transport-controlling mechanisms is made (e.g., attractive or repulsive interaction energies predicted by DLVO theory).
- Columns are packed using the site porous medium, and a tracer test is run on each column to determine porosity and dispersivity; a few transport tests under well-defined conditions are then performed at different injection rates (typically three or four).
- Transport coefficients (i.e., attachment/detachment coefficients and other specific parameters of the considered mechanism) are fitted against experimental results; if model fitting is unsatisfactory, the initial guess on transport mechanisms is updated via an iterative procedure (e.g., comparison of model and experimental data suggests that physical filtration may dominate over blocking).
- The trend of $k_a$ and $k_d$ with flow velocity and/or other relevant parameters (e.g., IS) is verified against the theoretical model formulations (Eqs. 5.8 and 5.9). The corresponding coefficients are determined.
- One-dimensional radial simulations are run in the forward mode, assuming the transport mechanisms and parameters as determined in the previous steps. Simulations are run assuming different injection rates and injection durations,

thus deriving the expected radius of influence, particle spatial distribution, and pressure buildup for each set of operative conditions (see also Ref. [17] for an example of application).

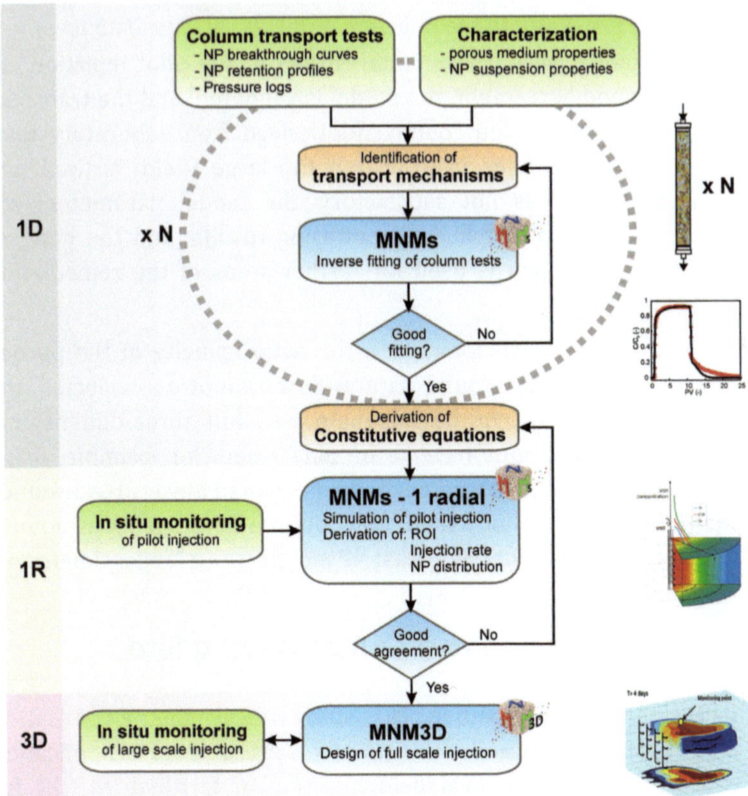


**Figure 5.5** Procedure for the interpretation of column tests to derive information for the simulation at larger scales. Green boxes refer to experimental data; blue boxes refer to transport modeling; orange boxes refer to conceptual relationships.

- Based on the model results, the preferred injection scenario for the pilot test is implemented. After the pilot test is performed, the particle distribution around the injection point and radius of influence predicted by the model are compared to those obtained in the field. To this aim, monitoring results are needed. In particular, it is crucial to have information on the NP mobility during the injection (which can be obtained, for example, by collecting water samples over time in monitoring piezometers around the injection point), and the final particle distribution after injection (to be reconstructed, for example, by analyzing the concentration of NPs in soil cores collected in the proximity to the injection point, at various distances).

- If model and experimental results of the pilot injection are in good agreement, the model is validated, and the transport mechanisms and coefficients derived from laboratory tests can be assumed valid also at the large (field) scale. If the agreement is not satisfactory, the model parameters are adjusted to match the monitoring results, and the new set of parameters is used for further steps of the remediation design.

In some cases, such as when the heterogeneity of the porous medium or the background natural flow cannot be neglected, the radial model may not be adequate and full three-dimensional numerical simulations have to be performed, for example, using MNM3D. A three-dimensional model is instead always required for the full-scale design of a field injection in a heterogeneous aquifer with multiple wells, or to forecast NP mobility on a longer timeframe.

## 5.5 Application to a Contaminated Site

In this paragraph, a synthetic case study concerning the application of the protocol proposed in this chapter is presented. The application concerns the hypothetical field injection of ferrihydrite NPs for the stimulation of the bioremediation of an aquifer contaminated by chlorinated hydrocarbons [15]. The experimental data used in this test case are obtained from a previous publication [39], which the reader can refer to for a detailed description of methods and experimental results of the column transport tests. Even though the tests in the cited study were performed using a quartz sand, and not soil collected at a specific site, we assume here that the transport behavior of the ferrihydrite NPs observed (and modeled) in the laboratory can be extrapolated to the full-scale application (i.e., to a real aquifer system). This is not, as a general rule, a procedure applicable to real cases. However, this assumption is made here since the application presented is a synthetic case, aimed at exemplifying the application of the modeling up-scaling procedure, rather than to present a real application.

### 5.5.1 Column Transport Tests

Ferrihydrite NPs, having an average radius of 106.7 ± 15.5 nm (dynamic light scattering measurement, Zetasizer Nano S90, Malvern Instruments Ltd, UK), were dispersed in DI water with the addition of NaCl at different concentrations (0.1 to 10 mM). Chromatographic columns (inner radius of 8 mm, length 11.3 cm) were wet-packed with quartz sand (Dorsilit 8, Dorfner, Germany; mean diameter $d_{50}$ = 0.194 mm). Tracer tests provided an average effective porosity of 0.447 (±0.021) and an average dispersivity of 2.82 (±0.62) × $10^{-4}$ m. The electrophoretic mobility measurements (Zetasizer Nano ZS, Malvern Instruments, Malvern, UK) evidenced a negative surface charge for both particles and sand, under the entire range of IS explored. The estimation of particle–collector interactions using the DLVO theory indicated repulsion for all explored IS, thus suggesting that particle deposition in column tests likely follows a blocking kinetics.

Ferrihydrite transport tests were performed injecting the particles dispersed in NaCl solutions at an NP concentration of 7.5 mg/L. The protocol included a pre-flushing with particle-free solution, the injection of the NP suspension, and one or more post-flushing steps. The detailed protocol is reported in Tosco et al. [39]. The inlet and outlet concentration of salts and NPs was monitored online via optical density measurements (UV-Vis spectrophotometer Specord S600, Analytik Jena, Germany) equipped with flow-through cells.

The tests are here analyzed with the aim of assessing the influence of IS (IS equal to 0.1 mM, 1 mM, 3 mM, 5 mM, and 10 mM) and flow rate ($q$ equal to 7.76 × $10^{-5}$ m/s, 1.55 × $10^{-4}$ m/s, 2.33 × $10^{-4}$ m/s) on the transport and deposition of the particles when injected in the porous medium. Assessing the impact of flow rate on ferrihydrite NPs is important since, during the injection in the field through a screened well, the pore–water velocity decreases hyperbolically with increasing distance from the delivery point. As a consequence, it is important to determine the coefficients $C_{a,i}$ and $C_{d,i}$ in Eqs. 5.3 and 5.4 to model this dependence. On the other hand, the influence of IS is relevant since usually NPs are dispersed in tap water before the injection in the subsurface, and the IS of the tap water is typically different from the IS of the groundwater. As a consequence, the transport of the NPs during the injection and, more importantly, after injection, when the natural flow is restored in the injection area, may be significantly affected by the IS of the water used to prepare the suspension to be injected, and of the groundwater. This dependence is modeled by Eqs. 5.5–5.7, and the parameters needed to model the phenomenon are $k_{a,\infty}$, CDC, $\beta_a$ (for the attachment rate), $k_{d,0}$, CRC, $\beta_d$ (for the detachment rate), and $\gamma_s$ and $\beta_s$ (for the blocking function).

To highlight the abovementioned impacts and to derive the associated model parameters, two sets of experiments were run. For each test, the experimental breakthrough curve was least-squares fitted to model Eqs. 5.1 and 5.2, assuming two concurrent deposition mechanisms, one ($i$ = 1) following a reversible blocking, as suggested by the DLVO profiles, and the other one ($i$ = 2) following a linear irreversible deposition. As a consequence, the equations for this application are:

$$\frac{\partial}{\partial t}(\varepsilon c)+\sum_i\left(\rho_b\frac{\partial s_i}{\partial t}\right)+\frac{\partial}{\partial x}(qc)-\frac{\partial^2}{\partial x^2}(\varepsilon Dc)=0 \qquad (5.10a)$$

$$\rho_b\frac{\partial s_1}{\partial t}=\varepsilon k_{a,1}\left(1-\frac{s_1}{s_{\max,1}}\right)c-\rho_b k_{d,1}s_1 \qquad (5.10b)$$

$$\rho_b\frac{\partial s_2}{\partial t}=\varepsilon k_{a,2}c \qquad (5.10c)$$

The corresponding $k_{a,1}$, $k_{d,1}$, $s_{max,1}$, and $k_{a,2}$ were determined for each test.

A first set of column tests was performed at the same flow rate ($q$ = 7.76 × $10^{-5}$ m/s), while the IS changed from test to test, and the breakthrough curve was least-squares fitted using MNMs. In Fig. 5.6a, the experimental and modeled breakthrough curves are reported. The coefficients obtained from the inverse modeling of the experimental data are reported in Fig. 5.6b: the trend of the attachment/detachment coefficients with changing IS is consistent with those previously observed for other particles, and in particular for latex microspheres [22] and graphene oxide [29]: the attachment rates $k_{a,1}$ and $k_{a,2}$ increase with increasing IS, while the detachment rate $k_{d,1}$ decreases with a similar but opposite trend. Also the maximum concentration of particles retainable onto the solid phase $s_{max,1}$ increases with IS. In Fig. 5.6b, the solid lines representing these trends reproduce the model curves of Eqs. 5.5–5.7 for both deposition sites, which were obtained fitting the trends of the four coefficients. The parameters are reported in Table 5.2.

The second set of column tests was performed keeping the IS constant among the tests (5 mM or 10 mM), while the flow rate was changed from test to test. The experimental breakthrough curves were fitted using MNMs, similar to the previous set of experiments, to the model Eqs. 5.1 and 5.2. The trends of $k_{a,1}$, $k_{d,1}$, $s_{max,1}$, and $k_{a,2}$ as a function of pore-water velocity were analyzed following a procedure similar to the one described for IS-dependent tests (data not reported). The parameters $C_{a,1}$, $C_{d,1}$, and $C_{a,2}$ of Eqs. 5.3 and 5.4 were determined for the two IS values, respectively, equal to 0.11, 58.41, and 0.0031 for IS of 5 mM; equal to 0.21, 12.30, and 0.0011 for IS of 10 mM. No evident dependence of $s_{max,1}$ on the flow velocity was observed, similar to previous studies [29]. As a consequence, $s_{max,1}$ was assumed independent of flow velocity. Finally, the information on the dependence of deposition and release kinetics was combined, determining $C'_{a,1}$, $C'_{d,1}$, and $C'_{a,2}$ of Eqs. 5.8 and 5.9, which were obtained equal to 0.255, $5.54 \times 10^3$, and $4.81 \times 10^{-3}$, respectively. The fitted coefficients were then used for the predictive simulation of particle injection at larger scales and in more complex geometries.

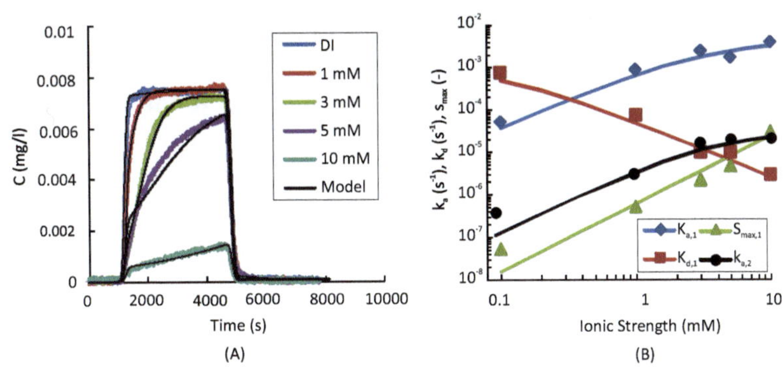


**Figure 5.6** (a) Experimental (color) and modeled (black) breakthrough curves for ferrihydrite transport tests performed at different ionic strength values, and (b) corresponding parameters $k_a$, $k_d$, and $s_{max}$. Modified from Ref. [39]. Copyright 2012 American Chemical Society.

**Table 5.2** Model parameters of Eqs. 5.5–5.7 for the two deposition sites

| | Parameter | Site 1 ($i$ = 1) | Site 2 ($i$ = 2) |
|---|---|---|---|
| Attachment rate $k_{a,i}$ (Eq. 5.5) | $k_{a\infty,i}$ ($s^{-1}$) | $4.1 \times 10^{-3}$ | $7.74 \times 10^{-5}$ |
| | $CDC_i$ (mM) | 3.7 | 3.7 |
| | $\beta_{a,i}$ (–) | 1.3 | 1.3 |
| Detachment rate $k_{d,i}$ (Eq. 5.6) | $k_{d0,i}$ ($s^{-1}$) | $1.0 \times 10^{-3}$ | — |
| | $CRC_i$ (mM) | 0.1 | — |
| | $\beta_{d,i}$ (–) | 1.3 | — |
| Maximum deposited concentration $s_{max,i}$ (Eq. 5.7) | $\gamma_{s,i}$ (–) | $6.0 \times 10^{-7}$ | — |
| | $\beta_{s,i}$ (–) | 1.6 | — |

**Note**: The model curves obtained from Eqs. (5.5–5.7) using the parameters reported in this table are reported as solid lines in Fig. 5.6.

### 5.5.2 Pilot Injection Modeling

A pilot injection of the ferrihydrite NPs through a screened well was then performed using the radial simulation tool of MNMs. The particles were dispersed in water with IS equal to 5 mM at a concentration of 0.5 g/L and injected following two injection protocols:

- Protocol 1: discharge rate of 60 L/min for 5 h;
- Protocol 2: discharge rate of 20 L/min for 15 h.

Both protocols consider the injection of 18 $m^3$ of NP suspension.

The aquifer material was assumed identical to the one used in the column tests, and consequently the NP transport mechanisms and parameters determined from the column tests were assumed valid. A porosity of 0.25 was adopted. The simulation results are reported in Fig. 5.7 in terms of total (deposited + suspended) NPs concentration. For comparison, the profile obtained for a tracer is also reported. The simulations evidence a radius of influence for the NPs close to the radius of influence of the tracer. The injection rate does not significantly affect the overall particle distribution, since the NP suspension is fairly stable and well transported in the porous medium. As a consequence, for a real application, the highest injection rate, 60 L/min, would be selected, thus limiting injection time, and consequently costs.

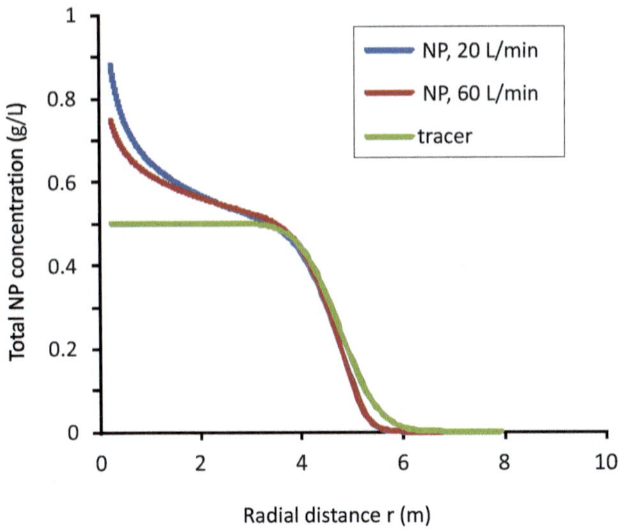


**Figure 5.7** Total (suspended + deposited) concentration of ferrihydrite NPs and of a tracer as a function of the radial distance *r* from the injection well.

### 5.5.3 Modeling Full-Scale Injection

A full-scale remediation of a contaminated site by injection of ferrihydrite NPs was finally simulated in MNM3D. The coupled IS-velocity-dependent model was used to take into account the concurrent influence of the pore–water velocity, mainly affecting the short-term transport, and groundwater salinity, governing the long-term behavior of the slurry. The site considered in this example is a heterogeneous multilayered aquifer contaminated by BTEX. The aquifer is composed of an upper sandy layer ($k = 5 \times 10^{-6}$ m/s), a lower gravel porous medium ($k = 5 \times 10^{-3}$ m/s) and a silty–clay lens ($k = 2 \times 10^{-8}$ m/s) interposed between the two main strata, for a total average depth of 16 m. The ferrihydrite particles were injected in the high permeability gravel layer, where the core of the contamination is located. Like for the radial simulation, a porosity of 0.25 was adopted for the gravel aquifer. The flow direction was S–N, with an average Darcy velocity of 0.5 m/day. Natural IS of 10 mM was assumed. The particle slurry was injected at a concentration of 1 g/L through three wells having a 1 m long screening, from 12 to 13 m b.g.l. The IS of the suspension was equal to 5 mM. The injection was 5 h long at a constant discharge rate of 60 L/min. The transport of nanoparticles under natural flow was also simulated to foresee the fate of the slurry 10 days after the injection.

Figure 5.8 reports the results of the simulation at the end of the 5 h long injection. The concentration distribution around the wells has a radial-like shape, since the natural flow velocity is much lower than the velocity field generated by the injection. However, the resulting particle plumes are slightly asymmetrical, because of the effect of the flow field, and strongly affected by the presence of the other wells. The total concentration profile along the section S–S, shown in Fig. 5.9a, reports a slightly lower concentration than the one simulated by the radial model (compare Fig. 5.7 and Fig. 5.9a). This is again due to the combined effect of the natural flow field and of the multi-well injection, which produced a lower attachment close to the well. The graph shows also the reduction of IS induced by the injection of a 5 mM suspension into a 10 mM groundwater, which is another factor promoting the wide spreading of the slurry. It is worth to highlight that, thanks to the high mobility of the particles considered in this example, only a small fraction of the particles is already deposed into the porous medium (green curve) at the end of the injection, while most part is still in suspension. It is, therefore, very important to be able to foresee the NP behavior on a longer timescale after the injection.

Figure 5.9b shows the concentration profile after 10 days from the end of the injection. Particles were transported just by natural flow, slowly mixing with the higher IS groundwater. The particle plume is still partly mobile after 10 days, even if the fraction of suspended NPs is constantly decreasing. The concentration profile shows a peak close to the injection well, and a constant concentration downstream, where the concentration of deposited particles reached $s_{max}$. Thanks to the good mobility of the ferrihydrite NPs, the natural flow contributes to create a wide area downstream the injection points where the NP concentration is pretty homogeneous, thus providing the optimal conditions for the stimulation of the bacterial activity, which is the final goal of the injection of ferrihydrite NPs.

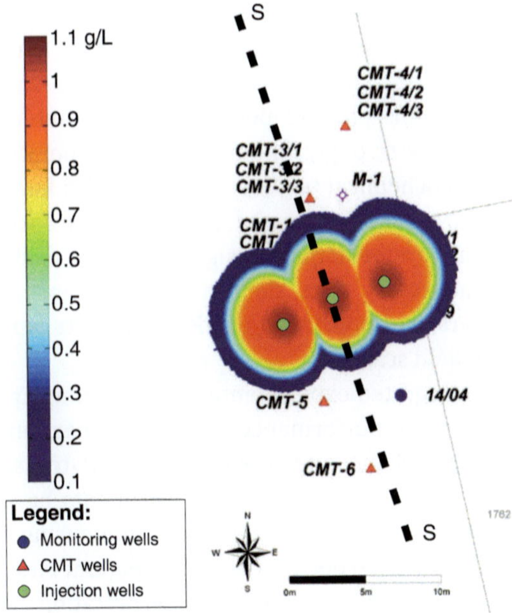


**Figure 5.8** Map of total (suspended + deposited) concentration of ferrihydrite NPs at the end of the injection.

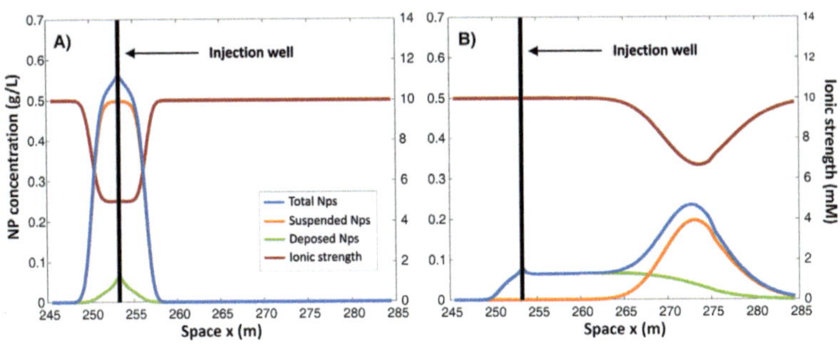


**Figure 5.9** Total (suspended + deposited) NP concentration, deposed NP concentration, and ionic strength profile along the section S–S at the end of the injection (A) and after 10 days (B).

## 5.6 Conclusion

In this chapter, the main mechanisms and driving forces governing the transport of nanoparticles in porous media have been presented, with a focus on iron-based NPs employed for groundwater remediation. The most common modeling approach for NP transport in groundwater is based on a modified advection–dispersion equation accounting for particle–porous medium interactions. These interactions can be strongly influenced by both operating (e.g., injection rate) and geochemical (e.g., IS) conditions. Understanding and modeling these phenomena is of pivotal importance in all the stages of a nanoremediation design, such as design and interpretation of laboratory tests, definition of the most suitable operating parameters, and prediction of NP fate and transport and effectiveness at the field scale.

To this aim, an integrated experimental and modeling approach was presented to foresee performances of an NP injection at the field scale, by characterizing and modeling the NP transport at the laboratory scale. Targeted laboratory experiments have to be performed to identify the main driving forces influencing the NP mobility and to derive site-specific transport parameters. These results are then supported by modeling interpretation and up-scaled to foresee the NP behavior at larger scales. Mathematical tools can be used to get a reliable estimation of several operative parameters, for example, particle distribution around the injection well, radius of influence for a target concentration, number of required injection wells. In this sense, the maodels are useful tools to test several alternative scenarios, allowing to easily span a wide range of conditions with a limited number of laboratory and pilot tests.

Two modeling tools were here proposed to deal with nanoparticle transport at the different scales of the problem: MNMs, for the assisted quantitative analysis of laboratory-scale column tests and preliminary design of pilot NP injections in a simplified geometry (radial one-dimensional simulations), and MNM3D, a full three-dimensional transport module for the simulation of particle injection in heterogeneous domains, and for the prediction of NP fate and transport at the field scale.

The combined experimental and modeling approach was here applied to the design of an iron oxide-based nanoremediation. The procedure proved to be effective in providing useful information for the design of a field-scale injection and for the prediction of the long-term fate of NPs introduced in groundwater. However, it is worth to highlight that in order to get reliable information, models require that a good conceptual model of the aquifer is formulated and that a suitable characterization and degree of knowledge of the hydrodynamics and geochemistry of the site is carried out. Even if it was not included in the example provided in this chapter, it is worth to recall here that the interplay between modeling and field monitoring data is a crucial step in the successful design of an NP-based remediation. Model forecasting should always go hand in hand with adequate monitoring and vice versa, model results can guide how, where and when to monitor, to prove that the expectations on NP placement—and remediation targets and safety—will be met. As a consequence, the use of models is not intended to supply unknown information of the site, but to support, strengthen, and extend known information, which still need to be retrieved by well-studied experimental tests and sampling campaign.

## References

1. Karn, B., Kuiken, T., and Otto, M. (2009). Nanotechnology and in situ remediation: A review of the benefits and potential risks, *Environ. Health Perspect.*, **117**(12), pp. 1823–1831.
2. Tosco, T., Petrangeli Papini, M., Cruz Viggi, C., and Sethi, R. (2014). Nanoscale iron particles for groundwater remediation: A review, *J. Clean. Prod.*, **77**, pp. 10–21.
3. O'Carroll, D., Sleep, B., Krol, M., Boparai, H., and Kocur, C. (2013). Nanoscale zero valent iron and bimetallic particles for contaminated site remediation, *Adv. Water Resour.*, **51**, pp. 104–122.
4. Yan, W., Lien, H.-L., Koel, B. E., and Zhang, W.-X. (2013). Iron nanoparticles for environmental clean-up: Recent developments and future outlook, *Environ. Sci. Process. Impacts*, **15**(1), pp. 63–77.
5. Zhang, W.-X. (2003). Nanoscale iron particles for environmental remediation: An overview, *J. Nanopart. Res.*, **5**(3–4), pp. 323–332.
6. Jegadeesan, G., Mondal, K., and Lalvani, S. B. (2005). Arsenate remediation using nanosized modified zerovalent iron particles, *Environ. Prog.*, **24**(3), pp. 289–296.
7. Ponder, S. M., Darab, J. G., and Mallouk, T. E. (2000). Remediation of Cr(VI) and Pb(II) aqueous solutions using supported, nanoscale zero-valent iron, *Environ. Sci. Technol.*, **34**(12), pp. 2564–2569.

8. Chang, M. C., Shu, H. Y., Hsieh, W. P., and Wang, M. C. (2005). Using nanoscale zero-valent iron for the remediation of polycyclic aromatic hydrocarbons contaminated soil, *J. Air Waste Manag. Assoc.*, **55**(8), pp. 1200–1207.
9. Joo, S. H., Feitz, A. J., and Waite, T. D. (2004). Oxidative degradation of the carbothioate herbicide, molinate, using nanoscale zero-valent iron, *Environ. Sci. Technol.*, **38**(7), pp. 2242–2247.
10. Mackenzie, K., Bleyl, S., Georgi, A., and Kopinke, F. D. (2012). Carbo-iron - An Fe/AC composite - As alternative to nano-iron forz groundwater treatment, *Water Res.*, **46**(12), pp. 3817–3826.
11. Elliott, D. W. and Zhang, W . X. (2001). Field assessment of nanoscale bimetallic particles for groundwater treatment, *Environ. Sci. Technol.*, **35**(24), pp. 4922–4926.
12. Hosseini, S. M. and Tosco, T. (2013). Transport and retention of high concentrated nano-Fe/Cu particles through highly flow-rated packed sand column, *Water Res.*, **47**(1), pp. 326–338.
13. Quinn, J., Geiger, C., Clausen, C., Brooks, K., Coon, C., O'Hara, S., Krug, T., Major, D., Yoon, W.-S., Gavaskar, A., and Holsworth, T. (2005). Field demonstration of DNAPL dehalogenation using emulsified zero-valent iron, *Environ. Sci. Technol.*, **39**(5), pp. 1309–1318.
14. Benjamin, M. M., Sletten, R. S., Bailey, R. P., and Bennett, T. (1996). Sorption and filtration of metals using iron-oxide-coated sand, *Water Res.*, **30**(11), pp. 2609–2620.
15. Bosch, J., Heister, K., Hofmann, T., and Meckenstock, R. U. (2010). Nanosized iron oxide colloids strongly enhance microbial iron reduction, *Appl. Environ. Microbiol.*, **76**(1), pp. 184–189.
16. Luna, M., Gastone, F., Tosco, T., Sethi, R., Velimirovic, M., Gemoets, J., Muyshond, R., Sapion, H., Klaas, N., and Bastiaens, L. (2015). Pressure-controlled injection of guar gum stabilized microscale zerovalent iron for groundwater remediation, *J. Contam. Hydrol.*, **181**, pp. 46–58.
17. Tosco, T., Gastone, F., and Sethi, R. (2014). Guar gum solutions for improved delivery of iron particles in porous media (Part 2): Iron transport tests and modeling in radial geometry, *J. Contam. Hydrol.*, **166**, pp. 34–51.
18. Velimirovic, M., Tosco, T., Uyttebroek, M., Luna, M., Gastone, F., De Boer, C., Klaas, N., Sapion, H., Eisenmann, H., Larsson, P.-O., Braun, J., Sethi, R., and Bastiaens, L. (2014). Field assessment of guar gum stabilized microscale zerovalent iron particles for in-situ remediation of 1,1,1-trichloroethane, *J. Contam. Hydrol.*, **164**, pp. 88–99.

19. Tufenkji, N. and Elimelech, M. (2004). Correlation equation for predicting single-collector efficiency in physicochemical filtration in saturated porous media, *Environ. Sci. Technol.*, **38**(2), pp. 529–536.

20. Saleh, N., Kim, H.-J., Phenrat, T., Matyjaszewski, K., Tilton, R. D., and Lowry, G. V. (2008). Ionic strength and composition affect the mobility of surface-modified $Fe^0$ nanoparticles in water-saturated sand columns, *Environ. Sci. Technol.*, **42**(9), pp. 3349–3355.

21. Tiraferri, A., Chen, K. L., Sethi, R., and Elimelech, M. (2008). Reduced aggregation and sedimentation of zero-valent iron nanoparticles in the presence of guar gum, *J. Colloid Interface Sci.*, **324**(1–2), pp. 71–79.

22. Tosco, T., Tiraferri, A., and Sethi, R. (2009). Ionic strength dependent transport of microparticles in saturated porous media: Modeling mobilization and immobilization phenomena under transient chemical conditions, *Environ. Sci. Technol.*, **43**(12), pp. 4425–4431.

23. Dalla Vecchia, E., Luna, M., and Sethi, R. (2009). Transport in porous media of highly concentrated iron micro- and nanoparticles in the presence of xanthan gum, *Environ. Sci. Technol.*, **43**(23), pp. 8942–8947.

24. Torkzaban, S., Wan, J., Tokunaga, T. K., and Bradford, S. A. (2012). Impacts of bridging complexation on the transport of surface-modified nanoparticles in saturated sand, *J. Contam. Hydrol.*, **136–137**, pp. 86–95.

25. Tosco, T. and Sethi, R. (2010). Transport of non-Newtonian suspensions of highly concentrated micro- and nanoscale iron particles in porous media: A modeling approach, *Environ. Sci. Technol.*, **44**(23), pp. 9062–9068.

26. Becker, M. D., Wang, Y., Paulsen, L., J., Song, Y. Q., Abriola, L. M., and Pennell, K. D. (2015). In situ measurement and simulation of nano-magnetite mobility in porous media subject to transient salinity, *Nanoscale*, **7**(3), pp. 1047–1057.

27. Katzourakis, V. E. and Chrysikopoulos, C. V. (2014). Mathematical modeling of colloid and virus cotransport in porous media: Application to experimental data, *Adv. Water Resour.*, **68**, pp. 62–73.

28. Šimunek, J., Van Genuchten, M. T., and Šejna, M. (2008). Development and applications of the HYDRUS and STANMOD software packages and related codes, *Vadose Zone J.*, **7**(2), pp. 587–600.

29. Bianco, C., Tosco, T., and Sethi, R. (2016). A 3-dimensional micro- and nanoparticle transport and filtration model (MNM3D) applied to the

migration of carbon-based nanomaterials in porous media, *J. Contam. Hydrol.*, **193**, pp. 10–20.

30. Boccardo, G., Marchisio, D. L., and Sethi, R. (2014). Microscale simulation of particle deposition in porous media, *J. Colloid Interf. Sci.*, **417**, pp. 227–237.

31. Crevacore, E., Tosco, T., Sethi, R., Boccardo, G., and Marchisio, D. L. (2016). Recirculation zones induce non-Fickian transport in three-dimensional periodic porous media, *Phys. Rev. E Stat. Nonlin. Soft Matter Phys.*, **94**(5), 053118.

32. Ko, C.-H. and Elimelech, M. (2000). The shadow effect in colloid transport and deposition dynamics in granular porous media: Measurements and mechanisms, *Environ. Sci. Technol.*, **34**, pp. 3681–3689.

33. Yao, K.-M., Habibian, M. T., and O'Melia, C. R. (1971). Water and waste water filtration. Concepts and applications, *Environ. Sci. Technol.*, **5**(11), pp. 1105–1112.

34. Messina, F., Marchisio, D. L., and Sethi, R. (2015). An extended and total flux normalized correlation equation for predicting single-collector efficiency, *J. Colloid Interface Sci.*, **446**, pp. 185–193.

35. Rajagopalan, R. and Tien, C. (1976). Trajectory analysis of deep-bed filtration with the sphere-in-cell porous media model, *Aiche J.*, **22**(3), pp. 523–533.

36. Gastone, F., Tosco, T., and Sethi, R. (2014). Guar gum solutions for improved delivery of iron particles in porous media (Part 1): Porous medium rheology and guar gum-induced clogging, *J. Contam. Hydrol.*, **166**, pp. 23–33.

37. Xue, D. and Sethi, R. (2012). Viscoelastic gels of guar and xanthan gum mixtures provide long-term stabilization of iron micro- and nanoparticles, *J. Nanopart. Res.*, **14**(11), 1239.

38. Clement, T. P., Sun, Y., Hooker, B. S., and Petersen, J. N. (1998). Modeling multispecies reactive transport in ground water, *Ground Water Monit. Rem.*, **18**(2), pp. 79–92.

39. Tosco, T., Bosch, J., Meckenstock, R. U., and Sethi, R. (2012). Transport of ferrihydrite nanoparticles in saturated porous media: Role of ionic strength and flow rate, *Environ. Sci. Technol.*, **46**(7), pp. 4008–4015.

## Chapter 6

# Nanoscale Zerovalent Iron Particles for Groundwater and Soil Treatment: Monitoring and Control of their Solid-State Synthesis, Stability, and Activity

**Jan Filip,[a] Jana Soukupová,[a] Josef Kašlík,[a] Jan Slunský,[b] and Radek Zbořil[a]**

[a]*Regional Centre of Advanced Technologies and Materials, Departments of Physical Chemistry and Experimental Physics, Palacký University, 17. Listopadu 12, CZ-77146 Olomouc, Czech Republic*

[b]*NANO IRON, s.r.o., Štefánikova 116, CZ-66461 Rajhrad, Czech Republic*

jan.filip@upol.cz

## 6.1 Introduction

In the last two decades, nZVI has been established as a multifunctional material exhibiting wide applicability in the large-scale remediation of surface water with microbial contamination [1, 2], groundwater in sediments, and soils at localities contaminated by various inorganic and organic hazardous substances [3–5]. Zerovalent iron nanoparticles exhibit extremely high efficiency and

*Iron Nanomaterials for Water and Soil Treatment*
Edited by Marta I. Litter, Natalia Quici, and Martín Meichtry

ISBN 978-981-4774-67-3 (Hardcover), 978-981-4669-49-8 (eBook)
www.panstanford.com

versatility of degradation (through effective reduction and catalysis) and/or removal of more than 70 environmental contaminants [6], and as such can be probably considered one of the most universal remediation materials [5, 7–9 and references therein]. The contaminants easily removable by nZVI particles involve halogenated hydrocarbons [10–13], metals (As, Cu, Co, Cr, Ni, Pb, Hg, etc. [14]), various anions [15], radionuclides, organic dyes [16], amoxicillin, ampicillin, pesticides, trinitrotoluene (TNT) [17], hexahydro-1,3,5-trinitro-1,3,5-triazine (RDX) [18], or chemical warfare agents [19]. This fact was successfully demonstrated in dozens of laboratory and large-scale studies [5, 8, 20]. The in situ technologies, utilizing nZVI particles, include construction of permeable reactive barriers (i.e., reactive zones) or injection of surface-modified nZVI dispersions into contaminated groundwater within saturated porous media such as soils and sediments [7, 8, 21]. The in situ applied nZVI particles thus represent one of the most widely investigated environmental reactive remediation technologies replacing usage of macroscopic (granular) or microsized zerovalent iron [9]. Recently, nZVI particles were successfully commercialized as a new generation of versatile and cost-effective environmental remediation technology [5]. Furthermore, they have shown even an ultimate ambition to replace conventional environmental technologies altogether.

However, the efficiency of nZVI particles is strongly dependent not only on their size (i.e., affecting specific surface area of nZVI particles), but also on the actual $Fe^0/Fe_{tot}$ ratio [4]. The best results were approached with powdered particles prepared using a thermally induced conversion of a suitable Fe(III)-based precursor under a reductive atmosphere [11, 22, 23]. The application of these nZVI particles, however, requires their transfer into an aqueous media in order to form a dense or diluted water slurry, which is then routinely transported, stored, and applied. However, most frequently, the usage of such a slurry requires intensive mixing or resuspension in either laboratory or for large-scale remediation of contaminated groundwater (i.e., poured or injected into treated wells at contaminated sites). The reactivity and subsurface mobility of nZVI particles are then predetermined by any surface modification [4, 24], which is frequently used in order to suppress massive aggregation of particles and, as such, retain maximal active surface area of the particles [6, 25]. However, even when stored in either surface-modified or surface-passivated form, nZVI particles spontaneously change to the form of core–shell particles Fe-@-(FeO, $Fe_3O_4$, FeOOH, $Fe(OH)_2$, $Fe(OH)_3$, etc.) because of their low selectivity to a variety of chemicals, including dissolved oxygen, sulfates, nitrates, and even water molecules.

Based on the above-mentioned statements, the real content of zerovalent iron ($Fe^0/Fe_{tot}$ ratio) is one of the crucial parameters predetermining the applicability of nZVI dispersion in any large-scale groundwater remediation [26]. As the application of nZVI dispersions represents a cost-demanding procedure, a detailed characterization of the used materials, including exact knowledge of the active content of zerovalent iron in the nZVI dispersion, is required prior to any laboratory and field experiments or site remediation. Although the content of active nZVI represents a crucial parameter, no paper has so far focused on the invention of a field method precisely evaluating the $Fe^0$ content in nZVI slurries. More importantly, numerous studies, dealing with nZVI reactivity published so far, utilize nZVI particles without their detailed characterization. Those results could be, therefore, negatively influenced by partial nZVI oxidation prior to their application or testing.

Although a great variety of techniques are available for the characterization of nZVI particles from the physical, structural, and chemical point of view [6, 22, 25, 27], most of them are accessible exclusively in highly specialized laboratories and research centers. Therefore, they are unavailable for routine controls of the phase composition of nZVI slurries used in remediation. In addition, only some of the techniques can reveal the degree of oxidation of nZVI particles and can accurately determine the required $Fe^0/Fe_{tot}$ ratio. The following standard techniques can detect even a thin layer of iron oxides on the surface of nZVI particles: XRD; transmission electron microscopy (TEM) in combination with electron diffraction, energy-dispersive X-ray spectroscopy or electron energy-loss spectroscopy; $^{57}Fe$ Mössbauer spectroscopy; X-ray photoelectron spectroscopy (XPS); Auger electron spectroscopy (AES); secondary ion mass spectroscopy (SIMS); and synchrotron-based X-ray absorption spectroscopy (XAS). Aside from the above-listed sophisticated techniques, the following techniques can also be applied: acid digestion accompanied by hydrogen evolution (the amount of hydrogen is typically measured by gas pressure or gas chromatography); temperature programmed reduction of iron oxides (to measure oxygen in nZVI); electrochemistry (i.e., voltammetry); magnetometry (vibrating-sample or SQUID magnetometers); Fourier transform infrared (FTIR), Raman, and electron paramagnetic resonance (EPR) spectroscopy; and chemical oxidation of metallic iron with copper (for the estimation of the oxide layer thickness).

Nevertheless, field applications require a quick, cheap, accurate, and easy-to-perform technique for the determination of the $Fe^0/Fe_{tot}$ ratio characterizing the reactivity of nZVI dispersions, which are delivered to the contaminated sites or prepared directly prior to or during the process of remediation. Alternatively, the same measurement technique would be useful for routine checks of nZVI dispersions in injection, as well as in monitoring wells during and after the performed remediation.

Following our previous extensive experience with synthesis, surface modification, detailed characterization, and application of nZVI particles in aqueous dispersions [11, 13, 14, 21, 28], we summarize in this chapter key techniques applicable for nZVI characterization, including the in situ monitoring of nZVI synthesis. Moreover, we present here comparison of two standard sophisticated laboratory techniques [29] with one, newly invented, technique to quantify the $Fe^0$ content in aqueous slurries. These three techniques were used to monitor the ageing process of nZVI dispersions stored under various conditions. The conditions were chosen with respect to known possible storage and/or transfer conditions preceding full-scale applications at the remediation sites. The ageing process of selected samples of nZVI particles, stored in their aqueous dispersions, was monitored using three key experimental techniques: XRD, $^{57}Fe$ Mössbauer spectroscopy, and optimized acid digestion accompanied by hydrogen evolution. It is well known that the first two techniques can provide the user with complementary quantitative information about the $Fe^0/Fe_{tot}$ ratio, but they are not available at the application sites. Therefore, for application in sites, we suggest the acid digestion method, which can be distinguished by its simplicity of experimental adjustment and performance and as such can be employed directly at the site as a necessary quality control of the nZVI dispersion prior to the in situ application.

## 6.2 Methods for in situ Monitoring the Mechanism and Kinetics of nZVI Preparation

There exist several ways for the preparation of highly reactive micro- to nanoscale zerovalent iron particles. Thermally induced solid-state syntheses represent simple and cost-effective methods to prepare zerovalent iron nanoparticles in a large scale. This is proved by the fact that the two most important producers of nZVI material (i.e., NANO IRON and TODA companies) utilize the process of high-temperature reduction of suitable iron oxide precursors under hydrogen. In this chapter, we will summarize the techniques for in situ monitoring of the iron oxide powder reduction in hydrogen leading to the preparation of micro- to nanoscale zerovalent iron particles.

Generally, the thermally induced solid-state transformation from iron oxide powder to metallic iron particles can be described by the simple Eq. (6.1) [11, 23]:

$$Fe_2O_3 + 3H_2 \rightarrow 2Fe^0 + 3H_2O \tag{6.1}$$

A similar mechanism is known in the case of ferric oxyhydroxide polymorphs:

$$FeOOH + 1.5H_2 \rightarrow Fe^0 + 2H_2O \tag{6.2}$$

However, as it will be shown later, the exact mechanisms are more complex, including the initial reduction of $Fe_2O_3$ to magnetite followed by its subsequent reduction to metallic iron nanoparticles [30]. This thermally induced solid-state approach allows only to precisely monitor the formation of metallic iron particles using several complementary approaches, but also to easily control the process leading to the preparation of nZVI particles with tailored properties.

### 6.2.1 High-Temperature X-Ray Powder Diffraction

High-temperature X-ray diffraction (HT-XRD) plays a significant role in the study of the transformation and synthesis processes of materials. The main goal of this technique is to in situ monitor the mechanism and kinetics of structural and phase changes during the thermal treatment of various solid materials in air, inert or reaction gases (reductive or oxidative) under variable conditions (e.g., temperature, gas pressure, duration of the treatment, etc.). The knowledge of the exact reaction kinetics is a basic prerequisite for nZVI preparation at a large scale. The in situ monitoring also allows monitoring of the crystallinity of all observed phases at any certain temperature [23]. The crystallinity, defined as the mean length of X-ray coherent domains, is proportionally related to particle size of given phase(s). HT-XRD thus represents a unique tool for monitoring the thermal reduction of an iron oxide precursor as it records structural characteristics of the treated sample at any defined temperature [30]. The set of collected XRD patterns, together with the knowledge of the exact experimental conditions, is then used to estimate and describe the solid-state reactions. The observed reaction mechanism in hydrogen depicted in Fig. 6.1 includes the initial reduction of $Fe_2O_3$ to magnetite ($Fe_3O_4$) between 250 and 350 °C, followed by its reduction to metallic $\alpha$-Fe (the reduction starts at around 300 °C). From the changes in broadening of diffraction peaks (i.e., from the calculated mean coherence lengths), one can control the resulting particle size and/or the degree of sintering, leading to a controlled specific surface area (and thus controlling the nanoparticle reactivity).

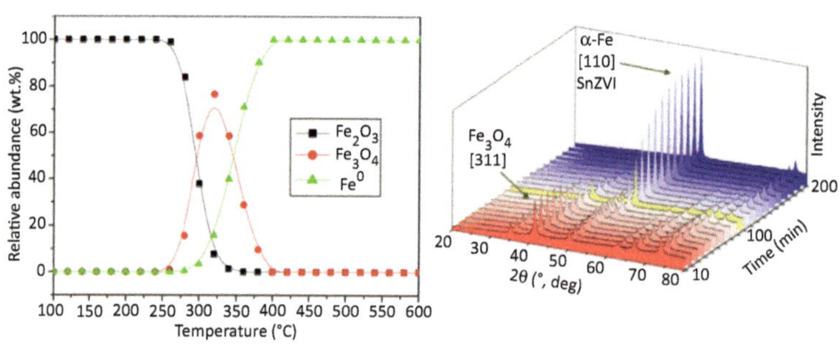


**Figure 6.1** Phase-composition plot derived from the dynamic HT-XRD monitoring of hematite reduction to nZVI in $H_2$ (left); XRD patterns showing the evolution from magnetite precursor to hollow-sphere nZVI particles (data recorded at 300 °C every 10 min). Adapted with permission from Ref. [30]. Copyright 2016 American Chemical Society.

Besides experiments in dynamic conditions (Fig. 6.2, right), the isothermal experiments are designed to evaluate the kinetics of the reactions (Figs. 6.2, left and 6.1, right). However, it is always advised to correlate results from HT-XRD with other techniques monitoring in situ the reaction pathway (see below) in order to avoid overlooking some hidden processes, including surface defects and formation of amorphous intermediates.

By changing the experimental conditions and/or using variable precursors, one can finely tune the properties of resulting nZVI particles (particle size, particle shape, specific surface area, some surface characteristics including passivation by additional oxide layer, and, therefore, reactivity of the prepared material) [19, 31].

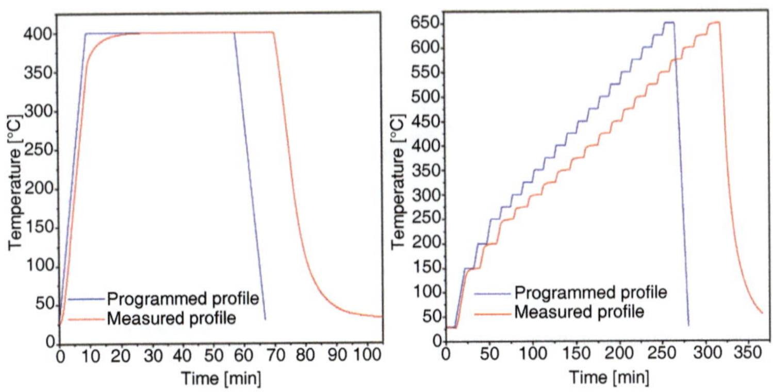


**Figure 6.2** Design of isothermal (left) and dynamic (right) HT-XRD experiments.

## 6.2.2 Thermoanalytical Methods

The process of iron oxide reduction could be further monitored using thermal analysis methods, employing either calorimetric methods (like differential scanning calorimetry, DSC) or thermogravimetry (TG) to directly monitor temperature-induced reactions under various atmospheres. The simultaneous TG and DSC measurement is an effective tool for understanding the complex investigated process on modern thermoanalytical instruments. The further advantage of this technique is a possible coupling of TG/DSC with evolved gas analysis (EGA) based on mass spectrometer or FTIR spectroscopy for the identification of volatile decomposition products.

The thermal reduction of nanocrystalline iron oxides (identical to sample from Ref. [11]) in a mixture of hydrogen and nitrogen gases (10% of $H_2$ in 90% of $N_2$) starts at about 400 °C, and the reduction is completed above 500 °C (Fig. 6.3). The process evidently comprises two steps, and it is consistent with Eq. (6.2) with a weight loss of 31%. In addition to the observed weight changes due to the reductive transformations of precursors, kinetic analysis can also be routinely performed using TG. A multivariate nonlinear regression of kinetic data could be used for the calculation of activation energy for the monitored process.

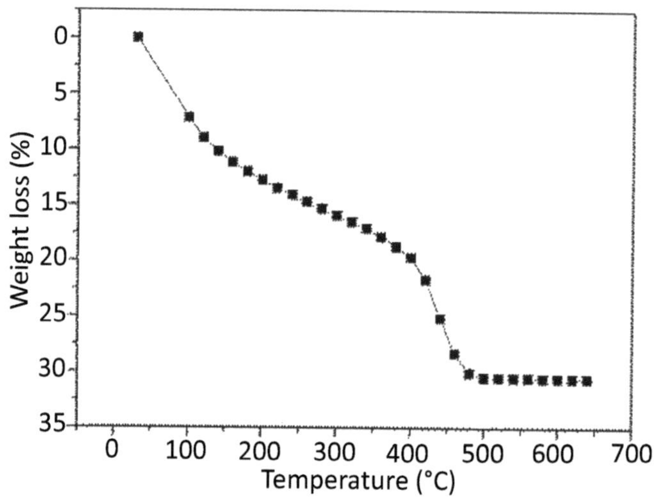


**Figure 6.3** The thermal reduction of nanocrystalline iron oxides (identical to sample from Ref. [11]).

The methods of thermal analysis can also be employed to evaluate the thermal stability of prepared nZVI particles. The as-prepared nZVI particles are highly pyrophoric, and they have to be stored either in a dry state under protective gas (typically $N_2$) or surface passivated for easy transport and storage on air. Especially for the storage and transport of surface-passivated nZVI particles, the evaluation of the thermal stability is critical. Figure 6.4 demonstrates the thermal stability of nZVI with a 4 nm thick oxide shell as monitored by TG.

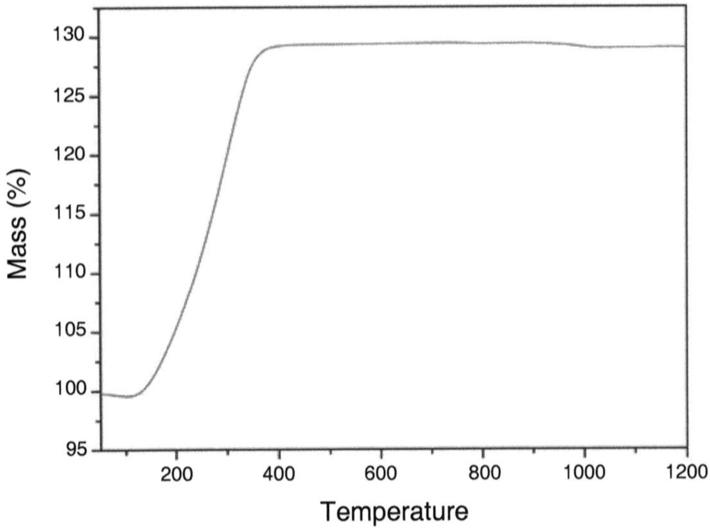


**Figure 6.4** Thermal stability of surface-passivated nZVI as viewed by TG.

### 6.2.3 Thermomagnetic Analysis

Another way to monitor the process of iron oxide reduction in hydrogen is to measure changes in magnetic properties of the sample during the temperature increase. Measurement of temperature dependence of the magnetic moment (i.e., thermomagnetic curves) of powdered materials is performed on a vibrating-sample magnetometer (VSM) equipped with a suitable furnace. The exact temperatures of transformation can be then estimated from the temperature dependences of the magnetic moment. Additionally, the observed Curie temperature of particular phases can help to identify the phases formed during the thermal treatment. The example shown in Fig. 6.5 illustrates the thermal reduction of ferrihydrite ($Fe_5HO_8{\cdot}4H_2O$) powder in hydrogen. From the evaluation of the thermomagnetic curve, we can deduce the two-step reduction of the iron oxide precursor toward metallic iron. At ~350 °C, the ferrihydrite is transformed to magnetite with Curie temperature at 585 °C. Then the subsequent growth of magnetization unambiguously indicates magnetite reduction to ferromagnetic $\alpha$-Fe nanoparticles [23].

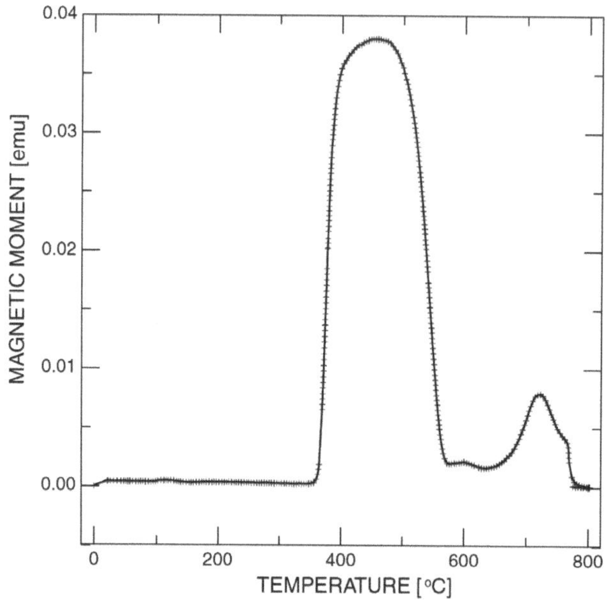


**Figure 6.5** The illustration of temperature dependence of the magnetic moment of ferrihydrite (sample identical to Ref. [11]) reduction in hydrogen. Adapted with permission from Ref. [23], copyright 2011 Springer Nature.

### 6.2.4 Temperature Programmed Reduction Using Gas Adsorption Technique

The temperature programmed reduction (TPR) method is a complementary technique to the above-mentioned tools to monitor the reduction of iron oxides in hydrogen under conditions of dynamic gas flow. The heating rates are carefully controlled for samples placed in a U-tube cell made of quartz glass. The hydrogen is diluted with an inert carrier gas (typically nitrogen), allowed to purge the sample cell before and during the heating, and the amount of consumed reactive gas is measured by a thermal conductivity detector (TCD) placed before and after the sample cell. Alternatively, the apparatus can be connected to a mass spectrometer to obtain the gas composition according to the molecular weight of the outgoing gas. The results are closely comparable with TG, where the same mixture of gases is used. The observed changes in the measured characteristics also support the two-step mechanism, including the initial reduction of $Fe_2O_3$ to $Fe_3O_4$, followed by magnetite reduction to metallic iron (Fig. 6.6).

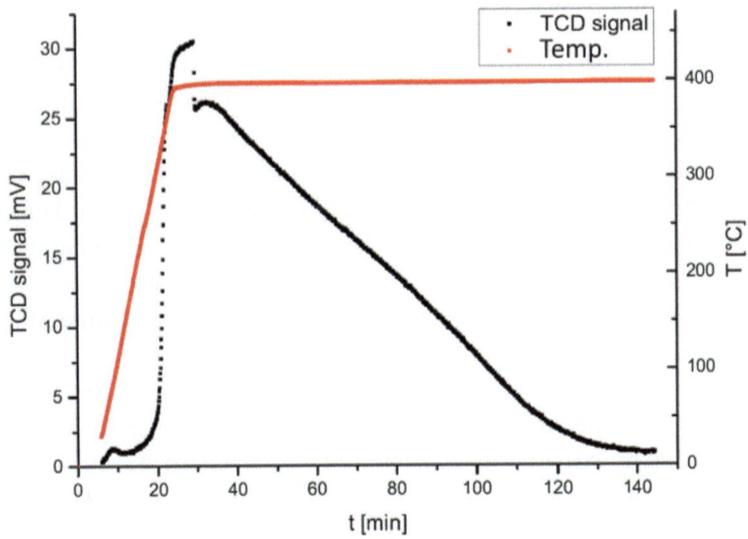


**Figure 6.6** Temperature programmed reduction of hematite precursor in hydrogen (isothermal measurement at 300 °C).

## 6.3 Methods for Detailed Characterization of As-Prepared, Surface-Modified and Reacted nZVI Particles

In order to demonstrate nZVI instability upon storage and to prepare nZVI samples with variable $Fe^0/Fe_{tot}$ ratio, several ageing experiments were conducted. Solid nZVI particles (synthesis protected by patent no. WO 2008/125068 A2) were purchased from NANO IRON Company (commercially available as NANOFER 25P). The powdered nZVI material (specific surface area of 27 $m^2/g$; mean particle size of 65 nm) was kept in a sealable metal container under protective nitrogen atmosphere at room temperature. The particles were left without coating (i.e., bare particles) and dispersed in deionized water under an inert gas and continuous intensive mechanical stirring for 10 min. All of the aqueous dispersions of the nZVI particles were relatively highly concentrated, i.e., 20% (w/w). All the experiments were performed with deionized water (18 MΩ/cm, Millipore), and all chemicals were used without any further purification.

The dispersions of the bare nZVI particles were sampled into 60 mL screw-cap glass vials (200 mg of nZVI in 50 mL of distilled water). All handling was performed under a protective atmosphere of nitrogen (i.e., under oxygen- and humidity-free conditions) in a sophisticated glove box (P[box]; JACOMEX, France). One-third of the vials were closed and stored in the glove box at room temperature, periodically shaken and regularly opened to allow them depressurize. Another third of vials were kept aerated at room temperature, half-closed and permanently shaken in order to simulate random opening of the containers containing delivered nZVI slurry during laboratory and/or field experiments. The last third of the vials were also tightly sealed and kept both at room temperature and 4 °C. Periodically, a set of these vials was opened and nZVI particles were directly analyzed in order to eliminate effects of their unwanted oxidation, which typically proceeds immediately when the particles are exposed to air.

In all of the experiments, the $Fe^0$ content was monitored by $^{57}Fe$ Mössbauer spectroscopy, XRD, and acid digestion method. These results were consequently mutually evaluated and compared.

### 6.3.1 X-Ray Powder Diffraction

X-ray diffraction patterns were recorded on an X′Pert PRO (PANalytical, The Netherlands) instrument in Bragg–Brentano geometry with iron-filtered $CoK_\alpha$ radiation ($\lambda$ = 0.178901 nm; 40 kV and 30 mA), equipped with a fast X′Celerator detector and programmable divergence and diffracted beam anti-scatter slits. Magnetically pre-concentrated slurries were inserted into a conventional cavity sample holder made of stainless steel and repeatedly scanned in the $2\theta$ range of 20–105° at ambient conditions (eight fast continuous scans per hour at a scan speed of about 11.5° $2\theta$/min and resolution of 0.017°, during which the sample was partially dried). Alternatively, the dry nZVI powder was immersed in silicon-based vacuum grease, spread between two Mylar foils and sealed (all handling was made in an inert atmosphere of $N_2$ in the glove box); such samples were immediately measured in the transmission mode utilizing an incident-beam hybrid monochromator. Commercial standards SRM640 (Si) and SRM660 ($LaB_6$) from NIST were used for the evaluation of the line positions and instrumental line broadening, respectively. The acquired patterns were processed (i.e., phase analysis and Rietveld refinement) using X′Pert HighScore Plus software (PANalytical, The Netherlands), PDF-4+ and ICSD databases (ICSD collection codes are $\alpha$-Fe—631729, magnetite—20596, hematite—82137). Peak shapes were modeled using the pseudo-Voigt function, separately refining the Caglioti parameters ($u$, $v$, $w$), unit cell parameters, and scale factor for each phase.

### 6.3.2 Transmission $^{57}$Fe Mössbauer Spectroscopy

The transmission $^{57}$Fe Mössbauer spectra were collected at a constant acceleration mode with a $^{57}$Co(Rh) source (1.85 GBq). The magnetically pre-concentrated slurries were fast frozen in liquid nitrogen bath, and measurements were carried out at 250 K and in external magnetic field of 0 T for 1 day per sample. The temperature of 250 K is selected as the most suitable to keep the samples frozen, but it is high enough ensuring that the sample is above the Verwey transition temperature of magnetite, as below Verwey transition, the spectral components of magnetite turn to a much more complicated feature. Alternatively, a dry nZVI powder was prepared under protective $N_2$ (i.e., in a glove box) into the form of conventional absorber (~5 mg Fe per cm$^2$), and measured at room temperature using a spectrometer located directly into the glove box. In both cases, the isomer shift values were calibrated against an $\alpha$-Fe foil at room temperature. The spectra were fitted with Lorentz functions using the computer program CONFIT2000 [32]. The effects of non-ideal absorber thickness and variable recoil-free fractions for iron atoms in non-equivalent structural sites of different phases were expected to be within experimental errors (hyperfine parameters ±0.02 mm/s, relative spectral area ±3%).

### 6.3.3 Acid Digestion

A simple technique for the measurement of $Fe^0$ content in nZVI-containing slurry was developed optimizing the well-known process of acid digestion of metals accompanied by hydrogen gas evolution (see also Ref. [33]). The method is based on the measurement of the volume of the evolved hydrogen gas during fast digestion of nZVI in a concentrated acid. The volume of evolved gaseous hydrogen is directly proportional to the amount of nZVI in the measured slurry. The iron oxides do not influence the volume of the generated hydrogen and produce no hydrogen under these conditions. The reaction of nZVI with either potassium bisulfate ($KHSO_4$) or hydrochloric acid (HCl) proceeds according to the following reactions:

$$2KHSO_4(aq) + Fe^0(s) \rightarrow FeSO_4(aq) + K_2SO_4(aq) + H_2(g) \quad (6.3)$$

or

$$6HCl(aq) + 2Fe^0(s) \rightarrow 2FeCl_3(aq) + 3H_2(g) \quad (6.4)$$

The weight and the concentration of nZVI in the slurry (either $Fe^0/Fe_{tot}$ ratio, or percentage of $Fe^0$ in slurry, i.e., w/w content of "active" metallic iron) can be consequently calculated from the volume of the evolved hydrogen. Based on Eq. (6.3), 55.85 g of zerovalent iron can evolve 22.41 $dm^3$ of hydrogen (see Appendix A: Practical example of acid digestion measurement).

The apparatus, enabling the reaction of nZVI with acids, was suggested to consist of a standard reaction glass bottle capped with septum (for injection of the nZVI slurry) and filled with saturated acid, connected to either a graduated U-tube or a graduated cylinder placed vertically and further connected to a storage vessel (Fig. 6.7). No leaks of evolved hydrogen from this test apparatus were detected using portable GS1 gas sniffer (Wöhler GmbH, Germany). The evaluation of $Fe^0$ content then proceeded as follows. The weighted amount of nZVI slurry is injected through the septum using a calibrated syringe (no need to evaluate volume of void space) at laboratory temperature. After completed the reaction (typically a few minutes in the case of nZVI particles well below 500 nm), the hydrogen volume is directly read from the graduation on the wall of the used glassware. To determine the accurate $Fe^0/Fe_{tot}$ ratio, it is needed to know the exact mass of the dispersed phase (i.e., nZVI particles) in the studied slurry. The nZVI can be weighed prior mixing with deionized water, or the air-dried solid sample can be annealed at 900 °C for 1 h in air in order to oxidize all iron into hematite ($\alpha$-$Fe_2O_3$) and then weighed. The correctness and limitations of this method were extensively tested on a series of nZVI samples with various initial concentrations of $Fe^0$.

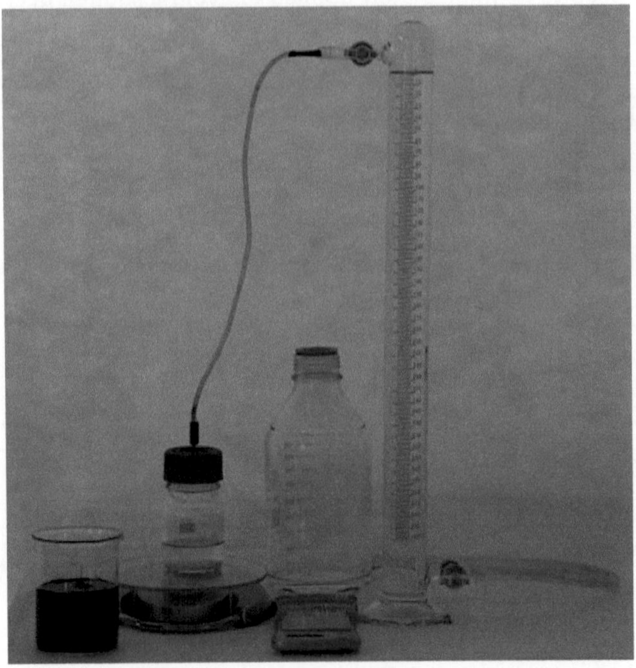

**Figure 6.7** Illustration of the suggested design of apparatus enabling the reaction of nZVI with acids.

### 6.3.4 Stability of nZVI Stored under Various Conditions

Under inert atmosphere (i.e., under gaseous nitrogen), the dry nZVI powder exhibits satisfactorily long stability, and no decrease in $Fe^0$ concentration was observed during the detailed monitoring for more than 300 days by XRD (Fig. 6.8, $A_1$). The nZVI samples, collected at various periods of ageing and analyzed with combination of XRD, $^{57}Fe$ Mössbauer spectroscopy, and acid digestion, yielded between 98 and 100 wt.% of $Fe^0$ relative to $Fe_{tot}$ (Fig. 6.8, $A_1$, $A_2$, and $A_3$). Therefore, the storage of dry nZVI powder under inert atmosphere was evaluated as the ideal choice for the long-term preservation of its original highly reactive state. However, prior any environmental application, an aqueous slurry of the nZVI particles has to be prepared transferring nZVI powder into a dispersing media under protective atmosphere. This way of preparing an nZVI slurry is then supposed to be immediately applied to avoid a sudden decrease in the $Fe^0$ content (see below).

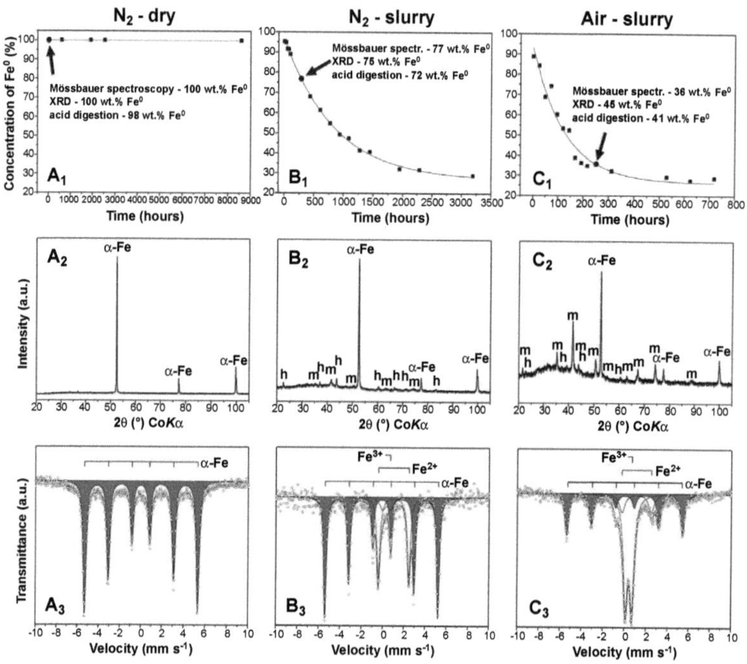


**Figure 6.8** Stability of nZVI in air and $N_2$ conditions in sequence of decreasing nZVI stability. Weigh percentages of measured $Fe^0$ (i.e., $\alpha$-Fe) as a function of the exposure time of nZVI to its storage environment. Characteristic X-ray powder diffraction patterns (m—magnetite/maghemite; h—hematite) and $^{57}$Fe Mössbauer spectra of quality sufficient for nZVI evaluation and collected at reasonable time for routine analysis. Data partly adapted with permission from Ref. [28]. Copyright 2014 American Chemical Society.

The aqueous nZVI slurries kept in tightly sealed vials at room temperature and 4 °C are relatively stable for more than 30 and 60 days, respectively (Fig. 6.9). However, progressive spontaneous reaction of nZVI with water molecules at room temperature typically causes the gradual evolution of gaseous hydrogen leading to overpressure of the vessels, occasionally ending in "explosion" and complete loss of the sample. On the contrary, the nZVI slurries stored at room temperature both in air or under nitrogen and with the allowed continuous release of the evolved hydrogen were nearly completely oxidized by reaction with water and/or dissolved oxygen under ambient conditions. The fractions of $Fe^0$ in the aged nZVI samples, derived from detailed $^{57}Fe$ Mössbauer spectroscopy measurements, are plotted in Fig. 6.8 ($B_1$ and $C_1$). For both series of the samples aged in water, the overall trends are very consistent in the exhibition of an exponential decrease in the $Fe^0$ content as a function of time [28]. Nevertheless, when the nZVI slurry is stored in air, the kinetics of its degradation is about seven times faster than in the case of the sample stored under oxygen-free atmosphere.

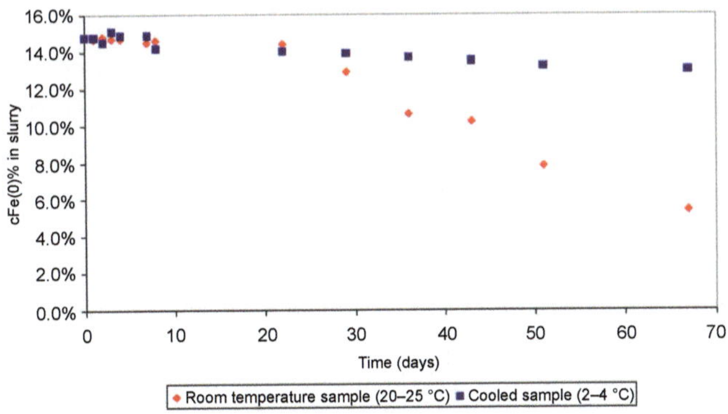


**Figure 6.9** Stability of aqueous nZVI slurries kept in tightly sealed vessels at room temperature and 4 °C; nZVI quantification is based on acid digestion.

The decrease in the relative $Fe^0$ content to the limiting value of approximately 30% was determined after ~300 h when stored in air (50% decrease after ~130 h); under a protective atmosphere of nitrogen, the same values of $Fe^0$ contents were reached after ~2000 and 900 h of ageing, respectively (Fig. 6.8). The validity of this model was independently proved by XRD and acid digestion method. Other parameters, such as particle size and morphology or specific surface area, do not change so dramatically [34]. The details on mechanisms and kinetics of nZVI ageing under various conditions are not further discussed in this chapter and are presented elsewhere [28]. The results clearly demonstrate the necessity to characterize any commercial nZVI slurries directly prior their application because the concentration of the "active" elemental iron could be significantly different to the declared value even after several days of its storage. This undesired property of the nZVI slurries side by side with the choice of a proper analytical technique represents a critical prerequisite if reliable results are expected from the targeted remediation processes. Moreover, the same experimental approach can be applied when the fate of highly reactive nZVI particles is monitored in groundwater (i.e., in series of wells), soils, or wastewater [35].

### 6.3.5 Comparison and Evaluation of Key Analytical Techniques

In this part, we will focus on the comparison and detailed evaluation of the results of the cheapest, most accessible, and simplest methods for evaluating the nZVI content in aqueous slurries and giving reproducible and accurate data at the same time. Special attention will also be paid to the aspect of possible applicability of the particular method directly at the treated sites (i.e., onsite measurement). Certainly, an inclusive multi-analytical approach, including a combination of XRD, $^{57}Fe$ Mössbauer spectroscopy, TEM, XPS, XAS, and several other methods, would give a nearly complete characterization of the nZVI particles intended for remediation, namely the extent and nature of the nZVI surface oxidation prior to its application. Nevertheless, such sophisticated techniques (as XAS, XPS, and TEM) for characterization of nanoparticles are not involved in this discussion because their availability is exclusively limited to highly specialized laboratories or even to synchrotron sources. In addition, these sophisticated laboratory techniques also have the following advantages/disadvantages. First, TEM is highly impractical for routine analysis of nZVI related to their large-scale application and cannot precisely distinguish the $Fe^0/Fe_{tot}$ ratio in the bulk sample. Second, the preparation of the samples for the TEM analysis requires only a small volume of the slurry and, therefore, because of the quite common inhomogeneity of such slurries, the resulting images can be weighed down by errors [36]. In turn, XPS is not a common technique, and like TEM, it does not give the average (i.e., bulk) $Fe^0/Fe_{tot}$ ratio as it analyzes a thin surface layer up to 10 nm. Finally, XAS requires synchrotron sources [37].

From each of the above-described experimental series, the nZVI samples containing predominantly about 80 wt.% of $Fe^0$ (i.e., a typical value of freshly opened commercial aqueous slurries used in remediation) were selected to demonstrate the consistency of the selected analytical techniques (data presented in Fig. 6.10). Additionally, several other samples representing a wide concentration range were measured in order to support the validity of the suggested acid digestion technique also for the highly concentrated nZVI slurries with high and low concentrations of $Fe^0$.

#### 6.3.5.1 X-ray powder diffraction

X-ray powder diffraction is an essential and generally accepted tool for the standard characterization of the phase composition (i.e., the crystal structure and quantitative phase analysis) of crystalline solids, including nZVI [38]. It provides also illustration of particle size according to the broadening of diffraction peaks or when combined with small-angle X-ray scattering (SAXS) [39]. Moreover, XRD is a widespread technique commonly available in numerous analytical laboratories, especially in those oriented on solid state, (nano)materials or environmental research. The procedure of sample preparation for the XRD measurement, data collection, and processing has been reviewed in detail many times [40].

However, when handling nZVI prior and during the measurement, several specific steps have to be followed, related mainly to the fast oxidation of nZVI when inappropriately exposed to air [28]. In order to maintain the reproducibility of the acquired data, two protocols were followed to allow measurement of either wet nZVI samples (i.e., magnetically pre-concentrated slurries) or dry nZVI powders. Moreover, successive short scans using a fast solid-state detector provide a unique insight into the possible process of nZVI oxidation during sample measurement (typically acquired within less than 1 h) accompanied by spontaneous drying of the slurry. For the subsequent quantitative phase analysis utilizing a full-profile fitting (i.e., Rietveld refinement), either a first fast scan collected just at the beginning of nZVI drying/oxidation can be used, or to sum up all of the collected scans when no phase changes took place in the course of successive measurements.

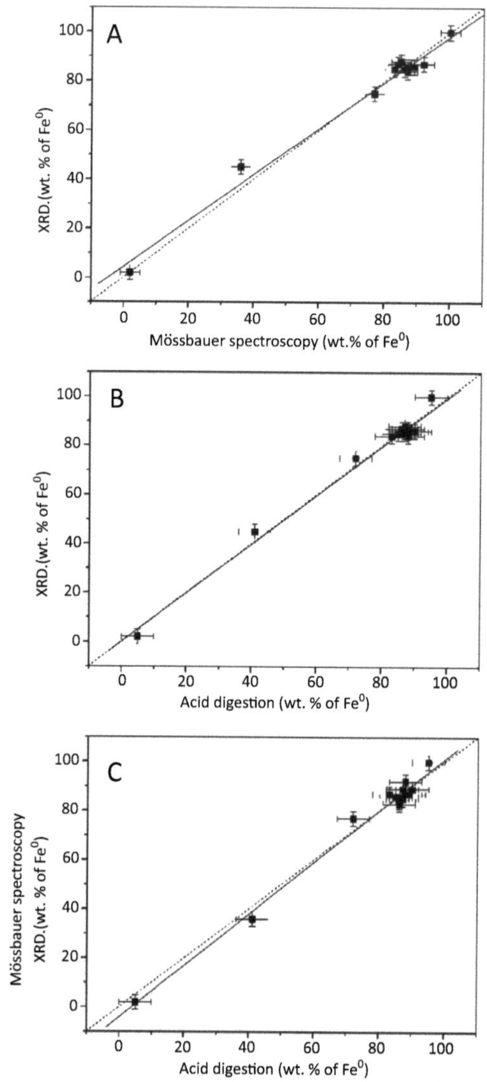


**Figure 6.10** Comparison of results of XRD, Mössbauer spectroscopy, and acid digestion measured on a broad range of $Fe^0/Fe_{tot}$ ratios. The dotted line represents the 1:1 ratio.

The typical XRD patterns of the aged nZVI particles are presented in Fig. 6.8 ($A_2$, $B_2$, and $C_2$). The major peak at ~2.027 Å (i.e., at ~52° $2\theta$ for Co$K_\alpha$ radiation or at ~44° $2\theta$ for Cu$K_\alpha$) is characteristic of metallic $\alpha$-Fe and represents the diffraction on the (110) plane.

Less intensive and broader peaks are ascribed to other crystalline nanosized phases resulting from the oxidation of $\alpha$-Fe. The hump typically centered at ca. 35° $2\theta$ (Fig. 6.8, $B_2$ and $C_2$) is caused by high humidity of the measured sample and it disappeared when the sample gets dry. Successive scans collected on each sample of nZVI slurry exposed to ambient air confirmed the sample stability in terms of nZVI oxidation (i.e., surface passivation of nZVI particles) during its drying as no change in phase composition was observed in the majority of cases. Hence, the slow drying of the nZVI slurry causes the effective passivation of nZVI particle surface, and the activity diminishes under air. The ratio of metallic iron to iron oxides in each sample was calculated from the Rietveld refinement [41] and is presented as weigh percent of the particular phase (Fig. 6.10). XRD measurements on samples that have been degraded under different conditions and for different periods of time are in a good agreement with the other two experimental techniques, namely with acid digestion (Fig. 6.10). It could be concluded that the sample preparation and the experimental conditions for XRD measurements lead to results highly comparable to other experimental techniques. No amorphous phase(s) was observed in quantities larger than the uncertainty of any employed methods. Exploiting Rietveld refinement, the accuracy of the XRD results are below 3% for the quantification of the $\alpha$-Fe phase.

### 6.3.5.2 Transmission $^{57}$Fe Mössbauer spectroscopy

Transmission $^{57}$Fe Mössbauer spectroscopy is another important tool for the characterization of nZVI (both in dry powders and aqueous slurries), and it represents a unique technique for probing the $Fe^0/Fe_{tot}$ ratio, yielding precise values of the relative $Fe^0$ content. It is a less common technique compared with XRD, and it is definitely more complicated than the acid digestion method. On the other hand, the Mössbauer spectrometer is usually compact, portable with some limitations (e.g., need of electric supply), and significantly cheaper compared to X-ray powder diffractometer. The main disadvantage of Mössbauer spectroscopy lies in the relatively long counting times (typically about 1 day per sample of nZVI) during which the measured sample could further oxidize. This drawback is compensated by the possibility of measuring nZVI samples under a protective atmosphere, as the whole spectrometer could be installed, for example, in a glove box [28]. Eventually, the pre-concentrated nZVI slurries could be fast frozen and measured at low temperatures (optimally at liquid nitrogen temperature) utilizing a special cryogenic system leading to significantly slowed down oxidation of nZVI [11]. The low-temperature measurements are further beneficial in the case when reaction products of nZVI are amorphous (therefore hard to be detected by XRD). The other very important benefit of this method is its entirely iron-selectivity [42]. Therefore, $^{57}Fe$ Mössbauer spectroscopy could be applied also for complicated subjects like for direct observation of nZVI oxidation in soil samples collected from either laboratory column experiments or field tests.

The $^{57}Fe$ Mössbauer spectra of representative samples measured at room temperature are depicted in Fig. 6.8 ($A_3$, $B_3$, and $C_3$). Spectra are characterized by one sextet of $\alpha$-Fe with theoretical hyperfine parameters: isomer shift $\delta = 0.00$ mm/s, quadrupole shift $\varepsilon_Q = 0.00$ mm/s, and hyperfine magnetic field $B_{hf} = 33$ T. Moreover, $Fe^{2+}$ and $Fe^{3+}$ components, clearly resolved as central (super)paramagnetic components or magnetically split hyperfine sextets, allow the possible identification even of amorphous and nanocrystalline phases. If necessary, low temperature and/or external magnetic field can be applied to induce magnetic splitting of doublet spectral lines of superparamagnetic and amorphous phases for their unambiguous identification (the sextets of respective phases differ mainly in effective magnetic fields) and quantification of their relative contents based on spectral areas [43]. Ignoring different recoil-free fractions for iron atoms in non-equivalent structural sites of different phases, the quantification of identified iron-bearing phases is basically done on relating spectral areas (i.e., Fe-atomic ratios) to known molar weights [44]. In full accordance with XRD, a dramatic change is visualized in $\alpha$-Fe spectral component in Fig. 6.8 ($A_3$, $B_3$, and $C_3$) during the course of nZVI ageing. The consistency of quantitative results derived from Mössbauer spectroscopy with XRD is illustrated in Fig. 6.10 (A).

#### 6.3.5.3 Acid digestion

The process of acid digestion of micron-sized iron particles is typically very slow and requires up to 20 days until the reaction is completed. Nevertheless, when working with particles in the diameter range well below 500 nm (i.e., as typical for all commercially available nZVI materials [5]), this process is significantly faster and all iron particles are usually dissolved within a few minutes. The suggested design of the newly introduced test apparatus for acid digestion of nZVI slurries is simple (Fig. 6.7) and represents the cheapest way for quantifying $Fe^0$ in aqueous dispersions. The apparatus can be constructed and easily operated at any chemical laboratory or at remediation sites where nZVI slurries are applied and/or monitored. Based on the simplicity of the method, also the interpretation of the acquired results is simple and clearly addressed (see Appendix A). Another aspect of this method is that it could bring not only quantification of the $Fe^0/Fe_{tot}$ ratio (similarly as Mössbauer spectroscopy and XRD), but also the actual content of $Fe^0$ in aqueous slurries. This property of the method is highlighted namely in relation to large-scale application of nZVI slurries at remediation sites and when studying their fate.

The comparison of quantitative results from acid digestion of selected nZVI samples with the results from XRD and Mössbauer spectroscopy is presented in Figs. 6.10(B) and 6.10(C), respectively. Regarding the simplicity of acid digestion, the acquired results are in surprisingly good agreement with conventional sophisticated analytical techniques. Test apparatus for hydrogen-volume measurement was also already tested with commercial nZVI slurry (TODA) prior and during laboratory column tests [33] and yielded results comparable to values declared by nZVI manufactures for fresh nZVI slurry, and to previously published results measured on the similar freshly delivered nZVI materials. The estimated errors are below 5% from four independent measurements. Therefore, the acid digestion of nZVI with quantification of evolved gaseous hydrogen volume turned out to be a simple method well applicable for routine characterization of nZVI slurries delivered to the site of application or collected from groundwater following their application. Nevertheless, this technique does not provide any information about the exact speciation of oxidized iron.

When the w/w concentration provided by the manufacturer is used as a starting parameter for the $Fe^0$ content calculation, the result represents an approximate value because the dispersion is never homogeneous and the particles spontaneously aggregate and sediment (it could be suitable for estimative in-field measurements, where the access to well-equipped laboratory tools is usually limited). For more precise results, the actual w/w value should be obtained, for example, by drying and subsequent annealing of the representative part of the slurry at 900 °C in air (see Appendix A). The precise data are obtained when the slurry is freshly prepared by weighting the solid nZVI under protective inert atmosphere into a known volume of deionized water.

## Acknowledgments

This work was supported by grants from the EU FP7 (project NANOREM), Technology Agency of the Czech Republic "Competence Centers" (project No. TE01020218), and the project LO1305 of the Ministry of Education, Youth and Sports of the Czech Republic. This work was further supported by Student Project IGA_PrF_2017_010 of Palacký University, Olomouc.

## References

1. Marsalek, B., Jancula, D., Marsalkova, E., Mashlan, M., Šafářová, K., Tuček, J., and Zbořil, R. (2012). Multimodal action and selective toxicity of zerovalent iron nanoparticles against cyanobacteria, *Environ. Sci. Technol.*, **46**, pp. 2316–2323.
2. Marková, Z., Machalová Šišková, K., Filip, J., Čuda, J., Kolář, M., Šafářová, K., Medřík, I., and Zbořil, R. (2013). Air stable magnetic bimetallic Fe-Ag nanoparticles for advanced antimicrobial treatment and phosphorus removal, *Environ. Sci. Technol.*, **47**, pp. 5285–5293.
3. Wei, Y.-T., Wu, S.-C., Chou, C.-M., Che, C.-H., Tsai, S.-M., and Lien, H.-L. (2010). Influence of nanoscale zero-valent iron on geochemical properties of groundwater and vinyl chloride degradation: A field case study, *Water Res.*, **44**, pp. 131–140.
4. Yan, W., Herzing, A. A., Kiely, C. J., and Zhang, W.-X. (2010). Nanoscale zero-valent iron (nZVI): Aspects of the core–shell structure and reactions with inorganic species in water, *J. Contam. Hydrol.*, **118**, pp. 96–104.
5. Fu, F., Dionysiou, D. D., and Liu, H. (2014). The use of zero-valent iron for groundwater remediation and wastewater treatment: A review, *J. Hazard. Mater.*, **267**, pp. 194–205.

6. Nurmi, J. T., Tratnyek, P. G., Sarathy, V., Baer, D. R., Amonette, J. E., Pecher, K., Wang, C., Linehan, J. C., Matson, D. W., Penn, R. L., and Driessen, M. D. (2005). Characterization and properties of metallic iron nanoparticles: Spectroscopy, electrochemistry, and kinetics, *Environ. Sci. Technol.*, **39**, pp. 1221–1230.
7. Zhang, W.-X. (2003). Nanoscale iron particles for environmental remediation: An overview, *J. Nanopart. Res.*, **5**, pp. 323–332.
8. Li, X.-Q., Elliott, D. W., and Zhang, W.-X. (2006). Zero-valent iron nanoparticles for abatement of environmental pollutants: Materials and engineering aspects, *Crit. Rev. Solid State Mater. Sci.*, **31**, pp. 111–122.
9. Tratnyek, P. G. and Johnson, R. L. (2006). Nanotechnologies for environmental cleanup, *Nano Today*, **1**, pp. 44–48.
10. Lien, H.-L. and Zhang, W.-X. (2001). Nanoscale iron particles for complete reduction of chlorinated ethenes, *Colloids Surf., A*, **191**, pp. 97–105.
11. Filip, J., Zbořil, R., Schneeweiss, O., Zeman, J., Černík, M., Kvapil, P., and Otyepka, M. (2007). Environmental applications of chemically pure natural ferrihydrite, *Environ. Sci. Technol.*, **41**, pp. 4367–4374.
12. Liu, Y., Phenrat, T., and Lowry, G. V. (2007). Effect of TCE concentration and dissolved groundwater solutes on NZVI-promoted TCE dechlorination and $H_2$ evolution, *Environ. Sci. Technol.*, **41**, pp. 7881–7887.
13. Šišková, K., Tuček, J., Machala, L., Otyepkova, E., Filip, J., Šafářová, K., Pechousek, J., and Zbořil, R. (2012). Air-stable nZVI formation mediated by glutamic acid: Solid-state storable material exhibiting 2D chain morphology and high reactivity in aqueous environment, *J. Nanopart. Res.*, **14**, 805.
14. Klimkova, S., Černík, M., Lacinova, L., Filip, J., Jancik, D., and Zbořil, R. (2011). Zero-valent iron nanoparticles in treatment of acid mine water from *in situ* uranium leaching, *Chemosphere*, **82**, pp. 1178–1184.
15. Sohn, K., Kang, S. W., Ahn, S., Woo, M., and Yang, S.-K. (2006). Fe(0) nanoparticles for nitrate reduction: Stability, reactivity, and transformation, *Environ. Sci. Technol.*, **40**, pp. 5514–5519.
16. Fan, J., Guo, Y., Wang, J., and Fan, M. (2009). Rapid decolorization of azo dye methyl orange in aqueous solution by nanoscale zerovalent iron particles, *J. Hazard. Mater.*, **166**, pp. 904–910.

17. Zhang, X., Lin, Y.-M., Shan, X.-Q., and Chen, Z.-L. (2010). Degradation of 2,4,6-trinitrotoluene (TNT) from explosive wastewater using nanoscale zero-valent iron, *Chem. Eng. J.*, **158**, pp. 566–570.
18. Zhang, Y., Li, Y., and Zheng, X. (2011). Removal of atrazine by nanoscale zero-valent iron supported on organobentonite, *Sci. Total Environ.*, **409**, pp. 625–630.
19. Zbořil, R., Andrle, M., Oplustil, F., Machala, L., Tuček, J., Filip, J., Marušák, Z., and Sharma, V. K. (2012). Treatment of chemical warfare agents by zero-valent iron nanoparticles and ferrate(VI)/(III) composite, *J. Hazard. Mater.*, **211–212**, pp. 126–130.
20. Cundy, A. B., Hopkinson, L., and Whitby, R. L. D. (2008). Use of iron-based technologies in contaminated land and groundwater remediation: A review, *Sci. Total Environ.*, **400**, pp. 42–51.
21. Soukupová, J., Zbořil, R., Medřik, I., Filip, J., Šafářová, K., Ledl, R., Mashlan, M., Nosek, J., and Černík, M. (2015). Highly concentrated, reactive and stable dispersion of zero-valent iron nanoparticles: Direct surface modification and site application, *Chem. Eng. J.*, **262**, pp. 813–822.
22. Wang, Q., Kanel, S. R., Park, H., Ryu, A., and Choi, H. (2009). Controllable synthesis, characterization, and magnetic properties of nanoscale zerovalent iron with specific high Brunauer–Emmett–Teller surface area, *J. Nanopar. Res.*, **11**, pp. 749–755.
23. Schneeweiss, O., Filip, J., David, B., Zbořil, R., and Mašláň, M. (2011). Iron nanoparticles prepared from natural ferrihydrite precursors: Kinetics and properties, *J. Nanopart. Res.*, **13**, pp. 5677–5684.
24. Martin, J. E., Herzing, A. A., Yan, W., Li, X.-Q., Koel, B. E., Kiely, C. J., and Zhang, W.-X. (2008). Determination of the oxide layer thickness in core–shell zerovalent iron nanoparticles, *Langmuir*, **24**, pp. 4329–4334.
25. Chun, C. L., Baer, D. R., Matson, D. W., Amonette, J. E., and Penn, R. L. (2010). Characterization and reactivity of iron nanoparticles prepared with added Cu, Pd, and Ni, *Environ. Sci. Technol.*, **44**, pp. 5079–5085.
26. Wang, C.-B. and Zhang, W.-X. (1997). Synthesizing nanoscale iron particles for rapid and complete dechlorination of TCE and PCBs, *Environ. Sci. Technol.*, **31**, pp. 2154–2156.
27. Sun, Y.-P., Li, X.-Q., Cao, J., Zhang, W.-X., and Wang, H. P. (2006). Characterization of zero-valent iron nanoparticles, *Adv. Colloid Interface Sci.*, **120**, pp. 47–56.

28. Filip, J., Karlický, F., Marušák, Z., Lazar, P., Černík, M., Otyepka, M., and Zbořil, R. (2014). Anaerobic reaction of nanoscale zerovalent iron with water: Mechanism and kinetics, *J. Phys. Chem. C*, **118**, pp. 13817–13825.

29. Bućko, M. S., Mattila, O.-P., Chrobak, A., Ziółkowski, G., Johanson, B., Čuda, J., Filip, J., Zbořil, R., Pesonen, L. J., and Leppäranta, M. (2013). Distribution of magnetic particulates in a roadside snowpack based on magnetic, microstructural and mineralogical analyses, *Geophys. J. Int.*, **195**, pp. 159–175.

30. Slovák, P., Malina, O., Kašlík, J., Tomanec, O., Tuček, J., Petr, M., Filip, J., Zoppellaro, G., and Zbořil, R. (2016). Zero-valent iron nanoparticles with unique spherical 3D architectures encode superior efficiency in copper entrapment, *ACS Sustainable Chem. Eng.*, **4**, pp. 2748–2753.

31. Hermankova, P., Hermanek, M., and Zbořil, R. (2010). Thermal decomposition of ferric oxalate tetrahydrate in oxidative and inert atmospheres: The role of ferrous oxalate as an intermediate, *Eur. J. Inorg. Chem.*, **2010**, pp. 1110–1118.

32. Žák, T. and Jirásková, Y. (2006). CONFIT: Mössbauer spectra fitting program, *Surf. Interface Anal.*, **38**, pp. 710–714.

33. Beneš, P., de Marco Rodrigo, M., Mašín, P., and Kubal, M. (2009). Nanoiron's activity measurement technique and soil enrichment possibilities, *Conference Proceedings 2009*, NANOCON 2009, pp. 217–223.

34. Liu, Y., Majetich, S. A., Tilton, R. D., Sholl, D. S., and Lowry, G. V. (2005). TCE dechlorination rates, pathways, and efficiency of nanoscale iron particles with different properties, *Environ. Sci. Technol.*, **39**, pp. 1338–1345.

35. Němeček, J., Pokorný, P., Lhotský, O., Knytl, V., Najmanová, P., Steinová, J., Černík, M., Filipová, A., Filip, J., and Cajthaml, T. (2016). Combined nano-biotechnology for in-situ remediation of mixed contamination of groundwater by hexavalent chromium and chlorinated solvents, *Sci. Total Environ.*, **563–564**, pp. 822–834.

36. Cao, G. (2004). *Nanostructures and Nanomaterials: Synthesis, Properties, and Applications*, 1st Ed. (Imperial College Press, London).

37. Gerson, A. R., Halfpenny, P. J., Pizzini, S., Ristíc, R., Roberts, K. J., Sheen, D. B., and Sherwood, J. N. (2007). *Materials Science and Technology: A Comprehensive Treatment, Vol. 2A, Characterization of Materials*, Cahn, R. W., Haasen, P., and Kramer, J. (eds.), Chapter 2 "Application

of synchrotron X-radiation to problems in materials science," (Wiley-VCH, Weinheim) pp. 551–617.

38. Clearfield, A., Reibenspies, J., and Bhuvanesh, N. (2008). *Principles and Applications of Powder Diffraction.* (Blackwell Publishing Ltd, Oxford).
39. Williams, C. E., May, R. P., and Guinier, A. (2006). *Materials Science and Technology: A Comprehensive Treatment, Vol. 2B, Characterization of Materials,* Cahn, R. W., Haasen, P., and Kramer, J. (eds.), "Small-angle scattering of X-rays and neutrons," (Wiley-VCH, Weinheim) pp. 611–656.
40. Buhrke, V. E., Jenkins, R., and Smith, D. K. (1998). *A Practical Guide for the Preparation of Specimens for X-Ray Fluorescence and X-Ray Diffraction Analysis.* (John Wiley & Sons, Inc., New York).
41. Young, R. A. (1993). *The Rietveld Method.* (Oxford University Press, New York).
42. Sharma, V. K., Klingelhöfer, G., and Nishida, T. (2013). *Mössbauer Spectroscopy: Applications in Chemistry, Biology, and Nanotechnology.* (John Wiley & Sons, Inc., New Jersey).
43. Prucek, R., Tuček, J., Kolařík, J., Filip, J., Marušák, Z., Sharma, V. K., and Zbořil, R. (2013). Ferrate(VI)-induced arsenite and arsenate removal by *in situ* structural incorporation into magnetic iron(III) oxide nanoparticles, *Environ. Sci. Technol.*, **47**, pp. 3283–3292.
44. Tuček, J., Zbořil, R., and Petridis, D. (2006). Maghemite nanoparticles by view of Mössbauer spectroscopy, *J. Nanosci. Nanotechnol.*, **6**, pp. 926–947.

## Appendix A. Practical example of acid digestion measurement

Brief description of the measurement procedure:

1. Set up the nZVI TESTER (see Fig. 6.7).
2. Take a sample of slurry with a syringe and transfer it to the reaction bottle containing acid.
3. Weigh the syringe involving the slurry and once again after the injection of the slurry into the acid-containing bottle (the difference is the actual amount of slurry injected into the reaction mixture).
4. Read the volume of the evolved hydrogen ($V_{H2}$ [mL]).
5. Calculate nZVI weight: $m_{nZVI} = (55.85 \times 101.325 \times V_{H2})/[8314.3 \times (273.15 + t)]$.
6. Calculate the content of the nZVI in the slurry: $c_{nZVI} = (m_{nZVI} \div m_{slurry}) \times 100$.

### Example of measurement

Sample: Nanofer 25S (20% of iron nanoparticles; 80% of water + stabilizer), age: 3 days.

Sample weight: $m_{slurry} = 4.95$ g.

Volume of the evolved hydrogen: $V_{H2} = 350$ mL.

Temperature = 20 °C

nZVI weight:

$m_{nZVI} = (55.85 \times 101.325 \times 350)/[8314.3 \times (273.15 + 20)] = 0.81304$ g.

nZVI in slurry content: $c_{nZVI} = (0.81304 \div 4.95) \bullet 100 = 16.43\%$.

Content of iron oxides: $c_{Fe(oxides)} = 20 - 16.43 = 3.57\%$

$c_{nZVI}$ to $c_{Fe(oxides)}$ ratio: 82% nZVI; 18% Fe(oxides).

If the Fe content in the slurry is unknown, its determination is necessary in order to use this information in the final calculations:

1. Take a sample of slurry by a syringe, and weigh the syringe with its content.
2. Inject the content into a porcelain crucible.
3. Weigh the syringe once again to get the information about the precise weight of the slurry.
4. Anneal the sample at 900 °C for 1 h in air in order to oxidize all iron into hematite ($\alpha$-$Fe_2O_3$) and then weigh it.
5. All Fe in Fe-containing compounds is oxidized to $Fe_2O_3$. From the weight of hematite, $Fe_{tot}$ content can be simply calculated.

Chapter 7

# Nanoiron for Site Remediation: Bench-Scale Assessment and Field Applications

**Gerardo Daniel López**
*National Technological University and Nanotek S.A., Güemes 3878 D 4, Santa Fe, S3002GHH, Argentina*
gerardo@santafe-conicet.gob.ar; gerardo.lopez@nanotek.ws

## 7.1 Introduction

Development and implementation of environmentally beneficial nanotechnologies for treatment and remediation have experienced noticeable growth in recent years. In terms of site remediation, the development and deployment of nanotechnology for contaminant destruction have already taken place. A multitude of nanotechnology applications for site remediation and wastewater treatment are currently in the research and development stages. Nanotechnology represents an extremely broad field, which encompasses a number of materials and technologies spanning multiple disciplines. Currently a wide variety of potential remedial tools employing nanotechnology are being examined at the bench scale for use in wastewater and

*Iron Nanomaterials for Water and Soil Treatment*
Edited by Marta I. Litter, Natalia Quici, and Martín Meichtry

ISBN 978-981-4774-67-3 (Hardcover), 978-981-4669-49-8 (eBook)
www.panstanford.com

soil remediation. But one emerging nanotechnology, nanosized zerovalent iron and its derivatives, has reached the commercial market for field-scale remediation. The scope of this chapter is to present a general overview of the technology and then focus on some relevant case stories based on our actual experience in applications of nanoiron for remediation of specific environmental impacts, including polychlorinated biphenyls (PCBs), hydrocarbons, heavy metals and metalloids, and the like. Results to be discussed originate from several levels of implementation (bench scale, pilot testing, and field-scale remediation) on different sites, including Argentina, Ecuador, Paraguay, and Canada.

Although most of the current commercial products that include nanomaterials in their formulation belong to manufacturing industries, there is a niche in the market for applications of this new transdiscipline in environmental care. Applications in this area can be conceptually distinguished into two criteria: avoiding contamination (ex ante technologies) and remediating affected sites once contamination has taken place (ex post technologies). This text focuses on ex post technologies, by reviewing mature processes in order to not only understand the technology but also take into account environmental, safety, and health issues associated with actual applications.

In the past few years, there has been a continuous growth in developments for mitigation of negative environmental impacts through the use of nanomaterials. Some of these developments have already left the laboratory and reached the field. The most successful of nanoparticles (NPs) used for these applications is iron in a diversity of configurations: nanozerovalent iron, core–shell structures, suspensions and emulsions of iron alone or bimetallic NPs, which are usually referred to as nanofluids or paramagnetic nanofluids.

By 2005, a review committed to the US Environmental Protection Agency (USEPA)[1] detected more than 15 field demonstration tests that involved iron NPs, as well as several projects at the laboratory level to develop techniques for treatment of contaminated sites, including soil and water matrixes.

Most affected sites show complex and variable situations, thus requiring that remediation procedures involve several specialties to interact with nanotechnologists, such as process engineers, geologists, chemists, and the like. That is why in actual field applications, solutions involving nanomaterials, the transdisciplinary nature of nanotechnology is fully revealed.

In order to assess the future and comprehend the potential of this emergent field of nanotechnology, it is paramount to understand the science underlying applications, for technologies currently in the market as well as for alternatives still under academic study. This assessment should balance efficacy of mitigation with eventual or collateral ecological impacts derived from nanomaterials, so as to avoid situations in which the cure could prove worse than the disease.

From a practical point of view, mature applications of nanoiron in environmental remediation can be classified into two broad types: use of NPs as a reactant and use of nanoiron as a catalyst. This classification is discussed in the following subsections.

### 7.1.1 Nanoiron as Reactant

By the end of the last century, the capacity of NPs (and particularly those based on iron) for chemical reduction–oxidation reactions was at the center of the academic community due to the potential to use these properties for the treatment of a wide range of contaminants found in effluent waters as well as subsurface water and soils impacted by spills of chemicals and petroleum derivatives. The usual form in which these nanomaterials were used was the creation of permeable reactive barriers designed to intercept and remediate contamination plumes. Unlike conventional systems based on extraction of contaminated material, its treatment and subsequent deployment on the site, these techniques for passive mitigation in situ proved to be less costly in operative terms than ex situ treatments. But the most important of the distinctive characteristics of this approach was the avoidance of health and environmental risks derived of dissemination of unwanted substances outside the original site due to eventual spills during transportation of contaminated material. These barriers have been used for the mitigation of different kinds of contaminants, including some recalcitrant to bioremediation such as chlorinated organic compounds, aromatic nitro compounds, PCBs, pesticides, and even metals and metalloids such as hexavalent chrome and arsenic.

The main limitations of these reactive barriers include the limitation in deep deployment of NPs, which becomes problematic if depth is higher than 15 m, the requirement of a very good previous characterization of the morphology and size of contaminated plume so as to avoid lixiviation of contaminants beyond the barrier and the decrease in nanoiron activity with time due to natural processes that take part underneath, such as precipitation as hydroxide or as carbonates, depending on soil composition.

Based on an extension of the concept of reactive barriers made form nanoiron, we developed a new generation of mitigation procedures technically efficient and commercially competitive to deal with environmental impacts that are not amenable to conventional technologies [2]. This statement can be better assessed through a practical example, such as the one summarized as case study 1.

## 7.1.2 Nanoiron as Catalyst

An alternative use of the high reactivity of nanoiron in the remediation of soils and waters contaminated with hydrocarbons and metals was developed and registered under the trade name nanocatox™ to represent the mechanism of nanocatalyzed oxidation [3]. It can be applied either as the so-called in situ chemical oxidation (ISCO) or as an onsite chemical oxidation (OSCO). During this process, not only hydrocarbons and other organics are destroyed, but also removal or immobilization of metals and metalloids takes place, as we have shown elsewhere for the case of arsenic [4]. The main advantages of this technology include relatively low cost in comparison with other alternatives, simplicity of operation and capability for scaling up. Practical application of ISCO procedures is summarized in this text as case study 2 and that for OSCO procedures as case study 3. Both of them are based on similar concepts, which are briefly described in the following paragraphs.

### 7.1.2.1 In situ nanocatalyzed chemical oxidation

The problem to be solved can be summarized as follows: Petroleum hydrocarbons such as gasoline, diesel, jet fuel, motor oil, BTEX (benzene, toluene, ethylbenzene, and xylene), and compounds comprised under the generic name volatile organic compounds

(VOCs) are commonly found at current and former fuel service stations and vehicle maintenance centers. Chlorinated solvents including tetrachlorethylene, trichloroethylene, and dichloroethylene (DCE) are commonly found at former industrial facilities and other locations. These kinds of contaminants occurring at industrial facilities, military bases, and other areas offer special challenges in locations where there is a high demand for the redevelopment of the former industrial area, for example for new suburban housing complexes. In cases such as these, a rapid and cost-effective method for the destruction of organic contaminants is needed, which can be applied without the commitment to years of costly operations and maintenance typically associated with soil vapor extraction or pump and treat systems currently employed for remediation of soil and groundwater containing hydrocarbons.

Hydrogen peroxide can chemically oxidize organics, forming carbon dioxide and water as end products if mineralization is complete. By using either a direct push method or a specialized lance system for the rapid delivery of treatment chemicals, in situ chemical oxidizers have the potential for rapidly treating soils contaminated with toxic and persistent organic wastes. Hydrogen peroxide, an ecofriendly reactant, when in contact with a metal catalyst such as iron, allows the generation of a powerful oxidizer, the hydroxyl radical ($OH^•$), in what is known as Fenton reaction. Procedures based on this reaction have been in use in water treatment plants for well over 50 years.

The oxidation of a contaminant by Fenton reaction involves complex mechanisms influenced by a number of variables, including pH, reaction time, temperature, catalysts, contaminant concentration, and hydrogen peroxide dosage. Based on this background, for nanocatox™ we have replaced commonly used iron salts for nanoiron as catalyst, thus avoiding the requirements for previous acidification of the site (usually by deploying sulfuric acid), which creates a collateral impact on an already contaminated site.

The theoretical model for this process when applied in the field involves a mechanism of interest, namely the mineralization of organics and the regeneration of the catalytic effect, which competes with mechanisms that are not of interest and results in excessive overall consumption of the oxidizer that has to deal with total organic carbon and free radical scavengers, most importantly, carbonate and bicarbonate alkalinity. Thus, careful evaluation of the chemistry of the soil and water is required prior to the start of any injection process.

The so-called mechanism of interest, due to its direct and intense decontamination effect, is the generation of hydroxyl radicals:

$$Fe^{\circ} + 2H_2O_2 \rightarrow 2^{\bullet}OH + 2HO^{-} + Fe^{2+} + \text{heat} \tag{7.1}$$

$$Fe^{2+} + H_2O_2 \rightarrow {}^{\bullet}OH + HO^{-} + Fe^{3+} + \text{heat} \tag{7.2}$$

These equations explain a basic advantage of nanocatox™ technology in relation to other advanced oxidation processes (AOP): Comparing the use of iron salts as it is usual in Fenton reaction, it can be seen that for each mole of dosed nanoiron, three hydroxyl radicals are generated instead of just one. The process also involves regeneration of catalyst (7.3 and 7.4) and mineralization of hydrocarbons (7.5).

$$Fe^{3+} + H_2O_2 \rightarrow 2H^{+} + 2O^{2-} \tag{7.3}$$

$$Fe^{3+} + 2O^{2-} \rightarrow Fe^{2+} + O_2 \tag{7.4}$$

$$^{\bullet}OH + C_xH_y \rightarrow H_2O + CO_2 + \text{heat} \tag{7.5}$$

Solving the generic Eq. (7.5) for specific hydrocarbons, the theoretical or stoichiometric demand of reactants for complete mineralization can be calculated [Eqs. (7.6 and 7.7)]. Based on this, an excess of reactant to take care of collateral or unwanted reactions can be estimated for the initial stages of injection.

$$\text{Benzene: } C_6H_6 + 15H_2O_2 \rightarrow 18H_2O + 6CO_2 \tag{7.6}$$

$$\text{Xylene: } C_8H_{10} + 21H_2O_2 \rightarrow 26H_2O + 8CO_2 \tag{7.7}$$

Taking into account experimental results obtained at the field after the first event of injection, which is based on the theoretical estimation, subsequent injections are performed until mitigation objectives defined by the client or by legal provisions are reached. Concurrently, conversion, inertization, and immobilization of metals take place due to the effect of injected nanoiron: adsorption and redox reactions that result in encapsulation and loss of solubility of metallic ions.

The basis for the efficiency of NPs for these functions is the high specific surface (up to 400 m$^2$ depending on average size), which maximizes interphase contact between contaminants and reactant. From a sustainability point of view, this process can be seen as ecofriendly because no previous acidification of the site is required (which usually leads to increased mobility of metals in groundwater), and the reactants are $H_2O_2$ (so that any excess of dosage will result in spontaneous decomposition into water and oxygen) and iron, which is the fourth most abundant element in the earth's crust.

In brief, in the current context where conventional techniques used to control environmental impacts, such as phytoremediation and bioremediation, are faced to practical limitations (manmade recalcitrant and non-biodegradable compounds) and economical restraints (occupation for years of land that results unavailable for other uses), nanotechnology allows the development of solutions easy and quick to operate and with minimal collateral impacts. Moreover, remnant hydrogen peroxide can also serve as an oxygen source for microbes in the subsurface to enhance biodegradation of contaminants. Therefore, many in situ chemical oxidation projects are designed to move into a second, longer-term bioremediation phase due to all the newly available oxygen in the subsurface.

Obviously, the remediation injection process (RIP), which is designed to handle high concentrations of hydrogen peroxide, requires significant safety training and planning of field procedures. The reaction time for hydrogen peroxide in the subsurface is usually seconds to minutes, with occasional reactions being completed within minutes to hours. Therefore, close spacing of the injection ports is generally required due to the short reaction period of hydrogen peroxide. During injections, the temperature of the site (soil and groundwater) is monitored not only to follow up the exothermic reactions, but also for safety reasons in order to avoid uncontrolled heating in an environment that contains VOCs. Based on field experience, temperatures in the range of 40–50°C indicate adequate rate of dosage of reactants. If the practical limit of 50°C is surpassed, injection is stopped until natural attenuation of the temperature occurs.

A typical nanocatox™ remediation task will include an initial review of existing physical and chemical data of the site, including pH, permeability, lithology, water depth, concentrations of VOCs, alkalinity, and similar data. Following this, a simple bench-scale test at the laboratory, performed with representative samples of the site, will allow preliminary adjustments of operation parameters before going to the actual field. After that, probably clients with previous experience in conventional methods will require a pilot test at some lingering hot spot in their facilities before committing the work, but usually the full-scale remediation program is implemented based on bench testing.

#### 7.1.2.2 Onsite nanocatalyzed chemical oxidation

The theoretical basis for onsite treatment is structured on the conceptual distinction into two stages in order to maximize extraction of contaminants from the solid matrix and have them available in a fluid aqueous-based matrix, so that controlling phenomena such as diffusion of reactants and temperature homogeneity are optimized and accelerated. In this way, the total lapse of time required for treatment can be substantially reduced in comparison to an in situ treatment for the same location. Efficacy and efficiency of the technique can be assessed by analysis of samples of treated soil, lixiviates, and effluent waters taken periodically during operation.

In the first instance, functionalized NPs are thoroughly mixed with the soil to be treated so as to make sure that hydrophilic and hydrophobic chains can be adsorbed on polar substances such as metallic salts as well as on nonpolar compounds such as hydrocarbons. In the second conceptual stage, organics are destroyed by the same mechanisms described in the previous subsection and metals, and metalloids adsorbed on the NPs are exposed to redox reactions that end with their insolubility and consequent immobilization.

## 7.2 Case Study 1: Full-Scale Remediation of PCBs in Concrete

Iron NPs are the basis for new-generation technologies used in remediation of contaminated sites. These NPs can provide innovative, technically efficient, and economically competitive solutions for the mitigation of negative environmental impacts whose complexity inhibits the application of conventional techniques due to their high surface/volume ratios, reactivity, and flexibility, zerovalent iron nanoparticles (nZVI) allow mitigation of persistent organic pollutants, such as halogenated organic compounds. This section describes the first case of scaling up to industrial level of a technology that previous literature has only described up to now through data at laboratory and pilot levels [5, 6]. This remediation is based on the application of nZVI to mitigate the contamination due to an explosion of electrical transformers at a power plant in Paraguay, in order to reach values below 10 μg/dm$^2$, following the suggestion of the USEPA about the maximum allowable PCBs level. The task included the definition of an initial isocontamination map of the facility, the non-destructive treatment of solid matrixes of concrete and structural masonry with an nZVI slurry, the assessment of the remnant isocontamination after each stage of treatment until the limits required in USEPA "*Decontamination standard for concrete*" were reached, and the final application of a reactive barrier under the form of an organic primer coating formulated with nanoiron to prevent eventual future bleeding of PCB from the porous substrate. The special interest of the case lays in an effective decontamination of concrete and masonry by dechlorination with nZVI, thus scaling up a technique that has previously been used only at experimental level with much less complicated porous substrates, such as natural soils. Besides, the use of locally manufactured nZVI was in fact the first effective export ever of nanotechnology from Argentina.

### 7.2.1 Background

Iron and nanoiron have been used for dechlorination of recalcitrant organic compounds, and their efficiency has been proved at laboratory and pilot scales. The technology described in this chapter is based on the chemistry reported in the bibliography, but it goes a step further in order to apply the process not only to relatively simple conditions such as soils and groundwater contaminated with persistent organic pollutants (POPs), but to extend the application of the technology to much more complex situations, such as solid matrixes of mortars and concrete. This demanding application has in fact been proved at full scale by decontamination, after obtaining an international tender from a hydroelectric facility. This technique is based on nZVI to mitigate the contamination effects of an explosion of two transformers containing high levels of PCBs at the engine house of generators 3 and 4 of the Hydroelectric Power Plant Acaray II, in Paraguay, as shown in Fig. 7.1. The closing of operations due to this accident affected half of the installed capacity of the facility.

**Figure 7.1** Explosion of transformers and initial PCBs spillage.

PCBs are a family of manmade chemicals that contain 209 individual compounds with various toxicity levels [7]. PCBs have been widely used as coolants and lubricants in transformers, capacitors, and other electrical equipment because of their insulating and nonflammable properties. Although PCBs are no longer manufactured, exposure still occurs due to their presence in old transformers and capacitors. Iron has been used to enhance de-chlorination of PCBs present in water by reducing the chlorinated compounds to their de-chlorinated form [8]. To get any significant PCB dechlorination in a reasonable lapse, the surface area of iron particles needs to be maximized, for example using iron NPs (1–100 nm). Also NPs can be transported effectively in porous media by the flow of a suitable carrier, such as water. Due to this attribute, the nanoparticle–water slurry can be injected under pressure and/or under capillary forces into solid matrixes where in situ treatment is needed. Direct subsurface injection, whether under gravity-fed or pressurized conditions, has already been shown to effectively transform chlorinated organic compounds [9].

The full-scale decontamination project was divided into stages that included an initial mapping of isocontamination areas, followed by a second stage of cleaning as a preparation for the application of an nZVI suspension. The decontamination advance was assessed by periodical sampling and analysis, until every value was below the allowable maximum established on the USEPA "Decontamination standard for concrete" document [10], namely PCBs $< 10\ \mu g/dm^2$.

Finally, all concrete surfaces were covered by an active barrier consisting of nZVI dispersed in a two-component polyurethane matrix. This barrier will prevent subsequent recontamination by reacting with remaining PCBs that may migrate out of the concrete due to porosity and capillary forces. The task was finished by conventional painting to fulfill mandatory color codes for each area. A more detailed description of the sequential stages of this project follows.

The magnitude of the remediation task refers to affected surfaces: concrete to be decontaminated amounted to a total of 1314 $m^2$, involving 100 $m^2$ of ceilings, 720 $m^2$ of walls, 192 $m^2$ of floors, 284 $m^2$ of staircases, and 18 $m^2$ of auxiliary structures.

According to EPA classification [11], the contamination was treated as bulk PCB remediation on porous surfaces, because the affected solid matrixes were either concrete or structural masonry. As the engine house is neither an area of permanent residence nor a zone of constant circulation of people, the site was classified as low occupancy area. Thus, required cleanup level for decontamination of concrete according to the aforementioned EPA reference was adopted, and maximum allowable limits for remnant contamination were established as less or equal to 10 $\mu g/dm^2$ as measured by the standard wipe test 40 CFR 761.123 [10], as shown in Fig. 7.2.

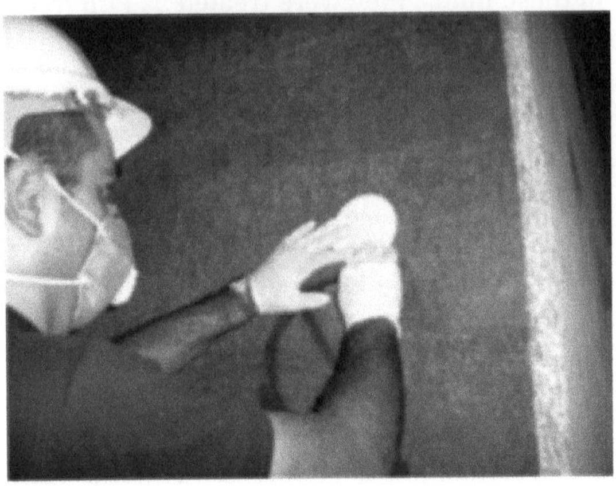

**Figure 7.2** Sampling of contamination by the wipe test technique.

An initial sampling of affected structures established a quantitative starting point, based on 600 individual samples. Of these, 53 showed moderate contaminations, in the range from 10 μg/dm$^2$ to 100 μg/dm$^2$, and 19 sites showed high contamination, meaning over 100 μg/dm$^2$, with a maximum of 829 μg/dm$^2$. After six months of work, a second sampling was made. From 259 inspected sites, 20 showed moderate contamination and only 4 were still highly contaminated, but the maximum value has lowered to 319 μg/dm$^2$. Further treatment with nZVI slurry was applied to these sites, and a month later a new assessment was made. On this occasion, all of the final samples showed PCB values lower than 10 μg/dm$^2$.

The aforementioned results were achieved by the following process. First, every surface to be treated was thoroughly cleaned up by a combination of mechanical brushing and solvent washing with isopropanol, in order to expose naked surfaces of concrete and masonry. This was done under the provisions described at section 761.61 PCB remediation waste, subpart a, instructions 1–6 of Ref. [10]. Decontamination solvent was used as required by the provisions of section 761.79 (d) of the same reference.

Cleanup began at the ceiling of the engine house and was continued downside up to the floor where generators are located, 60 m below. During the whole procedure, all liquids were collected by an ad hoc system of piping and stored in tankers until PCB levels on these fluids were analyzed. As no content of PCBs over 50 mg/L was detected in these fluids, they were treated as PCB free as provided by section 761.79 (g), "*Decontamination waste and residues*" of Ref. [10].

After mechanical and solvent cleaning, all surfaces were treated with slurry of nZVI in water, which penetrated the porous solid matrixes of masonry and concrete in order to react with the halogenated compounds, as shown in Fig. 7.3. This treatment was repeated several times before the second sampling mentioned before and was also repeated on the remnant contaminated spots between the second sampling and the final one.

After all the areas were shown to be cleaned up according to EPA criteria of <10 μg/dm$^2$ of PCBs, the mitigation task was completed. Nevertheless, as theoretical migration to the surface of minute quantities of deeply penetrated PCBs could not be overruled, it was decided that a long-range protection should take care of this eventual problem. Thus, an organic coating was prepared with a concentrated suspension of nZVI particles, which was applied as a suitable primer, 20 μm thick, and will act as a reactive barrier to any eventual exudation of halogenated compounds from the solid matrix.

**Figure 7.3** Application of nZVI slurry.

The last stage of the procedure was the application of a high traffic organic coating over the primer, such as provided in section 761.61(a)-7 "*Cap requirements*" of Ref. [11].

The specific interest of this case is that an effective decontamination of concrete and masonry structures by dehalogenation with nZVI was proved to be feasible at real-world scale. This experience extended the field of application of the technology, which previously has been used, at most, at the pilot level and with much more porous media (thus more amenable to nZVI migration into the solid matrix) such as natural soils. Moreover, the use of nanoiron manufactured at industrial scale by a local small and medium enterprises (SME) was indeed the first export of nanotechnology from Argentina.

As regards quality and results of the procedure, it must be emphasized that, PCBs being one of the compounds included in the so-called "*dirty dozen*," the whole work was audited by delegates of United Nation's Committee on Persistent Organic Pollutants, who finally issued the approval certificate (see Fig. 7.4) once the strict limits required by USEPA for PCBs mitigation were achieved over all the building, including walls, stairs, floors, elevator box, and the like.

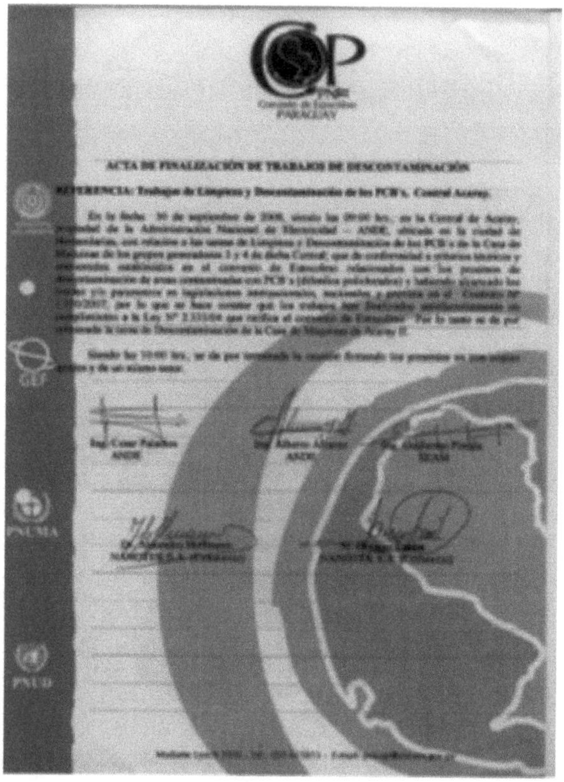

**Figure 7.4** Final certificate issued by UN Committee on POPs.

## 7.3 Case Study 2: ISCO of Hydrocarbons at an Urban Gas Station

The site is located in a highly populated zone of Adrogué, a province of Buenos Aires, Argentina, as shown in Fig. 7.5. A creek named "El Rey" runs 1 km southwest from the site. The gas station remained in activity during the treatment, showing that the technology is not disruptive for commercial operation of the site.

**Figure 7.5** Location of "Palo Verde" gas station.

Table 7.1 describes the composition and structure of the soil to be remediated.

**Table 7.1** Soil composition

| Depth (m below surface) | Description |
|---|---|
| 1 | Grey sand and coarse bits of hard clay |
| 2 | Brownish lime clayish soil and bits of hard clay |
| 3 to 6 | Brownish, reddish massive silt clay soil |
| 7 | Brownish, reddish massive silt clay soil with scarce chalky concretions |
| 8 | Brownish, reddish massive silt clay soil with medium chalky concretions |
| 9 | Light brownish lime clayish soil |
| 10 | Harder brownish lime clayish soil (Pampean II) |

Hydrogeological characterization of the site allowed definition of the topographic slope (biggest slope at shortest distance), so as to assess groundwater flow direction, which was southwest, pointing at the nearby creek. A hydraulic gradient of 0.009 was calculated from phreatic layers. The velocity of groundwater flow, calculated from data on permeability, hydraulic gradient, and effective porosity, was $1.48 \times 10^{-3}$ m/day. Vulnerability of the free aquifer to contamination was estimated by EKv method [12], resulting in a value of 6, indicating that the free aquifer has a natural capacity to resist contamination in the medium range. Once these data were assessed, the injection grid was designed and sampling wells were drilled in place with minimum disturbance to the gas station's daily operation, as shown in Fig. 7.6.

**Figure 7.6** Drilling of injection well.

Table 7.2 shows the initial level of contamination of groundwater before starting remediation.

**Table 7.2** Hydrocarbon contamination of groundwater (mg/L)

| TPH | GRO | DRO | VOCs | BTEX |
|---|---|---|---|---|
| 63.4 | 44.0 | 19.4 | 44.0 | 10.26 |

The radius of influence (ROI) was experimentally assessed at the field in order to estimate how far each injection well will distribute reactants underground. The value was 0.75 m, which in turn defines the geometry of the injection grid.

After all these initial preparations, the first event of injection took place, controlling the evolution of operative parameters such as pH, ORP, conductivity, and temperature. In all there were four injections, once per month, in order that reactions may take place underground, based only on natural diffusion because they cannot be accelerated by agitation. Table 7.3 shows the mitigation of contaminants after the fourth injection, as actual values of remnant hydrocarbons and as percentage of mitigation related to initial values shown in Table 7.2.

**Table 7.3** Mitigation of hydrocarbon contamination of groundwater (mg/L and %)

| TPH | GRO | DRO | VOCs | BTEX |
|---|---|---|---|---|
| 8.4 | 5.8 | 2.6 | 0.29 | 0.25 |
| 89% | 90% | 87% | 99.5% | 98.3% |

Figure 7.7 shows that dosage of reactants for ISCO was done without disturbing the normal operation of the gas station.

**Figure 7.7** Dosage of reactants at an injection well.

## 7.4 Case Study 3: OSCO of Hydrocarbons at an Industrial Treatment Site at Amazonia

A field demonstration of the viability of nanocatox™ as an onsite nanocatalyzed chemical oxidation was performed at a facility owned by Ecuador Petroleum Company, Petroamazonas EP (http://www.petroamazonas.gob.ec). The material to be treated was contaminated with hydrocarbons and metals. The site was identified as operations area of Sacha and was located in the Amazonian forest, some 30 min from the city of Coca in Ecuador.

Post-treatment analysis of soil and lixiviates showed that different levels of mitigation of contaminants can be reached, ranging from the most demanding classification "*sensitive ecosystems*" to the less demanding classification "*industrial use*" [13]. This allows the client to establish desired final mitigation based on the expected future use of the land and the costs involved for each alternative. Depending on each case, residence times for this type on nanoiron catalyzed OSCO go from a few hours to a day, which is an obvious advantage in comparison to biotreatments that may last months, and even years, depending on original contamination.

Three batches of contaminated soil were processed at the field, under the supervision of Amazonia Viva Management, who defined the representativeness of the samples. Figure 7.8 includes a general view and a close-up of the material to be treated. The batches were identified as CNT, E1T5, and E1T6. The main contamination of concern for each one was as follows. CNT contained an initial amount of 15,806 mg/kg of total petroleum hydrocarbons (TPH), with a high percentage of polycyclic aromatic hydrocarbons (PAHs), refractory to biodegradation. E1T5 contained an initial amount of 4204 mg/kg of TPH, plus 8.64 mg/kg of cadmium and 9.13 mg/kg of lead. E1T6 contained an initial amount of 4326 mg/kg of TPH, plus 8.64 mg/kg of cadmium and 9.13 mg/kg of lead.

**Figure 7.8** Soil for treatment: general view (left) and close-up (right).

The general procedure implemented for decontamination can be termed a combination of a chemical-enhanced soil washing with a nanocatalyzed oxidation of organics and immobilization of metals. The technique is a batch operation and includes recuperation and reuse of process water, which reduce the overall environmental impact of the technology because only a minimum amount of make-up water is required for each batch to account for losses and evaporation.

Scaling up of the procedure that was tested as the field demonstration scale will depend on the total volume of soil to be treated, space available for natural sedimentation and recuperation of process water, and urgency of the client to finish the treatment derived either from expected use or from legal enforcements.

Operations required adequate training for operators and technical identification of reactants at the field. Figure 7.9 depicts the use of safety equipment by operators involved in the procedure and mandatory identification of materials in the forms provided by the client.

**Figure 7.9** Safety garments for operators (left) and identification of materials (right).

The treatment took place in an open mixer that can deal with different proportions of solids and water and provides adequate homogenization capacity at the field-scale level: Figure 7.10 shows the operations of dosage of reactants and continuous control of temperature, which is an inference parameter for the evolution of the reaction and must be held below safe values to avoid eventual problems due to VOCs generation, while compounds of high molecular weight are being decomposed into lighter hydrocarbon chains.

**Figure 7.10** Dosage of reactants (left) and control of temperature (right).

Analysis of treated soil and effluent water was performed by a third-party laboratory chosen by the client. After treatment, each batch reached mitigation levels as described below.

CNT post-treatment characterization values: 997 mg/kg TPH (almost 94% of reduction in comparison to original value); <0.027 mg/kg de PAHs and <0.027 mg/kg cadmium. Based on these results, all parameters were below the strictest allowable limits required by local regulations. Thus, this treated material classified as adequate for sensitive ecosystems.

E1T5 post-treatment characterization values showed 1505 mg/kg TPH (a little over 64% of mitigation in comparison to original contamination) and 1.0 mg/kg cadmium. Thus, the dosage of reactants for this batch resulted in mitigation in the range of what regulations classify as agricultural use of land, as for instance <2500 mg/kg TPH.

E1T6 post-treatment characterization values showed 2969 mg/kg TPH and 1.0 mg/kg cadmium, which easily classifies this treated soil for industrial use, where the limit is <4000 mg/kg TPH, involving less consumption of reactants, which increases economic competitiveness.

Differences among results for each batch were not only due to reactants dosage and treatment times, but also due to different original conditions for each sample. Both E1T5 and E1T6 batches were taken from operating biopiles and thus included a higher load of organic carbons besides TPH, such as compost, humic and fulvic acids, and natural decomposition matter. On the other hand, CNT batch was contaminated material in its virgin form, and practically all of oxygen demand came from TPH and PAHs. Thus, dosage based on stoichiometric calculations allowed destruction of more than 90% of the contaminants.

These results were obtained in less than 12 h of treatment in a robust but non-optimized reactor that was adapted from a conventional concrete mixer. This means that there is still a great opportunity for optimization with adequate equipment and a change from manual operations to automatized procedures.

Cost comparisons with traditional methods such as biodigestion should be based not only on reactants and equipment costs, but also to hidden costs such as time (biopiles may take from months to years to achieve similar mitigation results), labor costs (biopiles require intensive manpower during the whole process), and last but not least, the possibility of effectively treating recalcitrant contaminants such as PAHs, not amenable to conventional biodigestion.

## 7.5 Case Study 4: Mature Fine Tailings Management—A Proof of Concept Test

Representing yet another turn of the screw, applications of nanoiron in environmental remediation do not need to be limited to exploitation of its potential as reactant or as catalyst, but may also take advantage of their paramagnetism. This concept will be described in this section, as applied to solving a problem in the oil-sands mining business.

The extraction of bitumen from sand by utilizing hot water processes results in the production of a slurry waste, which is stored in so-called "*tailings ponds*." Within these ponds, while fast settling sand particles segregate from the slurry in relatively short time, the fines fraction accumulates in the center of the pond and then settles, becoming mature fine tailings (MFT). Most of the water content of the pond is recycled back; however, around 86% of the volume of MFT consists of water [14]. It takes a few years after placement for MFT to settle to around 35% solids [15]. By 2008, there were about 750 million $m^3$ of MFT within the tailings ponds. Assuming that the tailing management remains the same, the amount of fluid tailings is expected to reach 2 billion $m^3$ by 2034 [16]. In 2009, it was estimated that there were around 130 $km^2$ of tailings ponds in the oil sand region of Canada. Thus, one of the most important environmental challenges regarding oil-sands mining is developing a process to separate water from the fine tailings within a reasonable timeframe, in order to allow for the reclamation of the site.

To illustrate the magnitude and economic importance of the problem, we refer to information published in the web. For example, the Canadian Association of Petroleum Producers estimated that by 2013, oil-sands operators have been investing more than $1 billion in tailings-reduction technology [17]. Suncor Energy, a company developing oil sands, mentions in its report on sustainability dated 2013 [18] that as of 2012, they have spent more than $1.3 billion to research, develop, and implement its TRO™ process, which allegedly will allow them to dry their mine tailings into a material solid enough to be reclaimed in a fraction of the time that earlier technologies require. This should, in turn, greatly accelerate their overall mine reclamation efforts because with the new drying process in place, they expect to reduce the time it takes from initial land disturbance to having a reclaimable surface to about 10 years, a third of what is now the industry standard.

Nevertheless, these technological promises seem to be unrealized to date. To put it in Antweiler's words [19]:

Tailings from bitumen mining are a mixture of water, sand, and fine clay, and also contain residual bitumen. These tailings are drained into settling ponds where sedimentation separates the ingredients. Water rises to the top, while heavier materials such as metals and minerals settle at the bottom of the pond. The difficulty arises with a middle layer of "mature fine tailings" (MFT), which take a very long time to settle, even decades. This makes tailings ponds inefficient for treating tailings. Speeding up the process has become a new challenge for oil companies, which are under a mandate from Alberta's provincial government to meet capture targets for fine tailings. Introduced in 2009 as Directive 74 by the Alberta Energy Regulator (AER), companies are required to capture more than 50% of fines (particles of 44 micrometers or less) since 2013. Companies are scrambling to meet this target. In 2013, oil-sands mines owned by Shell, Suncor, and Syncrude failed to meet targets.

Bearing in mind the problems described above, we initiated a preliminary applied research program in order to prove at bench scale the concept of speeding up sedimentation of MFT by means of NPs, and more specifically by using paramagnetic nanofluids manufactured with these NPs, based on the following conceptual framework.

The key factor for dealing with MFT is to break the stable multiphase emulsion. Chemical demulsifiers are amphiphilic compounds having both hydrophilic and hydrophobic properties, which allow them to adsorb and interact at the oil–water interface that surrounds inorganic particles of sand and the like. To break the stable oil–water interface around emulsified fine solids, chemical demulsifiers are required to be more surface active than the mechanisms acting as emulsion stabilizers. Magnetic nanoparticles (MNPs) are of great interest for this function because of their response to an external magnetic field for quick and easy isolation from the complex multiphase systems by magnetic separation. The interfacial activity of MNPs on the surface of micron and submicron particles has been found to allow them to be effectively attached to otherwise stable emulsified water droplets in diluted bitumen emulsions.

By working with hydrophobic MNPs, it should be possible to control the surface activity of the particles in mixed media. By tuning the wettability of the suspended particles, influence can be exercised on properties at the oil–water interface. This can be done by dispersion of MNPs into the stable emulsion, being the surface ligands number of the MNPs an important factor to affect the process, which is mainly governed by the increase in the interfacial viscoelasticity resulting from the attachment of MNPs.

The interfacial activity and high hydrophilic nature of polymer/ $Fe_nO_x$ MNPs (which we can denominate by the generic formula $Fe_3O_4$@) will allow them to be effectively attached to stable emulsified sand–water droplets in oil–water emulsion, and tagged water can be readily removed by an external magnetic field, which enhances the coalescence of magnetically tagged water droplets in an emulsion. The magnetic property of $Fe_3O_4$@ should allow this kind of MNPs to be readily recycled by magnetic separation and solvent washing. Polymer/$Fe_nO_x$ MNPs should be designed so as to retain their interfacial activity when recycled.

The conceptual mechanism is a combination of the following stages:

1. $Fe_3O_4$@ enters the oil–water interface surrounding inorganic solids micro-particles (MFT) and breaks up the interfacial film formed by surface-active components in this interface due to the more surface-active and hydrophilic nature of polymer/ $Fe_nO_x$ MNPs.
2. $Fe_3O_4$@ particles increase the partition coefficient and consequently the adsorption of demulsifier molecules at the oil–water interface.
3. $Fe_3O_4$@ particles aggregate water droplets through bridging flocculation leading to their coalescence.

This mechanism will obviously be influenced by parameters such as dosage, wettability (or water contact angle) of the NPs, and pH, at least. Demulsification efficiency (DE) should increase with increasing $Fe_3O_4$@ dosage. High wettability (i.e., water contact angle of ≥90°) will be desirable. Acid to neutral conditions should have little influence, but alkaline pH (≥9) should have adverse effects on DE. Probably a multistep operation will exhibit a higher DE than a single-step demulsification when the same amounts of $Fe_3O_4$@ are used.

To empirically prove this concept, tests were carried on with a sample of real MFT, as described below. The material was received from Canada provided by ALBERTA INNOVATES Technology Futures, which characterized the contents as follows:

Product: Mature fine tails

Use: By-product of the extraction of bitumen from oil sand

Composition: Thickened suspension of fine particles of clay and silt size minerals in water, formed by setting of process tailings from the separation of bitumen from oil sand.

Ingredients: Bitumen (CAS # 8052-42-4), 1 – 2 Wt. %

Naphta (CAS # 64742-48-9), < 0.1 Wt. %

Clay (N/A), 30-60 %

Water (CAS # 7732-18-5), balance

Based on the expertise of Nanotek S.A. and current range of own manufactured nanomagnetic particles as a starting point, several formulations were prepared to be assessed at lab scale in comparison to $AlSO_4$ flocculants. As shown in Fig. 7.11, after the first week of sedimentation, it was evident that conventional flocculants were not eligible, because instead of promoting a clear separation between phases, they generated a stratification of the MFT sample, with an upper floating phase with high solids content, a liquid phase of clarified water at the middle, and another agglomeration of material rich in solids at the bottom. This behavior will prevent the design of a sensible treatment at field scale.

On the other hand, the three most promising formulations of $Fe_3O_4$@ particles achieved, in the first 435 h of testing, respectively, 27, 28, and 31% of clarification (defined as the ratio between upper clear water phase and lower high solids phase). These 18 days behavior compared very favorably with the years that take natural sedimentation. Nevertheless, tests were continued for six months and additional clarification (up to 60% for the best performance) took place when aided by a static magnetic field. Figure 7.12 shows behavior of different formulations after 6 months of magnetically assisted sedimentation.

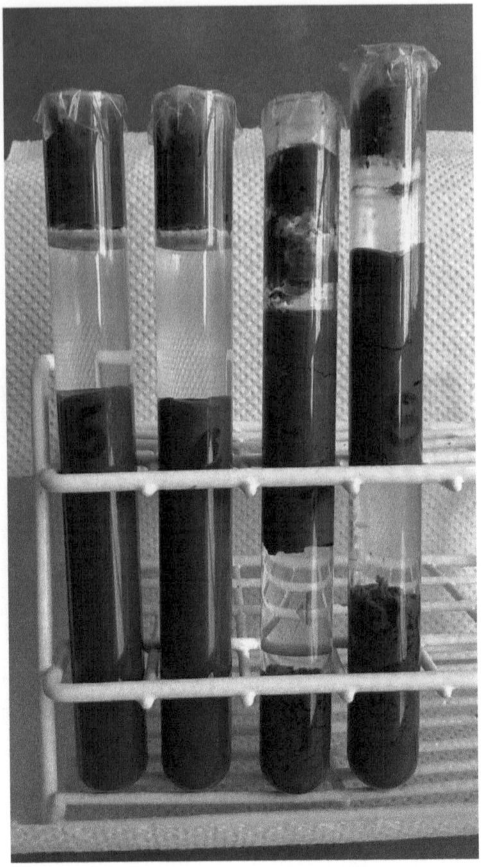

**Figure 7.11** Behavior of selected MNPs (left) versus $AlSO_4$ flocculant (right).

These preliminary series of tests prove that the concept of enhancing settling of MFT by means of paramagnetic nanofluids is feasible. But further steps are required in order to reach the market with a competitive formulation based on functionalized nanoiron. Optimization of dosage at bench scale should be tested as well as assessment of alternatives amenable to treatment at field scale, such as direct mixing into the slurry of existing tailings, mixing prior to deployment into the tailings pond, optimization of external magnetic fields (intensities and geometries), and the like. Scaling up of the production process for the functionalized NPs will be the final step in developing this technology at the commercial level.

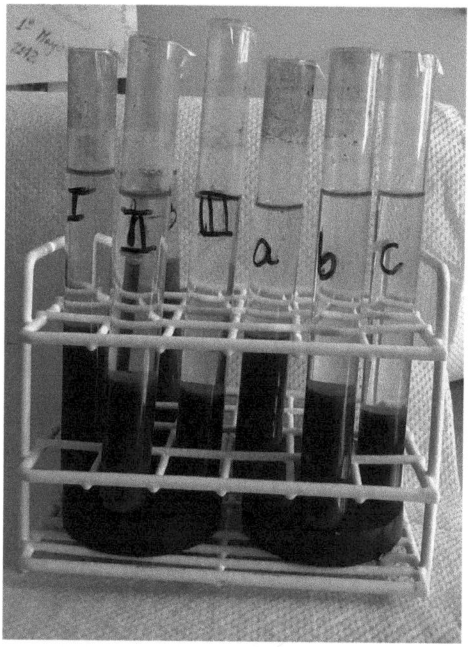


**Figure 7.12** Demulsification after 6 months (magnetically assisted).

## 7.6 Concluding Remarks

The case studies detailed in this chapter provide smoking gun evidence of the use of iron and iron oxides NPs to solve a wide variety of problematic situations regarding environment in the real world. These applications rely on different characteristics of nanoiron depending on the case, including its high chemical reactivity due to the great specific surface, its role as site for electrons interchange when used as catalyst, and its paramagnetic properties.

The description of cases also emphasized the need of interaction among different kinds of expertise when developing a procedure for mitigation of contaminants at field scale, including, but not limited to, chemists, engineers, geologists, and mechanics. Thus, nanotechnology applied to environmental remediation is a good example of what has been defined as transdiscipline.

Last, but not least, the cases demonstrate that environmental solutions need to be tailor made. The characteristics of the nanomaterials to be used are just the starting point, but field application requires solving specific logistics (it is quite different to work in a urban environment than in the Amazonian forest), extensive previous testing at lab and bench scale to define effective but simple control systems that can be used by any qualified operator (such as controlling evolution of hydrocarbons mineralization, and consequently dosage of reactants, through temperature alone), and pilot testing of procedures at real-world conditions, especially for large jobs or "not off the shelf" solutions, in order to optimize operative parameters.

## References

1. Watlington, K. (2005). *Emerging Nanotechnologies for Site Remediation and Wastewater Treatment* (USEPA, Washington, USA).
2. López, G. D. (2014). Nanohierro cerovalente para remediación in situ de compuestos organoclorados recalcitrantes: estudio de caso, *Proc. SAM-CONAMET 2014*; Santa Fe, Argentina.
3. López, G. D., Pagano, G., and Tobías, H. (2007). Remediación nanocatalizada de suelos con hidrocarburos, *Ing. Sanit. y Amb.*, **93**, pp. 39–50.
4. Morgada, M. E., Levy, I. K., Salomone, V., Farías, S. S., López, G., and Litter, M. I. (2009). Arsenic (V) removal with nanoparticulate zerovalent iron: Effect of UV light and humic acids, *Catal. Today*, **143**, pp. 261–268.
5. Tratnyek, P. G. and Johnson, R. L. (2006). Nanotechnologies for environmental cleanup, *Nanotoday*, **1**, pp. 44–48.
6. Zhang, W. X. (2003). Nanoscale iron particles for environmental remediation: An overview, *J. Nanopar. Res.*, **5**, pp. 323–332.
7. United States Environmental Protection Agency. (1987). *National Air Toxics Information Report: Qualitative and Quantitative Carcinogenic Risk Assessment.* (EPA, Washington DC, USA) pp. 11/1–6.
8. Wang C. B. and Zhang W. X. (1997). Synthesizing nanoscale iron particles for rapid and complete dechlorination of TCE and PCBs, *Environ. Sci. Technol.*, **31**, pp. 2154–2156.

9. Glazier, R., Venkatakrishnan, R., Gheorghiu, F., Walata, L., Nash, R., and Zhang, W. (2003). Nanotechnology takes root, *Civil Eng.*, **73**, pp. 64–69.
10. 40 CFR 761.79 - Decontamination standards and procedures, http://www.law.cornell.edu/cfr/text/40/761.79, Cornell University Law School.
11. http://www.epa.gov/osw/hazard/tsd/pcbs/pubs/2005-761.pdf.
12. Auge, M. (2003). *Vulnerabilidad de Acuíferos. Conceptos y Métodos*, Ebook: 1-38, RedIRIS, Spain.
13. Decreto Ejecutivo 1215 - RAHOE , table 6, http://www.miliarium.com/Paginas/Leyes/Internacional/Ecuador/Contaminacion/Decreto1215-01.pdf
14. Chalaturnyk, J., Scott, D., and Ozum, B. (2002). Management of oil sands tailings, *Petroleum Sci. Technol.*, **20**, pp. 1025–1046,
15. Beier, N., Alostaz, M., and Sego, D. (2009). Natural dewatering strategies for oil sands fine tailings, *Proc. 13th Internat. Conf. on Tailings and Mine Waste*, Banff, Alberta, Canada, pp. 845–858
16. Houlihan, R. and Haneef, M. (2008). Oil sands tailings: Regulatory perspective, *Proc. 1st Internat. Oil Sands Tailing Conf.*, Edmonton, Alberta, Canada, pp. 250–264
17. Canadian Association of Petroleum Producers, OIL SANDS TODAY, http://www.homerdixon.com/wp-content/uploads/2013/04/tar-sands-70-sq-mi.pdf, accessed in November 2016.
18. Suncor Energy, REPORT ON SUSTAINABILITY 2013, http://sustainability.suncor.com/2013/en/environment/tailings.aspx, Accessed in November 2016.
19. Antweiler, W., WERNER'S BLOG — OPINION, ANALYSIS, COMMENTARY, http://wernerantweiler.ca/blog.php?item=2014-09-02 , Accessed in November 2016.

# Chapter 8

# Use of Nanoparticulated Iron Materials for Chromium, Arsenic, and Uranium Removal from Water

**Natalia Quici,[a,b] Martín Meichtry,[a,b] and Víctor Nahuel Montesinos[a]**

[a]*Comisión Nacional de Energía Atómica, CONICET, Gerencia Química, Av. Gral. Paz 1499, San Martín, Prov. de Buenos Aires, 1650, Argentina*

[b]*Departamento de Ingeniería Química, Facultad Regional Buenos Aires, Universidad Tecnológica Nacional, Av. Medrano 951, Ciudad Autónoma de Buenos Aires, 1179, Argentina*

nquici@cnea.gov.ar

## 8.1 Introduction

Metals and metalloids in water have infinite lifetimes representing nowadays one of the most important environmental problems. Water treatment to remove these pollutants are carried out, generally, by precipitation, electrolysis, chemical oxidation, ozonation, adsorption, or chelation, all of them presenting drawbacks or being economically prohibitive [1].

The use of iron-based NPs (nFe), as zerovalent iron (nZVI), or iron oxides (nFeOx) has proved to be a promising technology for

*Iron Nanomaterials for Water and Soil Treatment*
Edited by Marta I. Litter, Natalia Quici, and Martín Meichtry

ISBN 978-981-4774-67-3 (Hardcover), 978-981-4669-49-8 (eBook)
www.panstanford.com

the removal of a wide range of pollutants, including metals and metalloids in water [2–7]. Their use combines properties such as chemical affinity to targeted oxyions, surface charge, and redox potential that, together with their stability and low cost [8], enable their use for diverse systems [9]. Figure 8.1 shows the possible mechanisms involving metals and metalloids removal on the nFe surface.

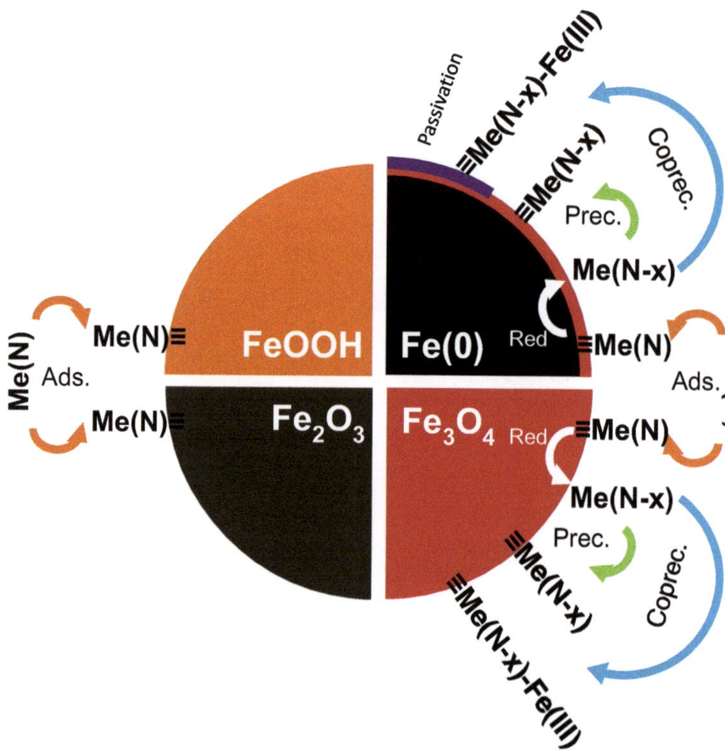


**Figure 8.1** Proposed removal mechanisms of metals and metalloids taking place on nFe surfaces (based on Ref. [9]). M: metal or metalloid; N: oxidation state.

Only for nZVI, more than 12 reviews on the synthesis and use of nZVI are available ([10] and references therein). Here, only a brief summary of the main removal mechanisms (see also Sections 3.2 and 3.3), material capacity, and reaction kinetics for Cr(VI), inorganic As(V)/As(III), and U(VI) are described.

## 8.2 Chromium

Among all chromium species, hexavalent chromium presents the highest environmental threat. Cr(VI) has a high mobility in water [11] and is present as a series of oxoanions capable to easily generate oxidizing radicals with a high carcinogenic potential [12]. Cr(III) is the thermodynamically stable Cr species in moderately oxidizing and reducing conditions, has low mobility at $5 < pH < 13.5$, is not toxic, and is an essential nutrient [13]. Due to its toxicity, the World Health Organization (WHO) established a maximum level of Cr(VI) in drinking water of 50 μg/L [14].

Iron-based nanomaterials constitute one of the most promising technologies for Cr(VI) removal from water [15]. A wide variety of nanoscale iron-containing agents for Cr(VI) removal are reported in the specific literature [3, 14, 16–18], and in the following pages, the most relevant materials for this purpose will be summarized.

The removal of hexavalent chromium by zerovalent iron in water is thermodynamically driven by the favorable reduction potential of Cr(VI) to Cr(III) against Fe(0)/Fe(II) ($\Delta E^0{}_{298\ K} = 1.77$ V, Eq. 8.1) and also against the formed aqueous Fe(II)/Fe(III) couples ($\Delta E^0{}_{298\ K} = 0.56$ V, Eq. 8.2) [19]. These reactions proceed with a high rate, and complete removal of Cr(VI) can be achieved in minutes in laboratory experiments [20–25].

$$Cr_2O_7{}^{2-} + 3Fe^0 + 14H^+ \rightarrow 2Cr^{3+} + 3Fe^{2+} + 7H_2O \qquad (8.1)$$

$$Cr_2O_7{}^{2-} + 3Fe^{2+} + 14H^+ \rightarrow 2Cr^{3+} + 3Fe^{3+} + 4H_2O \qquad (8.2)$$

In the case of nZVI, Fe(0) is unstable in aqueous media and readily reacts, giving rise to corrosion process; for this reason, its surface becomes usually covered by a protective layer of iron (oxy) hydroxides, with an exact composition depending on the synthesis process, particle size, and storage conditions [26–29].

It is accepted that the removal of Cr(VI) consists of a heterogeneous reaction where the contaminant is firstly adsorbed onto the external oxide layer and then reduced by the Fe(0) preserved inside the nanoparticle (NP) [7]. The surface reaction is the limiting step after an instantaneous adsorption of Cr(VI), and this explains why the time-dependent [Cr(VI)] reasonably fits the first- or pseudo-first order kinetics in most works on the subject [30].

The actual electron transfer mechanism is still under debate. On one hand, the electrons from the Fe(0) inside the NP could reach the Cr(VI) on the surface through localized states in the band structure of the external oxide layer [31, 32]. On the other hand, several authors propose that Fe(II) on the surface of the NP is the actual reducing agent ($0.34 \text{ V} < E^0_{Fe^{3+}(sup)/Fe^{2+}(sup)} < 0.65 \text{ V}$)[1] [33, 34]. In this model, Cr(VI) reacts with surface-bonded Fe(II) (≡Fe(II)) generating ≡Fe(III), which is continuously regenerated by reaction with Fe(0) [7, 35, 36]. In this regard, Mu et al. [20] prepared core–shell Fe@ $Fe_2O_3$ nanowires sequentially increasing the $Fe_2O_3$ shell and studied how the thickness of the external oxide layer affected the anoxic removal efficiency of Cr(VI). They found that a higher removal of Cr(VI) correlates with a higher superficial concentration of Fe(II) in the fresh material. Experiments where Fe(II) was scavenged by 1,10-phenantroline further confirmed its key role in Cr(VI) reduction.

According to Eqs. 8.1 and 8.2, the overall reduction process increases the pH of the reaction solution after the treatment. This is a beneficial effect given that, at pH > 5, Cr(III) precipitates as amorphous $Cr(OH)_3$ [37], and at pH > 4 in the presence of Fe(III), it forms very insoluble mixed Fe(III)–Cr(III) (oxy)hydroxides [2, 38] (Eqs. 8.3 to 8.5). Li et al. [38] analyzed the chromium deposition on the nZVI surface using high-resolution X-ray photoelectron spectroscopy (HR-XPS), finding the stoichiometry of the chromites to be approximately Cr/Fe 2:1, i.e., $(Cr_{0.67}–Fe_{0.33})(OH)_3$ or $Cr_{0.67}Fe_{0.33}OOH$. Later, Montesinos et al. [21] studied the in depth composition of the nZVI after Cr(VI) removal by $Ar^+$ sputtering cycles followed by XPS analysis of the exposed surface. It was found that the external layer has a complex composition with a decreasing concentration of Cr(III) and Fe(III) toward the core. This result was further confirmed by Ling et al. [39] through corrected scanning transmission electron microscopy (Cs-STEM) analysis.

$$Cr^{3+} + 3H_2O \rightarrow Cr(OH)_3\ (s) + 3H^+ \quad (8.3)$$

$$xCr^{3+} + (1 - x)\ Fe^{3+} + 3H_2O \rightarrow (Cr_xFe_{1-x})(OH)_3\ (s) + 3H^+ \quad (8.4)$$

[1]All reduction potentials given in this work are standard values versus SHE; therefore, the values correspond to pH 0 unless a different condition is specified. Values correspond to those in homogeneous solutions, although reactions at the interface can be somewhat different.

$$xCr^{3+} + (1-x)\ Fe^{3+} + 2H_2O \rightarrow Cr_xFe_{1-x}OOH(s) + 3H^+ \quad (8.5)$$

This highly stable Fe(III)–Cr(III) (oxy)hydroxide layer [40] passivates the surface of nZVI, and even a thin layer can inhibit the electron transfer from Fe(0) in the core to the remaining Cr(VI) in solution [21]. To overcome the nZVI surface passivation, four main strategies were explored: (1) addition of organic and inorganic Fe(III) and Cr(III) complexing agents [36, 41–43]; (2) combination with ultrasound [44]; (3) Fe(0) regeneration by dithionite [45]; and (4) photocatalytic activation of the external oxide layer by visible light irradiation [46].

The reduction potential of Cr(VI) to Cr(III) decreases with increasing pH, condition that also hinders reactions 8.1 and 8.2. At pH < 8, the nZVI surface is positively charged, favoring the interaction with $CrO_4^{2-}$ and/or $HCrO_4^-$ [19, 47]. However, according to Hu et al. [48], maghemite ($\gamma$-$Fe_2O_3$), usually reported as part of the nZVI external oxide layer, has a higher affinity for $CrO_4^{2-}$ than for $OH^-$ at pH > 8. This alternative way of sorption would explain the poor effect of the surface charge on the rate constants when increasing the pH. Additionally, at pH < 5.5, the iron oxidative dissolution occurs exponentially, increasing when the pH decreases [7], preventing the formation of the Cr(III)–Fe(III) layer and also consuming part of Fe(0) by the action of $O_2$ or $H^+$. Thus, the resulting removal efficiency of Cr(VI) in acidic media for different experimental setups will strongly depend on the balance between these opposed effects, controlled by the Cr(VI)/nZVI initial ratio, morphology, size, and surface area of the NPs employed.

Dissolved oxygen (DO) can compete with Cr(VI) for the Fe(0) electrons, but the reduction of the latter is thermodynamically and kinetically favored ($E^0_{HCrO4-/Cr3+} = 1.36$ V and $E^0_{O2/H2O} = 1.23$ V) as $O_2$ reduction presents a high overpotential [21]. Yu et al. observed that consumed DO increases with higher nZVI loading, as a parallel reaction, with no influence on the Cr(VI) removal efficiency [25]. Yoon et al. [49] and Qin et al. [50] found that the presence of $O_2$ accelerates Cr(VI) removal by regeneration of Fe(II) on the surface of microsized ZVI after oxidation of Fe(0). However, very recently, Mu et al. [51] showed a slight negative effect of DO, reaching 100 and 81.4% Cr(VI) removal with core–shell Fe@$Fe_2O_3$ nanowires in 60 min under anoxic and oxic conditions, respectively. As with other contaminants, no general conclusions have been achieved to date [47].

Iron oxide NPs act mainly as adsorbents for most heavy metals and constitute one of the most applied systems for their uptake from water. However, nanosized magnetite ($n Fe_3O_4$) and maghemite ($n\gamma$-$Fe_2O_3$) take the highest interest as they can act as magnetic carriers for Cr(VI) adsorbents and can be easily recoverable by means of a magnet [15]. A brief description of the mechanism of Cr(VI) removal by nFeOx is described in Section 3.2.

Particularly, $n Fe_3O_4$ can adsorb Cr(VI) and/or reduce it to Cr(III) but at a lower rate and extent than nZVI, as shown by Simeonidis et al. [9] in a comparative study of Cr(VI) removal with Fe, $Fe_3O_4$, and $Fe/Fe_3O_4$ NPs. Similar to nZVI, a higher initial pH derives in a lower Cr(VI) and total Cr removal efficiency, but also in a higher proportion of surface-bonded Cr(III) in the solid obtained after the experiments [15].

## 8.3 Arsenic

Chronic ingestion of arsenic present in groundwater can lead to different types of cancer, neurological and cardiovascular diseases, and perinatal conditions [52, 53]. More than 150 million people [54] consume water containing arsenic above the 10 μg/L limit set by the WHO [55].

The presence of arsenic in groundwater is mainly natural, coming from geological processes [56, 57]. In aqueous phase, inorganic species of As(III) and As(V) are the most relevant , being As(III) the most toxic and mobile [54, 58]. At pH < 9, neutral arsenite ($HAsO_2/H_3AsO_3$) and the bi- or monoprotonated arsenate ($H_2AsO_4^-$ or $HAsO_4^{2-}$) are the main species [53]. Organic As species are less common, and their toxicity is much lower than that of the inorganic species [53].

Iron (oxy)hydroxides, either alone or on the nZVI shell, can serve as strong adsorbents for both As(V) and As(III) in the pH range typical of groundwaters [59–62]. The use of crystalline nFeOx instead of amorphous (oxy)hydroxides is justified due to the tendency of these amorphous materials to form low-surface-area crystalline iron oxides, which greatly reduces their As removal capacity [63]. A brief description of the mechanism of As(III) and As(V) removal by nFeOx and nZVI are described in Sections 3.2 and 3.3, respectively.

Some studies report that under anoxic conditions, nZVI can also remove As(V) and As(III) by reduction of As(V) to As(III) [64, 65], and also by reduction of both As(V) and As(III) to insoluble As(0) [66–68]. A concentration gradient of As species was observed: As(V) and As(III) species are exclusively retained in the outer nFeOx layer, while As(0) is detected in the limits between the nFeOx shell and the Fe(0) core. The reaction mechanism involves arsenic diffusion into the interior of the NPs and/or arsenic capture by the growing nFeOx layer instead of being retained as surface-bound species. It should be emphasized that As(0) formation is a rather slow process, taking more than 24 h to be accomplished, and that this nZVI-promoted As(III)/As(V) redox removal mechanism has only been observed at relatively high As concentrations ([As] > 10 mg/L) [66, 68].

nFeOx have high removal capacity for both As(V) and As(III) [69], but the removal capacity per surface area unit is almost independent of the particle size [70]. However, when the NP diameter is smaller than 20 nm, a threefold increase in the capacity of As(III) adsorption [57] over magnetite [71] and maghemite [72] has been reported. This increase was ascribed to (1) the presence of new adsorption sites [72], (2) surface precipitation [71], related with partial Fe(II) dissolution followed by the formation of an As(III)-rich precipitate [57], and/or (3) to surface polymerization of adsorbed arsenite [73]. However, processes (2) and (3) take place only at As(III) concentrations far above the WHO limit [71]. There is also an increase in the adsorption driving force caused by a higher decrease in free surface energy after As(III) adsorption, as the NP diameter decreases, which can be related to As(III) replacement of tetrahedral Fe(III) in the crystal lattice of these nFeOx. The mechanisms of As adsorption over nFeOx can be described as surface adsorption by complexation/chemical adsorption, ion exchange, and ion precipitation [59], the involved forces being electrostatic attraction, hydrogen bonding, and configurational stabilization by interfacial water [57, 74]. As(III) [71–73] and As(V) [57, 74, 75] adsorption on nFeOx takes place by the formation of bidentate inner-sphere complexes, with contribution of monodentate inner-sphere and outer-sphere complexes; tridentate complexes were reported between As(III) and nanoparticulated magnetite and maghemite

[71, 72]. The adsorption equilibrium constants for both inner- and outer-sphere complexes decrease in the order $Fe_3O_4$ > hydrous ferric oxide > ferrihydrite > amorphous FeO > goethite [73]; these constants were found to be proportional to the dielectric constant of the solid.

The reaction rates of As(III) and As(V) removal with nFe can be fitted to pseudo-first [62, 64] or second-order kinetics [76]; for nZVI, a biexponential kinetics was also proposed for As(V) removal [77]. The removal capacity can be usually fitted with the Freundlich adsorption isotherm [54, 57, 64, 76, 78–80] or, less commonly, with the Langmuir model [54, 70, 81]. When different nFe are compared, As removal capacity is higher for the samples of higher Fe content, i.e., usually the most reduced materials: nZVI > $Fe_3O_4$ > $Fe_2O_3$ > FeOOH [68, 82]. However, nZVI may not be more advantageous over the other cheaper nFeOx [83]. As(V) adsorbs more strongly than As(III) on nFeOx surfaces [70], but the removal capacity is usually similar for both As species [70, 84].

Regarding the pH dependence, As(III) adsorption is usually higher at neutral pH [79, 84] due to an increase in the concentration of the $H_2AsO_3^-$ species, while iron (oxy)hydroxides have an isoelectric point at pH values from 6 to 8.5 [85, 86]; however, other studies report that the optimal As(III) adsorption on magnetite and maghemite NPs takes place at pH 2 [87]. As(V) adsorption is always higher under acidic conditions [79, 84, 87], as the anionic species are dominant even at pH 2.

In the presence of DO, arsenite adsorption over magnetite is followed by As(III) partial oxidation to As(V) by the action of reactive oxygen species (ROS) formed after Fe(II) oxidation by $O_2$ [88]; incomplete As(III) oxidation was ascribed to magnetite passivation.

Among the studied anions that can interfere with As(V) and As(III) removal with iron NPs, phosphate is the most relevant causing a strong adsorption inhibition [62, 64, 80, 83, 84, 89]. Carbonate/bicarbonate are also other significant interfering anions [80, 90], more noticeably for As(III) [91]. Sulfate has been reported to cause a slight interference with As(V) removal [80, 83], but a significant one toward As(III) with nFeOx [57]; however, for nZVI, some authors found a strong inhibition [83] while others found that the effect was negligible [90]. Regarding silicate, no definite position has been established: Some studies indicate a strong interference by competing adsorption and precipitation [62, 89, 92–94], whereas others report no influence [80, 95]. Nitrate and chloride [79, 80, 90] do not interfere with As removal with nFe. The effect of natural organic matter (NOM) is still under dispute, as some authors found no effect [80], while others report that NOM inhibits As removal by nZVI [77, 90], magnetite [70], and hematite [96] NPs. On the other hand, calcium was reported to enhance As(V) and As(III) removal with nZVI [90], and As(V) adsorption of iron hydroxides at high pH [57].

There are a few reports on As removal by green-synthesized iron NPs [76, 97], but the reported As removal capacity (up to 300 mg/g [76]) makes these low-cost materials very interesting. Regarding the combination of nFe with UV-Vis irradiation, there is only one work in the literature [77], indicating an enhanced As(V) removal (≈15% higher) with nZVI after 4 h of UV-Vis irradiation. Mixed metal oxide NPs, with Fe combined with other metals, have been also tested for As removal, but these materials will not be analyzed here and the reader is referred to Ref. [59].

## 8.4 Uranium

Uranium is present in the environment because of leaching from natural mineral deposits and as result of industrial activities, mainly nuclear applications [98, 99]. The main impact of uranium on health is its chemical toxicity [99, 100], associated with nephritis, high blood pressure, and bone dysfunction [101]. The WHO has established a guideline value of 30 μg/L in drinking water [99].

In aqueous media, uranium exists mostly in its hexavalent form, being the uranyl ion ($UO_2^{2+}$) the dominant species below pH 3.5 and in the presence of DO, while at higher pH and/or in the presence of different anions, other U(VI) complexes can be formed [102]. The tetravalent state (U(IV)) is the stable species in reducing environments. U(IV) can be found in very acid conditions as $U^{4+}$ (pH ≤ 1.2) and $U(OH)^{3+}$ (1.2 < pH < 1.8); at higher pH, it is present as $UO_2$, far less soluble, in equilibrium with $U(OH)_4$ (pKs = −7.3 [103]). Uranium species with +3 and +5 oxidation states are rather unstable in aqueous media, easily oxidized by $O_2$ [104, 105] and unstable even under anoxic conditions. U(II) existence in aqueous solution is still unclear [106], and U(0) has not been informed [106].

Generally, uranium remediation methods using iron-based materials can be classified as (1) not changing the uranium oxidation state, (2) changing the oxidation state, or (3) a combination of both [107]. It is accepted that U(VI) removal from solution using nZVI involves reduction, adsorption, and coprecipitation during the formation of Fe(III) hydroxides [108]. According to Yan et al. [109], Fe(0) originally present in the core of the NPs is the primary reduction agent, through electron transfer to U(VI) already adsorbed on the surface of the NPs, producing Fe(II) (Eq. 8.6).

$$\equiv UO_2^{2+} + Fe^0 \rightarrow UO_2 + Fe^{2+} \tag{8.6}$$

After that, surficial Fe(II) can rapidly reduce U(VI) (Eq. 8.7) while producing Fe(III) in a reaction favored under anoxic conditions.

$$\equiv UO_2^{2+} + 2 \equiv Fe^{2+} \rightarrow UO_2 + 2\ Fe^{3+} \tag{8.7}$$

Even when surface Fe(II) is considered a more powerful reducing agent than aqueous Fe(II), the latter can also reduce U(VI) but in a much slower process [110]. Fe(III) is a stronger oxidant than U(VI) ($E^0_{Fe(III)/Fe(II)} > E^0_{U(VI)/U(IV)} = 0.33$ V), and, therefore, the produced Fe(III) can oxidize Fe(0) (Eq. 8.8), followed by reduction of U(VI) by the resulting Fe(II):

$$Fe^{3+} + Fe^0 \rightarrow 2 \equiv Fe^{2+} \tag{8.8}$$

Li et al. [111] proposed that $UO_2^{2+}$ can be adsorbed by hydroxyl groups present in the oxide layer of the NPs through proton exchange. In that work, U(VI) hydrolysis and precipitation of $UO_2(OH)_2$ on the iron surface are also proposed, reducing the reactivity of the NPs due to the continuous precipitation of U(VI) and iron hydroxides on the surface.

Strong acid conditions are detrimental for U(VI) removal, and maximum removal is observed at pH around 5.5 with a rapid decline at pH values exceeding 6 [108, 112]. Under almost neutral pH and in the absence of complexing agents, near total reduction from U(VI) to U(IV) can be obtained [113].

The chemistry of uranium in water is highly affected by the presence of other ions in solutions. In the case of bicarbonate, uranyl carbonate complexes show an extremely high stability, and in solutions containing carbonates, reduction of U(VI) is arrested [114–116]. These U(VI)-carbonate ternary complexes were also found on the surface of hematite using EXAFS and electrophoretic mobility experiments [117]. If $HCO_3^-$ and $Ca^+$ are in solution, two ternary aqueous uranyl complexes are formed: $Ca_2UO_2(CO_3)_3$ and $CaUO_2(CO_3)_3^{2-}$. Crane et al. [118] showed that $Ca^+$, $Na^+$, and $HCO_3^-$ do not influence the efficiency of U removal for short-term experiments (30 min), but long-term experiments (48 h) demonstrated that the presence of $Ca^+$ significantly promotes U desorption in parallel to the re-establishment of chemically oxidizing conditions. This is in agreement with previous results of the group [115], which provided clear evidence that the corrosion lifespan and associated U(VI) removal efficacy are enhanced in the absence of dissolved oxygen. Additionally, it has been suggested that long-term U immobilization as U(IV) may require removal of nitrate [119]. Even when nitrate does not directly oxidize U(IV) at appreciable rates, nitrite slowly oxidizes U(IV) to U(VI) and the rate is speed up in the presence of Fe(II) ions [120].

Raman and XANES analysis of the NPs after treatment suggested the formation of $UO_2$ [121]. Studies carried out by XAS, XRD, and TEM showed that the structure of the reduced U solid phases is time dependent and largely influenced by the ratio of nZVI to U in the system with the formation of $UO_2$ at high U:Fe molar ratios (1:4) and U(IV) and U(V) surface complexes at lower molar ratios (1:21) [122]. In a recent work, Ling et al. [123] used Cs-TEM combined with X-ray energy dispersive spectroscopy (XEDS), electron energy-loss spectroscopy, and XPS in order to investigate diffusion and reaction processes for U(VI) on nZVI. It was shown that after 24 h, 87.3% of U is contained in the 50% central volume of the NPs, whereas after 1 h, almost 80% of the uranium is in the outer shell. The process was described as a sorption followed by time-dependent uranium penetration or diffusion across the oxide layer accelerated by its chemical reduction mediated by Fe(II) and/or Fe(0).

As expected, the uranium removal efficiency depends on the morphology and the composition of the NPs. Crane et al. [124] showed almost complete removal of U(VI) from mine water using nZVI and very low removal (15%) using $n\mathrm{Fe_3O_4}$. Vacuum-annealed nZVI was the only material able to prevent U re-release after 4-week experiments.

The removal of U(VI) using $n\mathrm{Fe_3O_4}$ is described through several simultaneous steps: (1) non-reductive adsorption on the NP surface, (2) U(VI) adsorption/incorporation on the Fe(III) (oxy)hydroxides, and (3) heterogeneous reduction to U(IV) by reaction with surficial or adsorbed Fe(II) [125, 126]. Evidences of monoelectronic U(VI) reduction with U(V) formation have been reported [127, 128]. It was shown that magnetite with Fe(II)/Fe(III) ratios ≥ 0.38 reduces U(VI) to $UO_2$, whereas with more oxidized magnetites (Fe(II)/Fe(III) ratio < 0.38) and maghemite (Fe(II)/Fe(III) ratio = 0), adsorbed U(VI) was the dominant phase observed [129].

Field applications for U removal with nZVI are scarce [118, 130]. Dickinson et al. [130] investigated the use of nZVI as a remediation strategy for a uranium-contaminated waste effluent from AWE (Atomic Weapons Establishment in Aldersmaston, UK). Experiments with a duration of 28 days showed 85% U removal in anoxic systems and 50% in oxic systems. Ex situ remediation of uranium-bearing mine water (Ciudanovita Uranium Mine, Banat, Romania) using nZVI and nZVI–nickel was tested, finding differential behavior when working at laboratory size (< 2 L) versus commercial-scale (> 1000 L) batch systems due to the significant technical challenge of homogeneous mixing/dispersion of NPs with the aqueous phase.

## 8.5 Conclusion

Iron-based NPs have been successfully used for Cr(VI), As(V)/As(III), and U(VI) removal from water with a vast number of studies published, mainly at the laboratory scale.

Parameters such as pH, DO, and presence of other dissolved ions are key to designing successful nFe-based treatments for any of these pollutants; nevertheless, particular issues have to be taken into account in each case. nZVI is the best material regarding Cr(VI) and U(VI) removal, as the reduction to Cr(III) and U(IV), respectively, is required and the Fe(0) core acts as an almost continuous source of electrons. On the other hand, nZVI surface passivation due to Cr(VI) adsorption/reaction is a major drawback for the system, as well as the presence of DO compromises the long-term stability of the removed U on the iron products. In the case of As(III) and As(V), both are removed mainly by adsorption processes; therefore, the best strategy is the use of nFeOx, which can be obtained through inexpensive procedures.

## Acknowledgments

Work performed as part of Agencia Nacional de la Promoción de la Ciencia y la Tecnología PICT 2011 0463, PICT 2013 2450 and PICT 2014 3460. NQ, MM, and VNM are members of CONICET.

## References

1. Litter, M. I. (2009). Treatment of chromium, mercury, lead, uranium, and arsenic in water by heterogeneous photocatalysis, *Adv. Chem. Eng.*, **36**, pp. 37–67.
2. Li, X. and Zhang, W. (2007). Sequestration of metal cations with zerovalent iron nanoparticles: A study with high resolution X-ray photoelectron spectroscopy (HR-XPS), *J. Phys. Chem. C*, **111**, pp. 6939–6946.
3. Zhang, W. X. (2003). Nanoscale iron particles for environmental remediation: An overview, *J. Nanoparticle Res.*, **5**, pp. 323–332.
4. Pradeep, T. A. (2009). Noble metal nanoparticles for water purification: A critical review, *Thin Solid Films*, **517**, pp. 6441–6478.
5. Yan, W., Herzing, A. A., Kiely, C. J., and Zhang, W.-X. (2010). Nanoscale zero-valent iron (nZVI): Aspects of the core–shell structure and reactions with inorganic species in water, *J. Contam. Hydrol.*, **118**, pp. 96–104.
6. Crane, R. A and Scott, T. B. (2012). Nanoscale zero-valent iron: Future prospects for an emerging water treatment technology, *J. Hazard. Mater.*, **211–212**, pp. 112–125.
7. Scott, T. B., Popescu, I. C., Crane, R. A., and Noubactep, C. (2011). Nano-scale metallic iron for the treatment of solutions containing multiple inorganic contaminants, *J. Hazard. Mater.*, **186**, pp. 280–287.
8. Adeleye, A. S., Conway, J. R., Garner, K., Huang, Y., Su, Y., and Keller, A. A. (2016). Engineered nanomaterials for water treatment and remediation: Costs, benefits, and applicability, *Chem. Eng. J.*, **286**, pp. 640–662.
9. Simeonidis, K., Tziomaki, M., Angelakeris, M., Martinez-Boubeta, C., Balcells, L., Monty, C., Mitrakas, M., Vourlias, G., and Andritsos, N. (2013). Development of iron-based nanoparticles for Cr(VI) removal from drinking water, *EPJ Web Conf.*, **40**, pp. 8007.
10. Zhao, X., Liu, W., Cao, Z., Han, B., Qian, T., and Zhao, D. (2016). An overview of preparation and applications of stabilized zero-valent iron

nanoparticles for soil and ground water remediation, *Wat. Res.*, **100**, pp. 245–266.

11. Fendorf, S. E. (1995). Surface reactions of chromium in soils and waters, *Geoderma*, **67**, pp. 55–71.
12. Saha, B. and Orvig, C. (2010). Biosorbents for hexavalent chromium elimination from industrial and municipal effluents, *Coord. Chem. Rev.*, **254**, pp. 2959–2972.
13. Anger, G., Halstenberg, J., Hochgeschwender, K., Scherhag, C., Korallus, U., Schmidt, H. K., and Ohlinger, M. (2012). *Ullmann's Encyclopedia of Industrial Chemistry* **9**, "Chromium Compounds" (Wiley-VCH, Weinheim), pp. 157–191.
14. WHO, Chromium in water, 2003 http://www.who.int/entity/water_sanitation_health/water-quality/guidelines/chemicals/chromium.pdf
15. Simeonidis, K., Mourdikoudis, S., Kaprara, E., Mitrakas, M., and Polavarapu, L. (2016). Inorganic engineered nanoparticles in drinking water treatment: A critical review, *Environ. Sci. Water Res. Technol.*, **2**, pp. 43–70.
16. Kharisov, B. I., Rasika Dias, H. V., Kharissova, O. V., Jimenez-Pérez, V. M., Olvera Pérez, B., and Muñoz Flores, B. (2012). Iron-containing nanomaterials: Synthesis, properties, and environmental applications, *RSC Adv.*, **2**, pp. 9325–9358.
17. Mehta, D., Mazumdar, S., and Singh, S. K. (2015). Magnetic adsorbents for the treatment of water/wastewater: A review, *J. Water Process Eng.*, **7**, pp. 244–265.
18. Tesh, S. J. and Scott, T. B. (2014). Nano-composites for water remediation: A review, *Adv. Mater.*, **26**, pp. 6056–6068.
19. Gheju, M. (2011). Hexavalent chromium reduction with zero-valent iron (ZVI) in aquatic systems, *Water Air Soil Pollut.*, **222**, pp. 103–148.
20. Mu, Y., Ai, Z., Zhang, L., and Song, F. (2015). Insight into core–shell dependent anoxic Cr(VI) removal with Fe@Fe2O3 nanowires: Indispensable role of surface bound Fe(II), *ACS Appl. Mater. Interfaces*, **7**, pp. 1997–2005.
21. Montesinos, V. N., Quici, N., Halac, B. E., Leyva, A. G., Custo, G., Bengio, S., Zampieri, G., and Litter, M. I. (2014). Highly efficient removal of Cr(VI) from water with nanoparticulated zerovalent iron: Understanding the Fe(III)–Cr(III) passive outer layer structure, *Chem. Eng. J.*, **244**, pp. 569–575.

22. Ponder, S. M., Darab, J. G., and Mallouk, T. E. (2000). Remediation of Cr(VI) and Pb(II) aqueous solutions using supported, nanoscale zero-valent iron, *Environ. Sci. Technol.*, **34**, pp. 2564–2569.

23. Wang, Q., Cissoko, N., Zhou, M., and Xu, X. (2011). Effects and mechanism of humic acid on chromium(VI) removal by zero-valent iron ($Fe^0$) nanoparticles, *Phys. Chem. Earth, Parts A/B/C*, **36**, pp. 442–446.

24. Wang, X., Le, L., Wang, A., Liu, H., Ma, J., and Li, M. (2016). Comparative study on properties, mechanisms of anionic dispersant modified nano zero-valent iron for removal of Cr(VI), *J. Taiwan Inst. Chem. Eng.*, **66**, pp. 115–125.

25. Yu, R. F., Chi, F. H., Cheng, W. P., and Chang, J. C. (2014). Application of pH, ORP, and DO monitoring to evaluate chromium(VI) removal from wastewater by the nanoscale zero-valent iron (nZVI) process, *Chem. Eng. J.*, **255**, pp. 568–576.

26. Pullin, H., Springell, R., Parry, S., and Scott, T. (2017). The effect of aqueous corrosion on the structure and reactivity of zero-valent iron nanoparticles, *Chem. Eng. J.*, **308**, pp. 568–577.

27. Nurmi, J. T., Tratnyek, P. G., Sarathy, V., Baer, D. R., Amonette, J. E., Pecher, K., Wang, C., Linehan, J. C., Matson, D. W., Penn, R. L., and Driessen, M. D. (2005). Characterization and properties of metallic iron nanoparticles: Spectroscopy, electrochemistry, and kinetics, *Environ. Sci. Technol.*, **39** , pp. 1221–1230.

28. Noubactep, C., Schöner, A., and Woafo, P. (2009). Metallic iron filters for universal access to safe drinking water, *CLEAN Soil, Air, Water*, **37**, pp. 930–937.

29. Sun, Y.-P., Li, X., Cao, J., Zhang, W., and Wang, H. P. (2006). Characterization of zero-valent iron nanoparticles, *Adv. Colloid Interface Sci.*, **120**, pp. 47–56.

30. Wang, X., Cong, S., Wang, P., Ma, J., Liu, H., and Ning, P. (2017). Novel green micelles Pluronic F-127 coating performance on nano zero-valent iron: Enhanced reactivity and innovative kinetics, *Sep. Purif. Technol.*, **174**, pp. 174–182.

31. Searson, P. C. and Latanision, R. M. (1988). Analysis of the photoelectrochemical response of the passive film on iron in neutral solutions, *J. Electrochem. Soc.*, **135**, pp. 1358–1363.

32. Gerischer, H. (1989). Remarks on the electronic structure of the oxide film on passive iron and the consequences for its electrode behaviour, *Corros. Sci.*, **29**, pp. 191–195.

33. Noubactep, C. (2008). A critical review on the process of contaminant removal in $Fe^0$-$H_2O$ systems, *Environ. Technol.*, **29**, pp. 909–920.

34. Scott, T. B., Allen, G. C., Heard, P. J., and Randell, M. G. (2005). Reduction of U(VI) to U(IV) on the surface of magnetite, *Geochim. Cosmochim. Acta*, **69**, pp. 5639–5646.

35. Dos Santos Coelho, F., Ardisson, J. D., Moura, F. C. C., Lago, R. M., Murad, E., and Fabris, J. D. (2008). Potential application of highly reactive Fe(0)/$Fe_3O_4$ composites for the reduction of Cr(VI) environmental contaminants, *Chemosphere*, **71**, pp. 90–96.

36. Zhou, H., He, Y., Lan, Y., Mao, J., and Chen, S. (2008). Influence of complex reagents on removal of chromium(VI) by zero-valent iron, *Chemosphere*, **72**, pp. 870–874.

37. Manning, B. A., Kiser, J. R., Kwon, H., and Kanel, S. R. (2007). Spectroscopic investigation of Cr(III)- and Cr(VI)-treated nanoscale zerovalent iron, *Environ. Sci. Technol.*, **41**, pp. 586–592.

38. Li, X., Cao, J., and Zhang, W. (2008). Stoichiometry of Cr(VI) immobilization using nanoscale zerovalent iron (nZVI): A study with high-resolution X-ray photoelectron spectroscopy (HR-XPS), *Ind. Eng. Chem. Res.*, **47**, pp. 2131–2139.

39. Ling, L. and Zhang, W. (2014). Mapping the reactions of hexavalent chromium [Cr(VI)] in iron nanoparticles using spherical aberration corrected scanning transmission electron microscopy (Cs-STEM), *Anal. Methods*, **6**, pp. 3211–3214.

40. Calderon, B. and Fullana, A. (2015). Heavy metal release due to aging effect during zero valent iron nanoparticles remediation, *Water Res.*, **83**, pp. 1–9.

41. Dong, H., Zeng, Y., Zeng, G., Huang, D., Liang, J., Zhao, F., He, Q., Xie, Y., and Wu, Y. (2016). EDDS-assisted reduction of Cr(VI) by nanoscale zero-valent iron, *Sep. Purif. Technol.*, **165**, pp. 86–91.

42. Rivero-Huguet, M. and Marshall, W. D. (2009). Reduction of hexavalent chromium mediated by micro- and nano-sized mixed metallic particles, *J. Hazard. Mater.*, **169**, pp. 1081–1087.

43. He, D., Ma, X., Jones, A. M., Ho, L., and Waite, T. D. (2016). Mechanistic and kinetic insights into the ligand-promoted depassivation of bimetallic zero-valent iron nanoparticles, *Environ. Sci. Nano*, **3**, pp. 737–744.

44. Zhou, X., Lv, B., Zhou, Z., Li, W., and Jing, G. (2015). Evaluation of highly active nanoscale zero-valent iron coupled with ultrasound for chromium(VI) removal, *Chem. Eng. J.*, **281**, pp. 155–163.

45. Xie, Y. and Cwiertny, D. M. (2010). Use of dithionite to extend the reactive lifetime of nanoscale zero-valent iron treatment systems, *Environ. Sci. Technol.*, **44**, pp. 8649–8655.

46. Montesinos, V. N., Quici, N., and Litter, M. I. (2014). Visible light enhanced Cr(VI) removal from aqueous solution by nanoparticulated zerovalent iron, *Catal. Commun.*, **46**, pp. 57–60.

47. Sun, Y., Li, J., Huang, T., and Guan, X. (2016). The influences of iron characteristics, operating conditions and solution chemistry on contaminants removal by zero-valent iron: a review, *Water Res.*, **100**, pp. 277–295.

48. Hu, J., Chen, G., and Lo, I. M. C. (2005). Removal and recovery of Cr(VI) from wastewater by maghemite nanoparticles, *Water Res.*, **39**, pp. 4528–4536.

49. Yoon, I. H., Bang, S., Chang, J. S., Gyu Kim, M., and Kim, K. W. (2011). Effects of pH and dissolved oxygen on Cr(VI) removal in Fe(0)/H2O systems, *J. Hazard. Mater.*, **186**, pp. 855–862.

50. Qin, H., Li, J., Bao, Q., Li, L., and Guan, X. (2016). Role of dissolved oxygen in metal(loid) removal by zerovalent iron at different pH: Its dependence on the removal mechanisms, *RSC Adv.*, **6**, pp. 50144–50152.

51. Mu, Y., Wu, H., and Ai, Z. (2015). Negative impact of oxygen molecular activation on Cr(VI) removal with core–shell Fe@$Fe_2O_3$ nanowires, *J. Hazard. Mater.*, **298**, pp. 1–10.

52. Bardach, A. E., Ciapponi, A., Soto, N., Chaparro, M. R., Calderon, M., Briatore, A., Cadoppi, N., Tassara, R., and Litter, M. I. (2015). Epidemiology of chronic disease related to arsenic in Argentina: A systematic review, *Sci. Total Environ.*, **538**, pp. 802–816.

53. Virender, K. S. and Sohn, M. (2009). Aquatic arsenic: Toxicity, speciation, transformations, and remediation, *Environ. Int.*, **35**, pp. 743–759.

54. Arpan, S. and Biswajit, P. (2016). The global menace of arsenic and its conventional remediation: A critical review, *Chemosphere*, **158**, pp. 37–49.

55. WHO. Arsenic in drinking-water, background document for development of WHO-Guidelines for drinking-water quality; WHO/SDE/WSH/03.04/75/Rev/1 2011.

56. Smedley, P. L. and Kinniburgh, D. G. (2013). *Essentials of Medical Geology*, 2nd Ed., eds.: Selinus, O., Alloway, B., Centeno, J. A., Finkelman, R. B., Fuge, R., Lindh, U., and Smedley, P. L., Chapter 12 "Arsenic in groundwater and the environment" (Springer), pp. 279–310.

57. Charlet, L., Morin, G., Rose, J., Wang, Y., Auffan, M., Burnol, A., and Fernandez-Martinez, A. (2011). Reactivity at (nano)particle-water interfaces, redox processes, and arsenic transport in the environment, *C. R. Geosci.*, **343**, pp. 123–139.

58. Bundschuh, J., Litter, M. I., Parvez, F., Román-Ross, G., Nicolli, H. B., Jean, J.-S., Liu, C.-W., López, D., Armienta, M. A., Guilherme, L. R. G., Gomez Cuevas, A., Cornejo, L., Cumbal, L., and Toujaguez, R. (2012). One century of arsenic exposure in Latin America: A review of history and occurrence from 14 countries, *Sci. Tot. Environ.*, **429**, pp. 2–35.

59. Lata, S. and Samadder, S. R. (2016). Removal of arsenic from water using nano adsorbents and challenges: A review, *J. Environ. Manag.*, **166**, pp. 387–406.

60. Su, C. and Puls, R. W. (2001). Arsenate and arsenite removal by zerovalent iron: Kinetics, redox transformation, and implications for in situ groundwater remediation, *Environ. Sci. Technol.*, **35**, pp. 1487–1492.

61. Sharma, Y. C., Srivastava, V., Singh, V. K., Kaul, S. N., and Weng, C. H. (2009). Nano-adsorbents for the removal of metallic pollutants from water and wastewater, *Environ. Technol.*, **30**, pp. 583–609.

62. Kanel, S. R., Manning, B., Charlet, L., and Choi, H. (2005). Removal of arsenic (III) from groundwater by nanoscale zero-valent iron, *Environ. Sci. Technol.*, **39 (5)**, pp. 1291–1298.

63. Mohan, D. and Pittman Jr., C. U. (2007). Arsenic removal from water/wastewater using adsorbents: A critical review, *J. Hazard. Mat.*, **142**, pp. 1–53.

64. Kanel, S. R., Che, J., Manning, B., Charlet, L., and Choi, H. (2006). Arsenic(V) removal from groundwater using nano scale zero-valent iron as a colloidal reactive barrier material, *Environ. Sci. Technol.*, **40**, pp. 2045–2050.

65. Rodová, A., Filip, J., and Černíki, M. (2015). Arsenic immobilization by nanoscale zero-valent iron, *Ecol. Chem. Eng. S.*, **22**, pp. 45–59.

66. Ramos, M. A. V., Yan, W., Li, Y.-Q., Koel, B. E., and Zhang, W.-X. (2009). Simultaneous oxidation and reduction of arsenic by zero-valent iron nanoparticles: Understanding the significance of the core-shell structure, *J. Phys. Chem. C*, **113**, pp. 14591–14594.

67. Yan, W., Vasic, R., Frenkel, A. I., and Koel, B. E. (2012). Intraparticle reduction of arsenite (As(III)) by nanoscale zerovalent iron (nZVI) investigated with in situ X-ray absorption spectroscopy, *Environ. Sci. Technol.*, **46**, pp. 7018–7026.

68. Yan, W., Ramos, M. A. V., Koel, B. E., and Zhang, W.-X. (2012). As(III) sequestration by iron nanoparticles: Study of solid-phase redox transformations with X-ray photoelectron spectroscopy, *J. Phys. Chem. C*, **116**, pp. 5303–5311.

69. Mayo, J. T., Yavuz, C., Yean, S., Cong, L., Shipley, H., Yu, W., Falkner, J., Kan, A., Tomson, M., and Colvin, V. L. (2007). The effect of nanocrystalline magnetite size on arsenic removal, *Sci. Technol. Adv. Mater.*, **8**, pp. 71–75.

70. Yean, S., Cong, L., Yavuz, C. T., Mayo, J. T., Yu, W. W., Kana, A. T., Colvin, V. L., and Tomson, M. B. (2005). Effect of magnetite particle size on adsorption and desorption of arsenite and arsenate, *J. Mater. Res.*, **20**, pp. 3255–3264.

71. Morin, G., Wang, Y. H., Ona-Nguema, G., Juillot, F., Calas, G., Menguy, N., Aubry, E., Bargar, J. R., and Brown, G. E. (2009). EXAFS and HRTEM evidence for As(III)-containing surface precipitates on nanocrystalline magnetite: Implications for As sequestration, *Langmuir*, **25**, pp. 9119–9128.

72. Auffan, M., Rose, J., Proux, O., Borschneck, D., Masion, A., Chaurand, P., Hazemann, J. L., Chaneac, C., Jolivet, J. P., Wiesner, M. R., Van Geen, A., and Bottero, J. Y. (2008). Enhanced adsorption of arsenic onto maghemites nanoparticles: As(III) as a probe of the surface structure and heterogeneity, *Langmuir*, **24**, pp. 3215–3222.

73. Sverjensky, D. A. and Fukushi, K. (2006). A predictive model (ETLM) for As(III) adsorption and surface speciation on oxides consistent with spectroscopic data, *Geoch. Cosm. Acta*, **70**, pp. 3778–3802.

74. Catalano, J. G., Park, C., Fenter, P., and Zhang, Z. (2008). Simultaneous inner- and outer-sphere arsenate adsorption on corundum and hematite, *Geochim. Cosmochim. Acta*, **72**, pp. 1986 –2004.

75. Wang, S. and Mulligan, C. N. (2008). Speciation and surface structure of inorganic arsenic in solid phases: A review, *Environ. Int.*, **34**, pp. 867–879.

76. Poguberović, S. S., Krčmar, D. M., Maletić, S. P., Kónya, Z., Pilipović, D. D. T., Kerkez, D. V., and Rončević, S. D. (2016). Removal of As(III) and Cr(VI) from aqueous solutions using "green" zero-valent iron nanoparticles produced by oak, mulberry and cherry leaf extracts, *Ecol. Eng.*, **90**, pp. 42–49.

77. Morgada, M. E., Levy, I. K., Salomone, V., Farías, S. S., López, G., and Litter, M. I. (2009). Arsenic (V) removal with nanoparticulate zerovalent iron: Effect of UV light and humic acids, *Catal. Today*, **143**, pp. 261–268.

78. Goswami, R., Deb, P., Thakur, R., Sarma, K. P., and Basumallick, A. (2010). Removal of As(III) from aqueous solution using functionalized ultrafine iron oxide nanoparticles, *Sep. Sci. Tech.*, **46**, pp. 1017–1022.

79. Lee, C. -G. and Kim, S.-B. (2016). Removal of arsenic and selenium from aqueous solutions using magnetic iron oxide nanoparticle/multi-walled carbon nanotube adsorbents, *Desal. Wat. Treat.*, **57**, pp. 28323–28339.

80. He, J., Bardelli, F., Gehin, A., Silvester, E., and Charlet, L. (2016). Novel chitosan goethite bionanocomposite beads for arsenic remediation, *Water Res.*, **101**, pp. 1–9.

81. Georgiou, Y., Mouzourakis, E., Bourlinos, A. B., Zboril, R., Karakassides, M. A., Douvalis, A. P., Bakas, Th., and Deligiannakis, Y. (2016). Surface decoration of amine-rich carbon nitride with iron nanoparticles for arsenite (As(III)) uptake: The evolution of the Fe-phases under ambient conditions, *J. Hazard. Mat.*, **312**, pp. 243–253.

82. Mamindy-Pajany, Y., Hurel, C., Marmier, C., and Romeo, M. (2011). Arsenic (V) adsorption from aqueous solution onto goethite, hematite, magnetite and zero-valent iron: Effects of pH, concentration and reversibility, *Desalination*, **281**, pp. 93–99.

83. Zhang, M., Wang, Y., Zhao, D., and Pan, G. (2010). Immobilization of arsenic in soils by stabilized nanoscale zero-valent iron, iron sulfide (FeS), and magnetite ($Fe_3O_4$) particles, *Chinese Sci. Bull.*, **55**, pp. 365–372.

84. Dixit, S. and Hering, J. G. (2003). Comparison of arsenic(V) and arsenic(III) sorption onto iron oxide minerals: Implications for arsenic mobility, *Environ. Sci. Technol.*, **37**, pp. 4182–4189.

85. Pang, S. C., Chin, S. F., and Anderson, M. A. (2007). Redox equilibria of iron oxides in aqueous based magnetite dispersions: Effect of pH and redox potential, *J. Colloid Interf. Sci.*, **311**, pp. 94–101.

86. Huang, S. H. and Chen, D. H. (2009). Rapid removal of heavy metal cations and anions from aqueous solutions by an amino-functionalized magnetic nano-adsorbent, *J. Hazard. Mater.*, **163**, pp. 174–179.

87. Chowdhury, S. R. and Yanful, E. K. (2010). Arsenic and chromium removal by mixed magnetite maghemite nanoparticles and the effect of phosphate on removal, *J. Environ. Manag.*, **91**, pp. 2238–2247.

88. Ona-Nguema, G., Morin, G., Wang, Y. H., Foster, A. L., Juillot, F., Galas, G., and Brown, G. E. (2010). XANES evidence for rapid arsenic(III) oxidation at magnetite and ferrihydrite surfaces by dissolved $O_2$ via $Fe^{2+}$-mediated reactions, *Environ. Sci. Technol.*, **44**, pp. 5416–5422.

89. Westerhoff, P., Highfield, D., Badruzzaman, M., and Yoon, Y. (2005). Rapid small-scale column tests for arsenate removal in iron oxide packed bed columns, *J. Environ. Eng.*, **131**, pp. 262–271.

90. Tanboonchuy, V., Grisdanuraka, N., and Liao, C.-H. (2012). Background species effect on aqueous arsenic removal by nano zero-valent iron using fractional factorial design, *J. Hazard. Mat.*, **205–206**, pp. 40–46.

91. Nicolli, H. B., Bundschuh, J., Blanco, M. del C., Tujchneider, O. C., Panarello, H. O., Dapeña, C., and Rusansky, J. E. (2012). Arsenic and associated trace-elements in groundwater from the Chaco-Pampean plain, Argentina: Results from 100 years of research, *Sci. Total Environ.*, **429**, pp. 36–56.

92. Goswami, R., Deb, P., Thakur, R., Sarma, K. P., and Basumallick, A. (2010). Removal of As(III) from aqueous solution using functionalized ultrafine iron oxide nanoparticles, *Sep. Sci. Tech.*, **46**, pp. 1017–1022.

93. Su, C. and Puls, R. W. (2001). Arsenate and arsenite removal by zerovalent iron: Kinetics, redox transformation, and implications for in situ groundwater remediation, *Environ. Sci. Technol.*, **35**, pp. 1487–1492.

94. Zhu, H., Jia, Y., Wu, X., and Wang, H. (2009). Removal of arsenic from water by supported nano zero-valent iron on activated carbon, *J. Hazard. Mater.*, **172**, pp. 1591–1596.

95. Savina, I. N., English, C. J., Whitby, R. L. D., Zheng, Y., Leistnerc, A., Mikhalovsky, S. V., and Cundy, A. B. (2011). High efficiency removal of dissolved As(III) using iron nanoparticle-embedded macroporous polymer composites, *J. Hazard. Mater.*, **192**, pp. 1002–1008.

96. Redman, A. D., Macalady, D. L., and Ahmann, D. (2002). Natural organic matter affects arsenic speciation and sorption onto hematite, *Environ. Sci. Technol.*, **36**, pp. 2889–2896.

97. Martínez-Cabanas, M., López-García, M., Barriada, J. L., Herrero, R., and Sastre de Vicente, M. E. (2016). Green synthesis of iron oxide nanoparticles. Development of magnetic hybrid materials for efficient As(V) removal, *Chem. Eng. J.*, **301**, pp. 83–91.

98. Bhalara, P. D., Punetha, D., and Balasubramanian, K. (2014). A review of potential remediation techniques for uranium(VI) ion retrieval from contaminated aqueous environment, *J. Environ. Chem. Eng.*, **2**, pp. 1621–1634.

99. WHO. Uranium in drinking-water, Background document for development of WHO-Guidelines for Drinking Water Quality, 2012: http://www.who.int/entity/water_sanitation_health/water-quality/guidelines/chemicals/background_uranium.pdf

100. Katz, S. (2014). The chemistry and toxicology of depleted uranium, *Toxics*, **2**, pp. 50–78.

101. Brugge, D., deLemos, J. L., and Oldmixon, B. (2005). Exposure pathways and health effects associated with chemical and radiological toxicity of natural uranium: A review, *Rev. Environ. Health*, **20**, pp. 1–2.

102. Amadelli, R., Maldotti, A., Sostero, S., and Carassiti, V. (1991). Photodeposition of uranium oxides onto $TiO_2$, from aqueous uranyl solutions, *J. Chem. Soc. Faraday Trans.*, **87**, pp. 3267–3273.

103. Casas, I., De Pablo, J., Giménez, J., Torrero, M. E., Bruno, J., Cera, E., Finch, R. J., and Ewing, R. C. (1998). The role of pe, pH, and carbonate on the solubility of $UO_2$ and uraninite under nominally reducing conditions, *Geochim. Cosmochim. Acta*, **62**, pp. 2223–2231.

104. Kulyukhin, S. A., Mikheev, N. B., Kamenskaya, A. N., Konovalova, N. A., and Rumer, I. A. (2006). Physicochemical properties of uranium in lower oxidation states, *Radiochemistry*, **48**, pp. 535–551.

105. Selbin, J. and Ortego, J. D. (1969). Chemistry of uranium (V), *Chem. Rev.*, **69**, pp. 657–671.

106. Grenthe, I., Droźdźyński, J., Fujino, T., Buck, E. C., Albrecht-Schmitt, T. E., and Wolf, S. F. (2011). *The Chemistry of the Actinide and Transactinide Elements*, eds.: Morss, L. R., Edelstein, N. M., and Fuger, J., "Uranium" (Springer), pp. 253–698.

107. Chen, A., Shang, C., Shao, J., Zhang, J., and Huang, H. (2016). The application of iron-based technologies in uranium remediation: A review, *Sci. Total Environ.*, **575**, pp. 1291–1306.

108. Li, X. Y., Zhang, M., Liu, Y. B., Li, X., Liu, Y. H., Hua, R., and He, C. T. (2013). Removal of U(VI) in aqueous solution by nanoscale zero-valent iron (nZVI), *Water Qual. Expo. Heal.*, **5**, pp. 31–40.

109. Yan, S., Chen, Y., Xiang, W., Bao, Z., Liu, C., and Deng, B. (2014). Chemosphere uranium (VI) reduction by nanoscale zero-valent iron in anoxic batch systems: The role of Fe(II) and Fe(III), *Chemosphere*, **117**, pp. 625–630.

110. Charlet, L., Liger, E., and Gerasimo, P. (1998). Decontamination of TCE- and U-rich waters by granular iron: Role of sorbed Fe(II), *J. Environ. Eng.*, **124**, pp. 25–30.

111. Li, Z. J., Wang, L., Yuan, L. Y., Xiao, C. L., Mei, L., Zheng, L. R., Zhang, J., Yang, J. H., Zhao, Y. L., Zhu, Z. T., Chai, Z. F., and Shi, W. Q. (2015). Efficient removal of uranium from aqueous solution by zero-valent iron nanoparticle and its graphene composite, *J. Hazard. Mater.*, **290**, pp. 26–33.

112. Jing, C., Li, Y. L., and Landsberger, S. (2016). Review of soluble uranium removal by nanoscale zero valent iron, *J. Environ. Radioact.,* **164**, pp. 65–72.

113. Riba, O., Scott, T. B., Vala Ragnarsdottir, K., and Allen, G. C. (2008). Reaction mechanism of uranyl in the presence of zero-valent iron nanoparticles, *Geochim. Cosmochim. Acta*, **72**, pp. 4047–4057.

114. Crane, R. A., Dickinson, M., Popescu, I. C., and Scott, T. B. (2011). Magnetite and zero-valent iron nanoparticles for the remediation of uranium contaminated environmental water, *Water Res.*, **45**, pp. 2931–2942.

115. Crane, R. A. and Scott, T. (2014). The removal of uranium onto carbon-supported nanoscale zero-valent iron particles, *J. Nanoparticle Res.*, **16**, pp. 2813.

116. Yan, S., Hua, B., Bao, Z., Yang, J., Liu, C., and Deng, B. (2010). Uranium(VI) removal by nanoscale zerovalent iron in anoxic batch systems, *Environ. Sci. Technol.*, **44**, pp. 7783–7789.

117. Bargar, J. R., Reitmeyer, R., Lenhart, J. J., and Davis, J. A. (2000). Characterization of U(VI)-carbonato ternary complexes on hematite: EXAFS and electrophoretic mobility measurements, *Geochim. Cosmochim. Acta*, **64**, pp. 2737–2749.

118. Crane, R. A., Pullin, H., and Scott, T. B. (2015). The influence of calcium, sodium and bicarbonate on the uptake of uranium onto nanoscale zero-valent iron particles, *Chem. Eng. J.*, **277**, pp. 252–259.

119. Wu, W. M., Carley, J., Green, S. J., Luo, J., Kelly, S. D., Van Nostrand, J., Lowe, K., Mehlhorn, T., Carroll, S., Boonchayanant, B., Löfller, F. E., Watson, D., Kemner, K. M., Zhou, J., Kitanidis, P. K., Kostka, J. E., Jardine, P. M., and Criddle, C. S. (2010). Effects of nitrate on the stability of uranium in a bioreduced region of the subsurface, *Environ. Sci. Technol.*, **44**, pp. 5104–5111.

120. Senko, J. M., Mohamed, Y., Dewers, T. A., and Krumholz, L. R. (2005). Role for Fe(III) minerals in nitrate-dependent microbial U(IV) oxidation, *Environ. Sci. Technol.*, **39**, pp. 2529–2536.

121. Crespi, J., Quici, N., Halac, E. B., Leyva, A. G., Ramos, C. P., Mizrahi, M., Requejo, F. G., and Litter, M. I. (2016). Removal of uranium(VI) with iron nanoparticles, *Chem. Eng. Trans.*, **47**, pp. 265–270.

122. Tsarev, S., Collins, R. N., Fahy, A., and Waite, T. D. (2016). Reduced uranium phases produced from anaerobic reaction with nanoscale zerovalent iron, *Environ. Sci. Technol.*, **50**, pp. 2595–2601.

123. Ling, L. and Zhang, W. X. (2015). Enrichment and encapsulation of uranium with iron nanoparticle, *J. Am. Chem. Soc.*, **137**, pp. 2788–2791.

124. Crane, R. A. and Scott, T. B. (2013). The effect of vacuum annealing of magnetite and zero-valent iron nanoparticles on the removal of aqueous uranium, *J. Nanotechnol.*, **2013**, article ID 173625.

125. Singer, D. M., Chatman, S. M., Ilton, E. S., Rosso, K. M., Banfield, J. F., and Waychunas, G. A. (2012). U(VI) sorption and reduction kinetics on the magnetite (111) surface, *Environ. Sci. Technol.*, **46(7)**, pp. 3821–3830.

126. Aamrani, S. E., Giménez, J., Rovira, M., Seco, F., Grivé, M., Bruno, J., Duro, L., and de Pablo, J. (2007). A spectroscopic study of uranium(VI) interaction with magnetite, *Appl. Surf. Sci.*, **253**, pp. 8794–8797.

127. Ilton, E. S., Ois Boily, J. -F., Buck, E. C., Skomurski, F. N., Rosso, K. M., Cahill, C. L., Bargar, J. R., and Felmy, A. R. (2010). Influence of dynamical conditions on the reduction of U(VI) at the magnetite–solution interface, *Environ. Sci. Technol.*, **44**, pp. 170–176.

128. Yuan, K., Ilton, E. S., Antonio, M. R., Li, Z., Cook, P. J., and Becker, U. (2015). Electrochemical and spectroscopic evidence on the one-electron reduction of U(VI) to U(V) on magnetite, *Environ. Sci. Technol.*, **49**, pp. 6206–6213.

129. Latta, D. E., Gorski, C., Boyanov, M. I., O'Loughlin, E. J., Kemner, K. M., and Scherer, M. M. (2012). Influence of magnetite stoichiometry on U(VI) reduction, *Environ. Sci. Technol.*, **46**, pp. 778–786.

130. Dickinson, M. and Scott, T. B. (2010). The application of zero-valent iron nanoparticles for the remediation of a uranium-contaminated waste effluent, *J. Hazard. Mater.*, **178**, pp. 171–179.

## Chapter 9

# Iron Nanoparticles for Cr(VI) Removal from Contaminated Soil

**Luca Di Palma, Elisabetta Petrucci, Nicola Verdone, and Giorgio Vilardi**
*Department of Chemical Engineering Materials Environment, Sapienza University, via Eudossiana 18, Rome, Italy*
luca.dipalma@uniroma1.it

## 9.1 Introduction

Heavy metals and metalloids contamination of soils is caused by the rapidly expanding industrial areas, mine tailings, land application of fertilizers, animal manures, sewage sludge, pesticides, wastewater irrigation, coal combustion residues, spillage of petrochemicals, and atmospheric deposition [1–4]. The geochemical cycles of heavy metals are strongly affected (influenced) by the environmental conditions (P, T, Eh, dissolved oxygen), soil characteristics (texture, organic content, humidity), the chemical speciation of the introduced metal and the presence of specific species of microorganisms in the soil microbial consortia. In fact, whereas organic pollutants are usually degraded and oxidized through biochemical and chemical

*Iron Nanomaterials for Water and Soil Treatment*
Edited by Marta I. Litter, Natalia Quici, and Martín Meichtry

ISBN 978-981-4774-67-3 (Hardcover), 978-981-4669-49-8 (eBook)
www.panstanford.com

processes, most metals and metalloids do not undergo chemical or microbial degradation [5] and their concentrations could persist for a long time in soil. The consequences of these phenomena could be the inhibition of enzymatic soil activity and organic pollutants degradation [6]. Indeed, the toxic effect of a metal depends on several factors (metal concentration, organism species, assumption pathway, presence of other chemicals that can increase or reduce the toxicity of the metal), but it may pose significant risks and hazards not only to the ecosystem but also to humans (contact with contaminated soil, through the food chain soil–plant–human or soil–plant–animal–human, contaminated groundwater drinking, reduction in food quality via phytotoxicity [7]). The most common heavy metals found in polluted soils are lead (Pb), zinc (Zn), cadmium (Cd), nickel (Ni), and chromium (Cr). In particular, the subject of this first part of the present chapter is the occurrence and characteristics of Cr in soil.

## 9.2 Chromium: Occurrence and Properties

Among the transition metal ions of the 3d series VIB group, Cr is the most widely studied because of its variability in oxidation state, coordination numbers, molecular structure, and physicochemical characteristics changes with its speciation [8]. It is the 21st most abundant metal in the earth's crust [9], and it has an atomic number of 24 and an atomic weight of 51.996; of the five known isotopes, the most common is $^{51}Cr$ (half-life 27.8 days). Chromium exists in nature as ore, and it was isolated for the first time by French chemist Nicolas Louis Vauquelin in 1797 from the mineral crocoite ($PbCrO_4$) [10]. The world production of Cr (as $FeCrO_4$) was increased from 9,570,000 t in 1994 [11] to 22,000,000 t in 2010 [12]. Cr can exist in oxidation states ranging from −2 to +6, but trivalent chromium Cr(III) (which is an essential nutrient), bivalent chromium Cr(II) (less stable than the trivalent form and present only in aqueous media), and hexavalent chromium Cr(VI) (which is carcinogenic for humans and other animals) are the most common forms observed, due to their stability in the environmental pH and redox range. The ionic radii are 0.052–0.053 nm for Cr(VI) and 0.064 nm for Cr(III) [13]; as a consequence, Cr(III) can readily substitute Fe in igneous rocks (the ionic ray of Fe(II) is 0.067 nm), whereas mafic and ultramafic rocks could be rich in both Cr(III) and Cr(VI). Thus, the natural occurrence of this metal in soil is due mainly to weathering phenomena, which involve the latter kind of rocks. However, the most stable chromium compounds in nature are the Cr(III)–Fe(II) minerals (known as chromites), which implies that the main Cr(VI) source comes from anthropogenic activities, related to several industrial processes such as metallurgical and mining activities, leather tanning, and timber preservation [10]. Chromium wastes generated by these activities are disposed of on land, wetlands, and/or landfills and increase the concentration of Cr(VI) and Cr(III) in the soils, subjacent groundwater, and nearby surface waters.

The average amount of chromium in various kinds of soils ranges from 0.02 to 58 mmol/g [14]. An increase in local Cr concentration originates from fallout and washout of atmospheric Cr-containing particles. Due to its presence in soil, Cr is usually ingested by humans through foods and beverages; regarding food, Cr was found in mashed potatoes, broccoli, and turkey breast [15], while regarding beverages, several well-known brandy beers are characterized by a mean Cr concentration (mainly Cr(III)) ranged between 0.1 and 0.4 mg/L [16]. Cr in soil is present mainly as Cr(III), as insoluble $Cr(OH)_3$, or as Cr(III) adsorbed to soil components (such as hydroxides and clay compounds), which prevents Cr leaching into the groundwater or its uptake by plants [17]. The dominant Cr form in soil depends on Eh and pH: In acidic soils (pH < 4), the dominant form is $Cr(H_2O)_6^{3+}$, whereas its hydrolysis products are dominant at pH < 5.5 [18]. In neutral-to-alkaline soils, Cr(VI) can exist in soluble (i.e., $Na_2CrO_4$) but also in moderately soluble chromates (i.e., $CaCrO_4$, $BaCrO_4$, $PbCrO_4$) [19]. In acidic soils (pH < 6), $HCrO_4^-$ becomes a dominant form. The relation between Cr(III) and Cr(VI) strongly depends on pH and oxidative conditions, and usually, the Cr(III) is the dominating species. To investigate the behavior of chromium in soil and to make some predictions about its speciation according to the environmental characteristics of the medium, the most stable forms are analyzed first separately and then together in a specific paragraph.

### 9.2.1 Trivalent and Bivalent Form

Cr(III) is the most stable oxidation state, while Cr(II) ions are strongly reducing and only stable in the absence of oxygen. When Cr(III) is present as soluble form in the soil solution, its speciation strongly depends on pH and Eh [20]. At pH < 4, Cr(III) is always an octahedral hexaaquo ion, $Cr(H_2O)_6^{3+}$. It tends to hydrolyze with increasing pH, resulting in the formation of polynuclear complexes containing $OH^-$ bridges due to the loss of a proton from coordinated water, followed by coordination of the $OH^-$ to a second cation. Insoluble $Cr(OH)_3$ is the end product of the hydrolysis. In Eq. (9.1), the pH equilibrium of Cr(III) is reported [21]:

$$Cr(H_2O)_6^{3+} \leftrightarrow [Cr(H_2O)_5OH]^{2+} \leftrightarrow [Cr(H_2O)_n(OH)_m]^{3-m} \leftrightarrow Cr(OH)_3 \quad (9.1)$$

$Cr(OH)_3$ exhibits amphoteric behavior and, at higher pH, is transformed into the readily soluble tetra-hydroxo complex, $Cr(OH)_4^-$ (pK = 15.4) [22]. At increased concentration of Cr(III) ([Cr(III)] > $10^6$ M), it is possible to observe other polynuclear hydrolytic products ($Cr_2(OH)_2^{4+}$, $Cr_3(OH)_4^{5+}$, $Cr_4(OH)_6^{6+}$). Besides the tendency to form hexacoordinate octahedral complexes with water and hydroxyl anions, Cr(III) forms other similar complexes with several ligands such as ammonia, urea, ethylenediamine, and other organic ligands containing oxygen, nitrogen, or sulfur donor atoms [23]. These complexes, except when the complex agents are humic acids or other organic macromolecules, are characterized always by a higher solubility and mobility in respect to that of Cr–OH complexes. Overall Cr(III) is non-mobile and non-bioavailable in soils.

Due to its instability, the aqueous chemistry of the Cr(II) cation has not been extensively studied. These ions are present in aqueous solution as octahedral high-spin hexaaquo ions, $Cr(H_2O)_6^{2+}$, and are unstable with respect to oxidation to Cr(III) [21] ($E_0$= −0.41 eV):

$$Cr(III) + e^- \leftrightarrow Cr(II) \quad (9.2)$$

Cr(III) is recognized as a nontoxic form of chromium by several researchers and an essential nutrient for humans and animals; its deficiency causes disturbance in the metabolism of glucose and lipids. Some other authors [24] stated that before recognizing Cr(III) as an essential nutrient for mammals, it is necessary to identify a Cr complex (biomolecule) with a biochemical role in the organism. Cr(III) seems to increase insulin sensitivity to favor the metabolism of lipids and glucose, and if it is delivered as chromium picolinate ($Cr(C_6H_4NO_2)_3$), it exhibits an antidepressant effect [25]. According to the U.S. Institute of Medicine, a daily intake of 25 and 35 μg of Cr(III) for females and males, respectively, is recommended [26]. The maximum accepted concentration for total Cr in drinking water is 100 μg/L in most countries of U.S. and Europe, although in California, Germany, UK, and Italy, it is lower (equal to 50 μg/L). The exposure limit for Cr(0), Cr(II), and Cr(III) recommended by the National Institute for Occupational Health and Safety (NIOSH) is 500 $\mu g/m^3$ for a 10 h workday, 40 h week [27].

### 9.2.2 Hexavalent Form

Hexavalent Cr is a strong oxidizing agent, and it exists in the environment as an oxyanion: chromate ($CrO_4^{2-}$), dichromate ($Cr_2O_7^{2-}$), hydrogen chromate ($HCrO_4^-$), or dihydrogen chromate ($H_2CrO_4$) [20]. At very low pH (pH < 0.1) and high Cr concentration (>1 M), it is possible to observe also hydrogen dichromate ($HCr_2O_7^{2-}$), trichromate ($Cr_3O_{10}^{2-}$), and tetrachromate ($Cr_4O_{13}^{2-}$). In general, the formation of more polymerized chromium oxide species occurs with an increase in the Cr concentration or with a decrease in the pH [33]. In particular, in the pH range 2–6, the following equilibrium is established:

$$2CrO_4^{2-} + 2H^+ \leftrightarrow Cr_2O_7^{2-} + H_2O \tag{9.3}$$

The redox potential of Cr(VI) strongly depends on pH. Considering acid media, it is possible to observe the following reaction ($E_0$ = 1.33–1.38 V):

$$Cr_2O_7^{2-} + 14H^+ + 6e^- \leftrightarrow 2Cr^{3+} + 7H_2O \tag{9.4}$$

whereas in basic media ($E_0$ = −0.12 V):

$$CrO_4^{2-} + 4H_2O + 3e^- \leftrightarrow Cr(OH)_3 + 5OH^- \tag{9.5}$$

Cr(VI) stable chemical forms are more soluble and mobile with respect to the trivalent forms, and also due to these characteristics, they have a significant toxicity for all organisms. In the scientific literature, several data regarding the exposure of workers to hexavalent chromium species airborne over an extensive period showed that the workers presented an increased risk of developing lung cancer [28]. Cr(VI) is also allergenic, corrosive, and mutagenic, in particular, if exposed to oral ingestion [28]. Several experiments were carried out on rats, and the conclusion of more than 20 years of research is that exposure to Cr(VI) causes cancer, based on the increased incidence of tumors in the small intestine of mice (within the duodenum, jejunum, and/or ileum) [29].

### 9.2.3 Behavior of Cr(VI) and Cr(III) in Soil

The metal retention mechanisms in soil involve sorption/desorption, precipitation/dissolution phenomena, redox and complexation reactions that depend on soil properties such as pH, redox potential, surface area, cation exchange capacity, organic matter content, clay content, Fe and Mn oxide content, and carbonate content [1, 13]. The Cr(III)/Cr(VI) system in soil is characterized by peculiar properties, starting from their large difference in mobility and solubility in soil solution. The most mobile forms of Cr in soil are $CrO_4^{2-}$ (predominant form of Cr(VI) in neutral soils, pH = 6.5–7.5) and $HCrO_4^-$ (predominant form of Cr(VI) in acidic soils). They can be assimilated by plants and can easily reach the deeper soil layers, causing groundwater pollution [19]. Only a minor amount of Cr(VI) is bound in soils and depends on the mineralogical composition and pH. Positively charged soil colloids such as aluminum and iron hydroxides (i.e., goethite, FeO(OH)) are generally involved in the adsorption phenomena of the hexavalent chromium species such as $CrO_4^{2-}$ ion, which, in more acidic soils, becomes $HCrO_4^-$, a more soluble species that usually tends to desorb from minerals and stay in the soil solution [30]. Hexavalent forms of chromium are the only metal species that are highly mobile in alkaline soils, and besides the variation of Eh and pH of the soil, only the presence of clay minerals containing free Fe and Mn oxides could retard their migration; other soil properties (i.e., cation exchange capacity), except for organic matter content, had no significant influence on Cr(VI) immobilization [31]. The sorption of Cr(VI) can occur simultaneously with its reduction to the trivalent form, and both are favored by acidic pH [1]. The sorption of Cr(VI) onto the positively charged surface of FeO(OH) and iron oxides seems to be nonspecific, as observed by

Stollenwerk and Grove [32] (the chromates adsorbed onto oxide coatings of alluvium aquifer materials desorbed immediately when uncontaminated groundwater was added). The reversibility of the process strongly depends on the properties of the soil solution (pH) and, in particular, on the presence of specific ligands ($SO_4^{2-}$ and $PO_4^{3-}$ tend to inhibit sorption respect to $Cl^-$ and $NO_3^-$, which have no significant effect on the process [33]). Among the possible Cr(VI) reduction mechanisms that can occur in soil, the only two that are characterized by a remarkable kinetics involve ferrous iron and organic matter [33]. The latter is the main electron donor and favors also the other reduction reactions carried out by $S^{2-}$, Fe(II) and some bacteria [1, 19, 30]. The reduction of Cr(VI) by ferrous iron can be written as [33]:

$$HCrO_4^- + 3Fe(II) + 7H^+ \leftrightarrow Cr(II) + 3Fe(III) + 4H_2O \qquad (9.6)$$

This reaction is very fast in acidic media (pH < 5), even in the presence of Fe-oxidizing species such as dissolved oxygen, and only when phosphate anion concentration is greater than 0.1 M or for pH > 10, the oxidation of Fe(II) to Fe(III) by oxygen becomes significant and can inhibit the process [34]. The most effective mineral for the reduction of Cr(VI) is pyrite (FeS), due to the presence of two electron donors (Fe(II) and $S^{2-}$). At pH < 4, a brown precipitate was observed, which was identified as a mixed Fe–Cr oxyhydroxide, according to the following reaction [34]:

$$x\text{Cr(III)} + (1-x)\text{Fe(III)} + 4H_2O \rightarrow (Cr_xFe_{1-x})(OH)_3 + 3H^+ \qquad (9.7)$$

where $x$ can vary from 0 to 1 and is 0.25 according to Eary and Rai [34]. The reaction can occur both in solution (after the dissolution of the ferrous iron from minerals or with Fe(II) already dissolved) or on the mineral surface. For instance, the Cr(VI) reduction performed by ferrous iron of biotite occurs in solution, and it is catalyzed by the presence of ferric ions, which can be reduced to Fe(II) on the biotite surface [34].

Humic and fulvic acids can reduce Cr(VI) to Cr(III) (which remains adsorbed on the macromolecule or precipitate as hydroxide) at moderate rates under environmental conditions; the process achieves significant rates only in acidic and anaerobic media, as already reported for iron reduction [35]. However, high concentrations of Cr(VI) could rapidly exhaust the available reducing capacity of the soil, whereas the presence of selected complexing agents (Al oxides, amorphous Al, hydroxides, and other soil components) could protect Cr(VI) from reduction, leading to the accumulation of the hexavalent species per several years [36] (in particular in sandy soils with a low organic matter content). The reduction of Cr(VI) is also possible through the biotic path, both in aerobic (bacteria) and anaerobic (bacteria and fungi) environments [33].

The trivalent form of chromium is naturally present in soils as hydroxide or as cation adsorbed to the soil minerals, thus reducing the mobility and bioavailability of the metal. $Cr(H_2O)_6^{3+}$ and $Cr(OH)^{2+}$ are the dominant forms in acidic soils, and both forms are easily sorbed by clay minerals (the process is favored by an increase in pH probably due to an increase in the negative charge of the surface of these minerals [1]). As the hexavalent form, the trivalent one can be also adsorbed by humic and fulvic acids and siliceous minerals [13]. However, the mobility of the obtained complex depends on pH and on the complexant; if it sorbs to humic acids, the product results immobile and non-reactive (the optimum pH range for this process is 2.7–4.5 [37]), whereas if it sorbs to fulvic acids or more soluble compounds as citric acid ($H_3C_4H_5O_7$), the product is more mobile and is susceptible of oxidation to Cr(VI). The fraction of Cr(III) dissolved in soil solution usually hydrolyzes to $Cr(OH)_3$ and precipitates, and in anaerobic environment such as deep sediments, it could remain immobilized as long as sediments remain stable [34]. In more oxidizing environments, dissolved oxygen and $MnO_2$ could oxidize Cr(III) to Cr(VI), and between these two oxidants, only the reaction carried out by manganese oxide seems to have an appreciable kinetic in soil systems [17, 30]. In fact, considering only the presence of dissolved oxygen, the oxidation becomes significant after two or three months: Palmer and Wittbrodt observed the increase in the Cr(VI) concentration in three different soils kept in moist state [38]. Richard and Bourge [39] proposed a three-step mechanism in aqueous solution for the Cr(III) oxidation mediated by manganese dioxide: Initially, the soluble Cr(III) sorbs to the mineral surface of the dioxide; subsequently, a fraction of Cr(III) is oxidized to Cr(VI) by Mn(IV) on the surface sites, and finally, Mn(II) and Cr(VI), as anions, are desorbed from the sites. In soil, this reaction (see Eq. 9.8) can occur both for $\delta$- and $\beta$-$MnO_2$ (pyrolusite), at a high rate (a significant Cr(III) oxidation was observed in less than 1 h) and seems to be catalyzed by acids (also organic acids) [40]:

$$CrOH^{2+} + 1.5\delta - MnO_2 \rightarrow HCrO_4^- + 1.5Mn(II) \quad (9.8)$$

This last phenomenon has important implications in Cr(VI) remediation strategy because it is well known that organic matter added to a Cr(VI)-polluted soil generally tends to reduce and immobilize hexavalent forms of chromium, while Mn speciation is generally less studied.

## 9.3 Hexavalent Chromium Reduction in Soil by Iron Nanoparticles

Several studies performed during the past two decades dealt with the experimental design and setup of immobilization technologies for the remediation of hexavalent chromium-contaminated soil [41, 42] and groundwater [43]. The most common alternative technologies, in fact, based on contaminant extraction processes, though successfully tested and applied, are often limited by several factors, such as the high cost of the extractant agent (chelant, surfactant), and its toxicity (though the use of biosurfactants is now widely adopted). In addition, the occurrence of side reactions during the extraction process, such as solubilization of organic matter, may induce a strong alteration of soil chemical and physical properties, thus also affecting chromium speciation [44, 45]. Furthermore, Cr(III) complexation by organic chelants could result in an increase in its availability for reoxidation to the hexavalent form [46], especially in the presence of a significant amount of Mn oxides [47].

Among the immobilization techniques, the in situ manipulation of redox status by chemical reduction using reactive solutions offers a promising alternative for the treatment of contaminated soil. This technique generally deals with the reduction to Cr(III), providing a source of electrons, followed by pH adjustment to neutrality to favor precipitation of $Cr(OH)_3$ or mixed oxyhydroxide phases [48]. To this aim, several studies in the recent years have investigated the use of nanozerovalent iron [49, 50] or bivalent iron [51, 52] as ferrous sulfate.

In particular, zerovalent iron (ZVI) is an excellent electron donor, which is used to transform via reduction or indirect oxidation many common contaminants in soil and groundwater. The large surface area of ZVI nanoparticles with respect to microsized ZVI strongly enhances the reactivity for the transformation of the recalcitrant environmental pollutants, due to the very high surface energy for the redox process [53, 54].

As widely reported in the recent literature, nZVI particles have been proved to be significantly effective in the removal of a wide range of organic pollutants from polluted water and wastewater, including halogenated solvents [55], chlorinated pesticides [56], organophosphates [57], dyes [58], nitroamines [59], nitroaromatics [60], and polyphenols [61] (in this last case, the obtained results were significantly better than those obtained by conventional processes [62]). Several literature experiences have also demonstrated that nZVI can be successfully used for the removal of hazardous inorganic species, such as nitrate [63, 64], metalloids [65], heavy metals [66], including uranium [67], as an alternative to microbial-assisted immobilization technologies [68, 69].

$Fe^0$ nanoparticles consists of a core–shell structure, with the $Fe^0$ core surrounded by an oxide/hydroxide shell, which grows thicker with the progress of iron oxidation due to the reaction with oxidizing species [70]. Reaction is limited by the passivation of the nanoparticle surface, which progressively reduces the available active sites for electron transfer.

The particles are usually injected as slurry directly into the subsurface environment to remediate contaminated groundwater plumes or contaminant source zones (thus avoiding the need for intrusive digging methods).

Pilot and field-scale remediation tests have been already performed or are under way, and the concentrations of approximately 4.5–300 g/L of nZVI slurry have been generally used [71]. The cost of the treatment depends on depth, local geology, type, and quantity of the contaminant [72].

### 9.3.1 Cr(VI) Reduction in Soil

The effectiveness of nZVI toward Cr(VI) reduction has been widely demonstrated by several studies in the literature. Several studies in the recent years have been carried out aiming at demonstrating the effectiveness of nZVI in soil remediation. Most of them were performed in spiked soil, and additional studies are required before implementing this technology on full scale [73]. According to the literature, nZVI can be successfully applied to soil treatment using slurry reactors under dynamic conditions [53, 74]. The reduction of Cr(VI) is time dependent and increases with the concentration of nZVI [75].

A comparison between ferrous sulfate and nZVI was performed by Di Palma et al. [76]. They showed that almost complete Cr(VI) removal from a polluted soil was achieved after several hours of treatment, in the presence of a large excess of Fe(II), providing a preliminary oxygen stripping from the slurry. At the same conditions, tests performed using colloidal nZVI nanoparticles showed that the reaction was faster than in the case of using Fe(II): A negligible Cr(VI) residue amount in the soil was achieved within 1 h of treatment, though a huge excess of nZVI with respect to the initial Cr(VI) content (30:1) was necessary. Tests conducted prolonging contact time up to 24 h did not show significant increase in the performance, due to progressive $Fe^0$ inactivation [75]. The typical product of this redox process is chromium hydroxide, $Cr(OH)_3$, with partial substitution of Cr by Fe; as a consequence, the most probable final product is a mixed insoluble compound described by $(Cr_{1-x}Fe_x)(OH)_3$, where "x" is typically around 0.33 [77]. This mechanism can be confirmed by performing sequential extraction tests on the treated soil, to investigate metal partitioning in different soil fractions. Due to Cr(VI) solubilization, reduction, and following Cr(III) precipitation, the reducing treatment resulted in an increase in the amount of chromium bound to the oxide–hydroxide fraction, thus confirming a mechanism of chromium–iron hydroxides precipitation [75].

## 9.3.2 Combination of nZVI and Dithionite

The use of nZVI in combination with dithionite ($Na_2S_2O_4$) inhibits oxidation, aggregation, and precipitation of ferrous iron, which mitigates the surface passivation of nZVI. The mechanism of Cr(VI) treatment in $Na_2S_2O_4$ applications involves the conversion of Fe(III) in soils to Fe(II) by $Na_2S_2O_4$, and the subsequent reduction of Cr(VI) to Cr(III) by Fe(II) to form the $Cr_xFe_{1-x}(OH)_3$ solid [78].

Sodium dithionite in water undergoes dissociation and disproportionation reactions to form primarily sulfoxyl radicals ($SO_2^{\bullet-}$), sulfites ($SO_3^{2-}$), and thiosulfates ($S_2O_3^{2-}$) via the following equations:

$$SO_4^{2-} \rightarrow 2SO_2^{*-} \tag{9.9}$$

$$4SO_2^{*-} + H_2O \rightarrow 2SO_3^{2-} + S_2O_3^{2-} + H^+ \tag{9.10}$$

During dissociation reactions, dithionite can reduce structural iron in clays and dissolve and reduce amorphous and some crystalline Fe(III) oxides to produce one or more Fe(II) species [79]:

$$SO_2^{*-} + Fe^{2+} + H_2O \rightarrow Fe^{3+} + SO_3^{2--} + 2H^+ \tag{9.11}$$

To limit the precipitation of $Fe(OH)_2$, which can inhibit the dissemination of Fe(II), acidic conditions must be assured, and the reduction of Cr(VI) by the Fe(II) can be written as:

$$HCrO_4^- + 3Fe^{2+} + 7H^+ \rightarrow Cr^{3+} + 3Fe^{3+} + 4H_2O \tag{9.12}$$

Laboratory-scale tests combining nZVI and sodium dithionite carried out at several molar ratios $Fe^0$/Cr(VI) and $Na_2S_2O_4$/Cr(VI) indicated that dithionite increases nZVI effectiveness in the long term, assuring an almost complete Cr(VI) reduction at lower nZVI dose under optimized conditions [75].

### 9.3.3 Toxicity of nZVI

Soil remediation by nZVI is generally considered a promising technology, but the environmental risks associated to the use of such nanoparticles are not yet fully understood.

Only a few studies have been performed so far to the evaluation of the environmental and biological impact of nZVI and its by-products injected during remediation processes. Studies were mainly devoted to the investigation of possible toxic effect on biological indicators [71, 80, 81].

The objective of a complete toxicity study must be the investigation of the persistence of nZVI and its by-products in the environment, as well as the nature and rate of iron transformation, and how soil and water chemistry may influence the partitioning of the iron species and distribution between solid and liquid phases. Few data are available about these crucial aspects, and additional studies are required to fully understand the potential impact of a remediation technology based on nanoparticles.

According to Keller et al. (2012), basing on the typical dose of nZVI injected in remediation tests, nZVI concentration in the aquatic environment impacted by a remediation site could range from μg/L to mg/L [71].

In tests carried out on selected commercial iron nanoparticles at those level, Keller et al. [71] showed that selected commercial nanoparticles exhibited toxicity in freshwater media at 0.5–1.0 mg/L to the freshwater phytoplankton *P. subcapitata* and the water flea *D. magna* and were also toxic for three species of marine phytoplankton at 0.3–3.0 mg/L [71].

The toxicity is mainly due to the product of iron oxidation, such as bivalent and trivalent iron cations and oxides, and may be reduced by dilution in the subsurface. Both iron species are chemical species that most organisms are well adapted to deal with, but the effects of a quick increase in their concentration in a relatively small area have been not fully investigated. In addition, it is well known that bivalent iron acts as a catalyst of hydrogen peroxide decomposition, thus generating hydroxyl radicals, which are highly reactive species. Furthermore, Fe ions are able to enter the cytoplasm of cells, thus inducing oxidative stress, which may damage cell membrane.

Nanoparticle coating prevents aggregation, thus prolonging the contact of nanoparticles with the contaminated media, but may also increase the risk of diffusion of nanoparticles in large areas, even reaching uncontaminated zones.

In a comprehensive study, Adeleye et al. [82] showed that large nZVI aggregates tend to be more persistent due to their decreased reactivity, and groundwater ions and natural organic materials may increase the persistence of nZVI in the subsurface via reduced reactivity and surface passivation. On the contrary, groundwater chemistry results impacted by nZVI injection, thus potentially influencing ecosystems [82].

## 9.4 Synthesis of Iron Nanoparticles

Several approaches have been proposed for the preparation of zerovalent iron at the nanoscale, including both physical and chemical technologies. However, the most commonly used method is the reduction of iron ions to elemental iron by aqueous reaction with sodium borohydride. During the synthesis, to prevent fast oxidation of the nanoparticles on contact with air, an inert atmosphere has to be provided. The pioneer paper by Wang and Zhang [83], where $FeCl_3$ was used as the zerovalent iron precursor, proposes the occurrence of the following reaction:

$$Fe(H_2O)_6^{3+} + 3BH_4^- + 3H_2O \rightarrow Fe^0 \downarrow + 3B(OH)_3 + 10.5H_2 \uparrow \quad (9.13)$$

and estimates that 90% of produced nanoparticles exhibit a size smaller than 100 nm with specific surface area of 33.5 $m^2/g$. More often, the reaction between ferric ions and the reducing agent is expressed as follows [84]:

$$4Fe^{3+} + 3BH_4^- + 9H_2O \rightarrow 4Fe^0 \downarrow + 3H_2BO_3^- + 12H^+ + 6H_2 \uparrow \quad (9.14)$$

Iron nanoparticles are composed of a metallic core coated by a passivating layer of iron and boron oxides and hydroxides that constitute the shell [85]. In addition, it has been verified that some operating conditions significantly affect the morphology and therein the reactivity of the nanoparticles. In particular, fast reaction and high concentration of reagents promote the formation of needle-shaped aggregates composed by particles endowed of small radii (9.5 nm) and large surface (45.4 $m^2/g$) [86]. It has long been known that zerovalent iron nanoparticles can be synthesized also by using ferrous [87] instead of ferric salts (mainly $FeSO_4$), with the advantage of consuming less $NaBH_4$, thus limiting both costs and hazard of the synthesis.

$$Fe(H_2O)_6^{2+} + 2BH_4^- + 3H_2O \rightarrow Fe^0 \downarrow + 2B(OH)_3 + 7H_2 \uparrow \quad (9.15)$$

In fact, a comparison of Eqs. (9.1) and (9.3) clearly indicates that the use of ferrous ions implies a more favorable molar ratio of zerovalent iron to $NaBH_4$ (from one-third to two-thirds). Moreover, the Fe(II) reaction reduces the released amount of gaseous hydrogen that passes from 10.5 to 7 mol per mole of metallic iron. Nonetheless, only a few studies report the use of Fe(II)-synthesized nZVI.

However, in spite of the stoichiometric coefficients, either starting from ferric or ferrous ions, excess of sodium borohydride is generally needed to enhance the reaction rate of synthesis [88] and to obtain highly reactive nanoparticles [86]. In particular, the stoichiometric excess of borohydride used with Fe(III) was 7.4 while with Fe(II) was 3.6 [89].

Although these bottom-up methods result in facile and relatively quick preparation, at ambient temperature, the release of gaseous hydrogen, with potential explosion hazard and the $NaBH_4$ toxicity, raises concerns about the process safety. Moreover, the high cost of $NaBH_4$ represents a further drawback and a serious deterrent to large-scale application. Therefore, the development of cost-effective and environmentally friendly methods represents a topic of great interest.

Environmentally benign approach to nZVI synthesis relies on the adoption of greener reductants. For example, sodium dithionite has been proposed to reduce ferrous chloride [90] and ferrous sulfate [91]. The reaction requires high pH but is not affected by the presence of oxygen; therefore, it can be conducted in air with significant simplification and cost reduction. Another alternative is the exploitation of the reducing power of dithionite for the chemical regeneration of passivated iron nanoparticles, thus assuring their long-term action [92]. Other green routes of nanoparticles synthesis include the use (as reducing and capping agents) of extracts from food scrap [93] or from plants such as coffee [90], sorghum [94], rosemary, melaleuca, eucalyptus [95], and tea [96]. Perennial and grass plants, typically less water and labor demanding and which can be cultivated worldwide, are of major interest. A size ranging from 5 to 100 nm is always reported, as well as enhanced stability and longevity of the nanoparticles, due to the capping ability of polyphenols released in the extracts. The nature and concentration of polyphenols do affect both size and structures of the nanoparticles [97]. The feasibility of producing nZVI from natural extracts, however, is a controversial topic, and conflicting results regarding safety, mechanism, and effectiveness can be found in the literature. In fact, a recent study verified an ecotoxicological impact on different aquatic organisms due to the onset of oxidative stress, due to the Fe(II)-catalyzed generation of reactive oxygen intermediates [98]. Finally, a low performance of green tea deriving nZVI in the treatment of Cr(VI)-contaminated soils is observed [98].

Another green approach to be considered is the use of an iron precursor effluent or industrial residue. For example, Wang et al. [99] proposed, for the in situ remediation of Cr(VI) in soils, nZVI synthesized by reaction of borohydride with steel pickling waste liquor.

As known, the enhanced reactivity of the metallic iron at the nanoscale and, therefore, its formidable performance in adsorbing/reducing a broad range of metals and organics can be attributed to its high surface-to-volume ratio. However, without stabilization, the ZVI nanoparticles undertake many different interactions driven by magnetic attraction and van der Waals bonds, resulting in the formation of aggregates and possible settling.

There is a consensus on the fact that, in the remediation of contaminated soils, nanoparticles agglomeration results in a decreased mobility in porous media and, therefore, in a difficult deliverability into soils, thus preventing the possibility of in situ applications. To improve nanoparticles dispersion, electrostatic and steric stabilizations are needed. To this aim, different approaches may be adopted [100] by modifying viscosity and surface characteristics or by introducing solid supports. Actions aimed at promoting the stability of the nanoparticles are generally differentiated into pre- and post-synthesis stabilization, depending on whether stabilizers are introduced during or after the nZVI production. Pre-synthesis methods exhibit better performance due to their superior reactivity and increased nanoparticles surface area [53].

Since stabilizing agents for in situ application should meet strict environmental safety requirements, the addition of surfactants is generally avoided, and a few studies investigate their use [101, 102]. Major concern is represented by their possible toxicity, also considering the concentration needed, and the tendency to mobilize other pollutants that may be present [99].

Stable and green suspensions of iron nanoparticles have been obtained by using stabilizers based on natural polymers such as xanthan [103] or guar gum [104]. He and Zhao [105] verified successfully the use of carboxymethyl cellulose (CMC) and postulated that it exerts an electrosteric stabilization of nanoparticles by forming a negatively charged layer on the ZVI nanoparticles. This depresses aggregation, accelerates iron nucleation, and results in the formation of small and dispersed particles. Moreover, it has been verified that the reduction in the particle size increased with increasing CMC to iron molar ratio. A study by Franco et al. [53] compared the reactivity of iron sulfate, granular iron, non-stabilized nZVI, and CMC-stabilized nZVI in the reduction of Cr(VI) during batch and semi-batch experiments, indicating that the latter is the most effective reductant [100]. This result has been confirmed in a more recent study by Di Palma et al. [76]. The advantage of using these polysaccharides is always their low cost.

A further approach is represented by immobilizing the nanoparticles on a support that promotes the access into porous media. In particular, for this purpose, different kinds of materials have been tested, including resins [77], hydrophilyzed carbon [106], silica fume [107], and electrospun PAA/PVA [108]. All syntheses involve the addition of the support to a ferric or ferrous ions solution prior to the drop-wise addition of borohydride. A promising result has been obtained by using smectite clay [109]. In fact, it not only acts as a support but also as a layer template where, due to the presence of negative charges, sparse iron ions are linked, thus resulting, after reduction, in metallic iron at sub-nanoscale. The immobilization on supports produces more transportable and stable nanoparticles but reduces the active surface and, in consequence, the reactivity. It is, therefore, necessary to reach a compromise by taking into account numerous site-specific parameters [106]. However, in spite of the encouraging results, deeper investigation and full scale-up data are needed to understand the interactions of stabilized nanoparticles with the medium they enter and the possible effect on the ecosystem [110].

## 9.5 Modeling Nanoparticles Behavior in Adsorption Processes

The particular and extremely low dimension of iron-based nanoparticles makes the proper modeling of their behavior in the adsorption processes a difficult task. The unit cell edge of the iron crystal is 0.287 nm, value not so much different from a reference diameter of the nanoparticle of 100 nm. Moreover, even if a porosity could be measured in the nanoparticles, the characteristic time of diffusion through the pores inside the particle is proportional to the ratio of the squared radius of the particle to the adsorbing species diffusion coefficient. For a particle diameter of 100 nm and considering the order of magnitude of the diffusion coefficient of species in a liquid medium of $10^{-5}$ [$cm^2$/s], the resulting characteristic diffusion time should be about $2.5\times10^{-6}$ s. These circumstances result in a difficult and perhaps unnecessary application of sophisticated models, such as the different formulations of the shrinking core model. As a consequence, in the literature, simplified and phenomenological models are, by far, preferred. For this reason, in the following description, reference is made only to this class of models.

### 9.5.1 Modeling Adsorption Equilibria

Adsorption isotherms explain the interaction that adsorbent and adsorbate exhibit with each other and assume a major importance in optimizing the use of adsorbents. In the adsorption processes, equilibrium can be described by several available adsorption isotherm models [111]; however, the following brief description is limited to the models more suitable or adopted in the literature to model adsorption in aqueous media. According to the majority of the reviewed papers, Langmuir, Freundlich, and Dubinin–Radushkevich isotherm models are best fitted models for Cr(VI) adsorption onto iron-based nanosorbents [112–114] and are briefly explained below.

#### 9.5.1.1 Langmuir adsorption isotherm model

Monolayer adsorption occurring at specific homogeneous sites on the adsorbent material can be successfully modeled by the Langmuir isotherm. All the available sites are considered identical and energetically equivalent so that once a site is occupied by an adsorbate molecule, no further adsorption can take place in the same site [111, 115]. The Langmuir model can be expressed in the nonlinear form as [116]:

$$q_e = q_m \times K_L \times \frac{C_e}{1 + K_L \times C_e} \qquad (9.16)$$

where $C_e$ is the concentration of the solution [mg/L], $q_e$ is the corresponding adsorption capacity [mg/L], and $q_m$ [mg/g] and $K_L$ [L/mg] are constants related to adsorption capacity and net enthalpy of adsorption, respectively. In the case of aqueous solutions, however, the theoretically derived Langmuir model is often not suitable in describing the experimental data, in particular when the basic assumptions of monolayer coverage of the adsorbent surface and energetic homogeneity of the adsorption sites are not fulfilled [117].

#### 9.5.1.2 Freundlich adsorption isotherm model

Freundlich isotherm model is applicable to heterogeneous adsorption surfaces and describes the non-ideal and reversible multilayer adsorption. The model is based on the hypotheses that several layers of adsorbate can be attached on the adsorbent; in this way, the adsorbate will continuously keep bound to the adsorbent, and the energy required for adsorption is not constant, but it varies according to an exponential distribution. The Freundlich model can be represented by the nonlinear form [118]:

$$q_e = K_F \times C_e^{1/n} \tag{9.17}$$

where the constants $K_F$ and $n$ measure the adsorption capacity and the energetic heterogeneity of the adsorbent surface, respectively. The unit of $K_F$ depends on the units used for $q_e$ and $C_e$ and includes the exponent $n$ (dimensionless). The value of $n$ describes a favorable adsorption condition when included in the range 1–10. The Freundlich model is not able to describe the linear range at very low concentrations, nor the saturation effect at very high concentrations, but in general, the intermediate concentration range is accurately modeled. In the literature, the model is widely adopted in describing the adsorption process in aqueous solution and is often applied to model equilibrium in the most kinetic and breakthrough curve interpretations [117].

#### 9.5.1.3 Dubinin–Radushkevich adsorption isotherm model

The Dubinin–Radushkevich isotherm model is an empirical model initially formulated to describe the adsorption process according to a pore-filling mechanism. The model is generally applied to infer the adsorption process occurring onto both homogeneous and heterogeneous surfaces. The nonlinear expression of the Dubinin–Radushkevich isotherm model can be represented by the following equations [119]:

$$q_e = q_s \times \exp(-\mathrm{K}_{DR} \times \varepsilon^2) \tag{9.18}$$

$$\varepsilon = R \times T \times \ln\left(1 + \frac{1}{C_e}\right) \tag{9.19}$$

where $q_s$ [mg/g] is a constant related to the adsorption capacity, $K_{DR}$ [mol$^2$/kJ$^2$] is a constant related to the mean free energy of adsorption, $R$ [J/mol K] is the gas constant, and T [K] is the absolute temperature.

## 9.5.2 Modeling Adsorption Kinetics

The heterogeneous nature of the adsorption process involving ZVI nanoparticles has encouraged many researchers to adopt mathematically complex models [120]. However, standing the considerations invoked at the beginning of this section, first-order kinetics is, at present, the most common model used to explain the experimental results from batch adsorption tests. This approach has been applied in a wide range of situations due to the basic relative simplicity of the mathematical procedure.

Assuming that in the current situation the first-order kinetics prevails, the rate of contaminant degradation is proportional to its concentration in solution, or in mathematical form:

$$\frac{dC}{dt} = -k \times C \tag{9.20}$$

where $C$ is the contaminant concentration [mol/L] at time $t$, and $k$ [t$^{-1}$] is the first-order rate constant.

However, when the adsorbent nanoparticles concentration strongly affects the adsorption process, this influence should appear in the kinetic model, and the first-order kinetic expression should be modified in the form:

$$\frac{dC}{dt} = -k \times C \times P \tag{9.21}$$

where $P$ is the concentration of the adsorbent [mol/L], and the rate of the contaminant degradation depends on both the contaminant and adsorbent concentrations. The reaction is now modeled as a second-order kinetics.

It is worth to note, however, that in practical situations, the concentration of the adsorbent into the solution medium is set appreciably larger than that of the contaminant, leading to a useful simplification in the modeling framework, which assumes that the adsorbent concentration does not change significantly in the course of the experiment with respect to the variation of the contaminant concentration. If this is the case, the rate law can be applied in the form:

$$\frac{dC}{dt} = -k_{obs} \times C \tag{9.22}$$

In this formulation, the observed first-order rate constant, $k_{obs}$, is related to the first-order rate constant $k$, according to $k_{obs} = k \times P$, and the previous equation is known as pseudo first-order model.

Despite its wide use in the literature, this form leads to the regression of $k_{obs}$ values different up to three orders of magnitude when applied to specific contaminants in experimental tests conducted in unlike conditions [120]. In an effort to avoid this discrepancy, the rate equation can be structured in the form:

$$\frac{dC}{dt} = -k_{SA} \times \rho_a \times C \tag{9.23}$$

where $k_{SA}$ is the surface-area-normalized rate constant [L/h m$^2$], and $\rho_a$ is the surface area concentration of the adsorbent [m$^2$/L] [121]. This last parameter can be related to the specific surface area of the adsorbent nanoparticle, $a_s$ [m$^2$/g], and its mass concentration, $\rho_m$ [g/L]:

$$\rho_a = a_s \times \rho_m \tag{9.24}$$

Finally, the relationship between $k_{obs}$ and $k_{SA}$ holds:

$$k_{obs} = k_{SA} \times a_s \times \rho_m \tag{9.25}$$

The successful application of this approach does clearly depend on the precision in measuring the physical characteristics of the nanoparticles.

## 9.6 Conclusion

Several studies in the last two decades proved the nZVI effectiveness in Cr(VI) reduction in polluted soil. The use of nZVI ensures high removal efficiency (almost quantitative) within a few hours, and nontoxic by-products are released at the end of the treatment. Nanoparticles are usually coupled to a stabilizing agent to prevent iron passivation in the reaction media. The simultaneous addition of a reducing agent such as dithionite was successfully tested to achieve a further treatment enhancement.

nZVI particles are typically produced by the borohydride reduction method, though innovative green synthesis have been recently proposed and developed, involving natural extract from agro-industrial wastes. Cr(VI) removal has been widely modeled by a pseudo first-order kinetics: Basing on the results of experiments reported in literature, this form leads to the regression of $k_{obs}$ values different up to three orders of magnitude when applied to specific contaminants. More complex models can be, therefore, developed, considering contaminant adsorption onto nanoparticle surface. However, due to the extremely low dimension of iron-based nanoparticles and characteristics time of diffusion, the application of sophisticated models, such as the different formulations of the shrinking core model, appears to be a quite difficult challenge.

## References

1. Sposito, G. (2008). *The Chemistry of Soils.* Oxford University Press.
2. Wuana, R. A. and Okieimen, F. E. (2011). Heavy metals in contaminated soils: A review of sources, chemistry, risks and best available strategies for remediation. *ISRN Ecology*, Article ID 402647, pp. 1–20.
3. Schroeder, W. H. and Munthe, J. (1998). Atmospheric mercury: An overview. *Atmos. Environ.*, **32**(5), pp. 809–822.
4. Zhang, M. K., Liu, Z. Y., and Wang, H. (2010). Use of single extraction methods to predict bioavailability of heavy metals in polluted soils to rice. *Commun. Soil Sci. Plan.*, **41**(7), pp. 820–831.
5. Kirpichtchikova, T. A., Manceau, A., Spadini, L., Panfili, F., Marcus, M. A., and Jacquet, T. (2006). Speciation and solubility of heavy metals in contaminated soil using X-ray microfluorescence, EXAFS spectroscopy, chemical extraction, and thermodynamic modeling. *Geochim. Cosmochim. Acta,* **70**(9), pp. 2163–2190.
6. Madejón, E., Burgos, P., López, R., and Cabrera, F. (2001). Soil enzymatic response to addition of heavy metals with organic residues. *Biol. Fert. Soils,* **34**(3), pp. 144–150.
7. McLaughlin, M. J., Zarcinas, B. A., Stevens, D. P., and Cook, N. (2000). Soil testing for heavy metals. *Commun. Soil Sci. Plan.*, **31**(11–14), pp. 1661–1700.
8. Greenwood, N. N. and Earnshaw, A. (2012). *Chemistry of the Elements.* Elsevier.

9. Barnhart, J. (1997). Chromium chemistry and implications for environmental fate and toxicity. *Soil Sediment Contam.*, **6**(6), pp. 561–568.

10. Enghag, P. (2008). *Encyclopedia of the Elements: Technical Data-History-Processing-Applications*. John Wiley & Sons.

11. Australian Government Department of Sustainability, Environment, Water, Population and Communities. (2000). National Pollutant Inventory: Organo-tin compounds: Overview.

12. U.S. Geological Survey. (2011). Mineral commodity summaries 2011: U.S. Geological Survey, 198 pp.

13. Bartlett, R. J. and James, B. R. (1988). Mobility and bioavailability of chromium in soils. In: J. O Nriagu and E. Nieboer, Eds.: *Chromium in Natural and Human Environments*, pp. 267–306, New York, Wiley Interscience.

14. Coleman, R. N. (1988). Chromium toxicity: Effects on microorganisms with special reference to the soil matrix. In: J. O Nriagu and E. Nieboer, Eds.: *Chromium in Natural and Human Environments*, pp. 335–350, New York, Wiley Interscience.

15. Anderson, R. A., Bryden, N. A., and Polansky, M. M. (1992). Dietary chromium intake. *Biol. Trace Elem. Res.*, **32**(1–3), pp. 117–121.

16. Iwegbue, C. (2010). Composition and daily intakes of some trace metals from canned beers in Nigeria. *J. I. Brewing*, **116**(3), pp. 312–315.

17. Bartlett, R. J. and Kimble, J. M. (1976). Behavior of chromium in soils: I. Trivalent forms. *J. Environ. Qual.*, **5**(4), pp. 379–383.

18. Ritchie, G. S. P. and Sposito, G. (1995). Speciation in soils. In: A. M. Ure and C. M. Davidson, Eds.: *Chemical Speciation in the Environment, Second Edition*, pp. 237–264, John Wiley & Sons.

19. Bartlett, R. J. and Kimble, J. M. (1976). Behavior of chromium in soils: II. Hexavalent forms. *J. Environ. Qual.*, **5**(4), pp. 383–386.

20. Brookins, D. G. (2012). *Eh-pH Diagrams for Geochemistry*. Springer Science & Business Media.

21. Weckhuysen, B. M., Wachs, I. E., and Schoonheydt, R. A. (1996). Surface chemistry and spectroscopy of chromium in inorganic oxides. *Chem. Rev.*, **96**(8), pp. 3327–3350.

22. Rai, D., Sass, B. M., and Moore, D. A. (1987). Chromium (III) hydrolysis constants and solubility of chromium (III) hydroxide. *Inorganic Chem.*, **26**(3), pp. 345–349.

23. Salem, F. Y., Parkerton, T. F., Lewis, R. V., Huang, J. H., and Dickson, K. L. (1989). Kinetics of chromium transformations in the environment. *Sci. Total Environ.*, **86**(1–2), pp. 25–41.

24. Vincent, J. B. (2010). Chromium: Celebrating 50 years as an essential element? *Dalton Trans.*, **39**(16), pp. 3787–3794.

25. Franklin, M. and Odontiadis, J. (2003). Effects of treatment with chromium picolinate on peripheral amino acid availability and brain monoamine function in the rat. *Pharmacopsychiatry*, **36**(05), pp. 176–180.

26. Shenkin, A. (2003). Dietary reference values for vitamin A, vitamin K, arsenic, boron, chromium, copper, iodine, iron, manganese, molybdenum, nickel, silicon, vanadium and zinc. *J. Hum. Nutr. Diet.*, **16**(3), pp. 199–200.

27. National Institute for Occupational Safety and Health (NIOSH). (1999). Online Pocket Guide to Chemical Hazards, NIOSH, http://www.cdc.gov/niosh/npg/npg.html.

28. Langard, S. (1990). One hundred years of chromium and cancer: A review of epidemiological evidence and selected case reports. *Am. J. Ind. Med.*, **17**(2), pp. 189–214.

29. National Toxicology Program. (2008). Toxicology and carcinogenesis studies of sodium dichromate dihydrate (Cas No. 7789-12-0) in F344/N rats and B6C3F1 mice (drinking water studies). *National Toxicology Program Technical Report Series*, (546), pp. 1–192.

30. James, B. R. and Bartlett, R. J. (1983). Behavior of chromium in soils: V. Fate of organically complexed Cr(III) added to soil. *J. Environ. Qual.*, **12**(2), pp. 169–172.

31. Korte, N. E., Skopp, J., Fuller, W. H., Niebla, E. E., and Alesii, B. A. (1976). Trace element movement in soils: Influence of soil physical and chemical properties. *Soil Sci.*, **122**(6), pp. 350–359.

32. Stollenwerk, K. G. and Grove, D. B. (1985). Adsorption and desorption of hexavalent chromium in an alluvial aquifer near Telluride, Colorado. *J. Environ. Qual.*, **14**(1), pp. 150–155.

33. Palmer, C. D. and Puls, R. W. *EPA Ground Water Issue*. EPA/540/5-94/505.

34. Eary, L. E. and Rai, D. (1988). Chromate removal from aqueous wastes by reduction with ferrous ion. *Environ. Sci. Technol.*, **22**(8), pp. 972–977.

35. Wittbrodt, P. R. and Palmer, C. D. (1995). Reduction of Cr (VI) in the presence of excess soil fulvic acid. *Environ. Sci. Technol.*, **29**(1), pp. 255–263.

36. Baron, D., Palmer, C. D., and Stanley, J. T. (1996). Identification of two iron-chromate precipitates in a Cr (VI)-contaminated soil. *Environ. Sci. Technol.*, **30**(3), pp. 964–968.

37. James, B. R. (1996). Peer reviewed: The challenge of remediating chromium-contaminated soil. *Environ. Sci. Technol.*, **30**(6), pp. 248A–251A.

38. Palmer, C. D. and Wittbrodt, P. R. (1990). Geochemical characterization of the United Chrome Products Site, Final Report. *IN: Stage, 2.*

39. Richard, F. C. and Bourg, A. C. (1991). Aqueous geochemistry of chromium: A review. *Water Res.*, **25**(7), pp. 807–816.

40. Fendorf, S. E. and Zasoski, R. J. (1992). Chromium (III) oxidation by delta-manganese oxide ($MnO_2$). 1. Characterization. *Environ. Sci. Technol.*, **26**(1), pp. 79–85.

41. Dermatas, D. and Meng, X. (2003). Utilization of fly ash for stabilization/solidification of heavy metal contaminated soils, *Eng. Geol.*, **70**, pp. 377–394.

42. Dermatas, D. and Moon, D. H. (2006). Chromium leaching and immobilization in treated soils, *Environ. Eng. Sci.*, **23**, pp. 75–85.

43. Mulligan, C. N., Yong, R. N., and Gibbs, B. F. (2001). Remediation technologies for metal-contaminated soils and groundwater: An evaluation, *Eng. Geol.*, **60**, pp. 193–201.

44. Manouchehri, N., Besancon, S., and Bermond, A. (2006). Major and trace metal extraction from soil by EDTA: Equilibrium and kinetic studies. *Anal. Chim. Acta,* **559**, pp. 105–112.

45. Di Palma, L. (2009). Influence of indigenous and added iron on copper extraction from soil. *J. Hazard. Mater.,* **170**, pp. 96–102.

46. Di Palma, L., Mancini, D., and Petrucci, E. (2012). Experimental assessment of chromium mobilization from polluted soil by washing. *Chem. Eng. Trans.,* **28**, pp. 145–150.

47. Negra, C., Ross, D. S., and Lanzirotti, A. (2005). Oxidizing behavior of soil manganese: Interactions among abundance, oxidation state, and pH, *Soil Sci. Soc. Am. J.*, **69**, pp. 87–95.

48. Bolan, N., Kunhikrishnan, A., Thangarajan, R., Kumpiene, J., Park, J., Makino, T., Kirkham, M. B., and Scheckel, K. (2014). Remediation of heavy metal(loid)s contaminated soils: To mobilize or to immobilize? *J. Hazard. Mater.,* **266**, pp. 141–166.

49. Du, J., Lu, J., Wu, Q., and Jing, C. (2012). Reduction and immobilization of chromate in chromite ore processing residue with nanoscale zero-valent iron. *J. Hazard. Mater.,* **215–216**, pp. 152–158.

50. Chrysochoou, M., Johnston, C. P., and Dahal, G. (2012). A comparative evaluation of hexavalent chromium treatment in contaminated soil by calcium polysulfide and green-tea nanoscale zero-valent iron. *J. Hazard. Mater.*, **201–202**, pp. 33–42.

51. Seaman, J. C., Bertsch, P. M., and Schwallie, L. (1999). In situ Cr(VI) reduction within coarse-textured, oxide-coated soil and aquifer systems using Fe(II) solutions. *Environ. Sci. Technol.*, **33**, pp. 938–944.

52. Kostarelos, K., Rao, E., Reale, D., and Moon, D. H. (2009). Reduction of Cr(VI) to Cr(III) in artificial, contaminated soil using ferrous sulfate heptahydrate and sodium thiosulfate. *Pract. Period. Hazard. Toxic Radioact. Waste Manag.*, **13**, pp. 135–139.

53. Franco, D. V., Da Silva, L. M., and Jardim, W. F. (2009). Reduction of hexavalent chromium in soil and ground water using zero-valent iron under batch and semi-batch conditions. *Water Air Soil Pollut.*, **197**, pp. 49–60.

54. Singh, R., Misra, V., and Singh, R. P. (2011). Synthesis, characterization and role of zero-valent iron nanoparticle in removal of hexavalent chromium from chromium-spiked soil. *J. Nanoparticle Res.*, **13**, pp. 4063–4073.

55. Liu, Y., Majetich, S. A., Tilton, R. D., Sholl, D. S., and Lowry, G. V. (2005). TCE dechlorination rates, pathways, and efficiency of nanoscale iron particles with different properties. *Environ. Sci. Technol.*, **39**(5), pp. 1338–1345.

56. Elliott, D. W., Lien, H. L., and Zhang, W. X. (2009). Degradation of lindane by zero-valent iron nanoparticles. *J. Environ. Eng.*, **135**(5), pp. 317–324.

57. Satapanajaru, T., Anurakpongsatorn, P., Pengthamkeerati, P., and Boparai, H. (2008). Remediation of atrazine-contaminated soil and water by nano zerovalent iron. *Water Air Soil Pollut.*, **192**(1–4), pp. 349–359.

58. Cao, J., Wei, L., Huang, Q., Wang, L., and Han, S. (1999). Reducing degradation of azo dye by zero-valent iron in aqueous solution. *Chemosphere*, **38**, pp. 565–571.

59. Naja, G., Halasz, A., Thiboutot, S., Ampleman, G., and Hawari, J. (2008). Degradation of hexahydro-1,3,5-trinitro-1,3,5-triazine (RDX) using zerovalent iron nanoparticles. *Environ. Sci. Technol.*, **42**(12), pp. 4364–4370.

60. Choe, S., Lee, S. H., Chang, Y. Y., Hwang, K. Y., and Khim, J. (2001). Rapid reductive destruction of hazardous organic compounds by nanoscale $Fe^0$. *Chemosphere*, **42**(4), pp. 367–372.

61. Lopes, D. V., Lobo, J., Santos, S., Martins, R. C., Quina, M. J., Gando-Ferreira, L. M., Dias-Ferreira, C., and Quinta-Ferreira, R. M. (2015). Treatment of olive mill solid wastes by chemical processes enhanced with iron rich materials. *Proceedings of the 3rd International Conference on Sustainable Solid Waste Management*, 2–4 July 2015, Tinos Island, Greece.

62. Stoller, M., Azizova, G., Mammadova, A., Vilardi, G., Di Palma, L., and Chianese, A. (2016). Treatment of olive oil processing wastewater by ultrafiltration, nanofiltration, reverse osmosis and biofiltration. *Chem. Eng. Trans.*, **47**, pp. 409–414.

63. Muradova, G. G., Gadjieva, S. R., Di Palma, L., and Vilardi, G. (2016). Nitrates removal by bimetallic nanoparticles in water. *Chem. Eng. Trans.*, **47**, pp. 205–210.

64. Vilardi, G. and Di Palma, L. (2017). Kinetic study of nitrate removal from aqueous solutions using copper-coated iron nanoparticles. *Bull. Environ. Contam. Toxicol.*, **98**, pp. 359–365.

65. Kanel, S. R., Manning, B., Charlet, L., and Choi, H. (2005). Removal of arsenic (III) from groundwater by nanoscale zero-valent iron. *Environ. Sci. Technol.*, **39**(5), pp. 1291–1298.

66. Wang, W., Hua, Y., Li, S., Yan, W., and Zhang, W.-X. (2016). Removal of Pb(II) and Zn(II) using lime and nanoscale zero-valent iron (nZVI): A comparative study. *Chem. Eng. J.*, **304**, pp. 79–88.

67. Jing, C., Li, Y. L., and Landsberger, S. (2016). Review of soluble uranium removal by nanoscale zero-valent iron. *J. Environ. Radioact.*, **164**, pp. 65–72.

68. Marsili, E., Beyenal, H., Di Palma, L., Merli, C., Dohnalkova, A., Amonette, J. E., and Lewandowski, Z. (2005). Uranium removal by sulfate-reducing biofilms in the presence of carbonate. *Water Sci. Technol.*, **52**(7), pp. 49–55.

69. Marsili, E., Beyenal, H., Di Palma, L., Merli, C., Dohnalkova, A., Amonette, J. E., and Lewandowski, Z. (2007). Uranium immobilization by sulfate-reducing biofilms grown on hematite, dolomite and calcite. *Environ. Sci. Technol.*, **41**(24), pp. 8349–8354.

70. Li, X. Q. and Zhang, W. X. (2006). Iron nanoparticles: The core-shell structure and unique properties for Ni(II) sequestration. *Langmuir*, **22**, pp. 4638–4642.

71. Keller, A. A., Garner, K., Miller, R. J., and Lenihan, H. S. (2012). Toxicity of nano-zero valent iron to freshwater and marine organisms. *PLoS ONE*, **7**(8), e43983. doi:10.1371/journal.pone.0043983.

72. Cundy, A. B., Hopkinson, L., and Whitby, R. L. D. (2008). Use of iron-based technologies in contaminated land and groundwater remediation: A review. *Sci. Total Environ.*, **400**, pp. 42–51.

73. Noubactep, C. (2015). Metallic iron for environmental remediation: A review of reviews. *Water Res.*, **85**, pp. 114–123.

74. Gueye, M. T., Di Palma, L., Allahverdeyeva, G., Bavasso, I., Petrucci, E., Stoller, M., and Vilardi, G. (2016). The influence of heavy metals and organic matter on hexavalent chromium reduction by nano zero-valent iron in soil. *Chem. Eng. Trans.*, **47**, pp. 289–294.

75. Thierno Gueye, M., Petrucci, E., and Di Palma, L. (2015). Chemical reduction of hexavalent chromium (vi) in soil slurry by nano zero-valent iron. *Chem. Eng. Trans.*, **43**, pp. 655–660.

76. Di Palma, L., Gueye, M. T., and Petrucci, E. (2015). Hexavalent chromium reduction in contaminated soil: A comparison between ferrous sulfate and nanoscale zero-valent iron. *J. Hazard. Mater.*, **281**, pp. 70–76.

77. Ponder, S. M., Darab, J. G., and Mallouk, T. E. (2000). Remediation of Cr(VI) and Pb(II) aqueous solutions using supported, nano-scale zero-valent iron. *Environ. Sci. Technol.*, **34**, pp. 2564–2569.

78. Paul, C. J., Khan, F. A., and Puls, R. W. (2002). In situ reduction of chromium contaminated groundwater, soils, and sediments by sodium dithionite. In: D. L. Morrison, S. J. Davis, J. A. Fuller, and C. C. Naftz, Eds.: *Handbook of Ground Water Remediation Using Permeable Reactive Barriers, Applications to Radionuclides, Trace Metals, and Nutrients*, pp. 465–493, San Diego, Academic Press.

79. Szecsody, J. E., Fruchter, J. S., Williams, M. D., Vermeul, V. R., and Sklarew, D. (2004). In situ chemical reduction of aquifer sediments: Enhancement of reactive iron phases and TCE dechlorination. *Environ. Sci. Technol.*, **38**, pp. 4656–4663.

80. Chen, P.-J., Su, C.-H., Tseng, C.-Y., Tan, S.-W., and Cheng, C.-H. (2011). Toxicity assessments of nanoscale zerovalent iron and its oxidation products in medaka (*Oryzias latipes*) fish. *Mar. Pollut. Bull.*, **63**(5–12), pp. 339–346.

81. Wang, H., Kou, X., Pei, Z., Xiao, J., Shan, X., and Xing, B. (2011). Physiological effects of magnetite ($Fe_3O_4$) nanoparticles on perennial ryegrass (*Lolium perenne* L.) and pumpkin (*Cucurbita mixta*) plants. *Nanotoxicology*, **5**(1), pp. 30–42.

82. Adeleye, A. S., Keller, A. A., Miller, R. J., and Lenihan, H. S. (2013). Persistence of commercial nanoscaled zero-valent iron (nZVI) and by-products. *J. Nanoparticle Res.*, **15**, pp. 1418–1435.

83. Wang, C.-H. and Zhang, W.-X. (1997). Synthesizing nanoscale iron particles for rapid and complete dechlorination of TCE and PCBs. *Environ. Sci. Technol.*, **31**, pp. 2154–2156.

84. Zhang, W.-X. (2003). Nanoscale iron particles for environmental remediation: An overview. *J. Nanoparticle Res.*, **5**, pp. 323–332.

85. Nurmi, J. T., Tratnyek, P. G., Sarathy, V., Baer, D. R., Amonette, J. E., Pecher, K., Wang, C., Linehan, J. C., Matson, D. W., Penn, R. L., and Driessen, M. D. (2005). Characterization and properties of metallic iron nanoparticles: Spectroscopy, electrochemistry, and kinetics. *Environ. Sci. Technol.*, **39**, pp. 1121–1230.

86. Hwanga, Y.-H., Kimb, D.-G., and Shina, H.-S. (2011). Effects of synthesis conditions on the characteristics and reactivity of nano-scale zero-valent iron. *Appl. Cat. B-Environ.*, **105**, pp. 144–150.

87. Glavee, G. N., Klabunde, K. J., Sorensen, C. M., and Hadjipanayisld, G. C. (1995). Chemistry of borohydride reduction of iron(II) and iron(III) ions in aqueous and nonaqueous media. Formation of nanoscale Fe, FeB, and $Fe_2B$ Powders. *Inorg. Chem.*, **34**, pp. 28–35.

88. Sun, Y.-P., Li, X.-Q., Cao, J., Zhang, W.-X., and Wang, H. P. (2006). Characterization of zero-valent iron nanoparticles. *Adv. Colloid Interfac Sci.*, **120**, pp. 47–56.

89. Zhang, X.-W. and Elliott, D. W. (2006). Applications of iron nanoparticles for groundwater remediation. *Remediation J.*, **16**, pp. 7–21.

90. Kozma, G., Rónavári, A., Kónya, Z., and Kukovecz, Á. (2016). Environmentally benign synthesis methods of zero-valent iron nanoparticles. *ACS Sust. Chem. Eng.*, **4**, pp. 291–297.

91. Sun, Q., Feitz, A. J., Guan, J., and Waite, T. D. (2008). Comparison of the reactivity of nanosized zero-valent iron (nZVI) particles produced by borohydride and dithionite of iron salts. *NANO Brief Reports Rev.*, **3**, pp. 341–349.

92. Xie, Y. and Cwiertny, D. M. (2010). Use of dithionite to extend the reactive lifetime of nanoscale zero-valent iron treatment systems. *Environ. Sci. Technol.*, **44**, pp. 8649–8655.

93. Wei, Y., Fang, Z., Zheng, L., Tan, L., and Tsang, E. P. (2016). Green synthesis of Fe nanoparticles using Citrus maxima peels aqueous extracts. *Mater. Lett.*, **185**, pp. 384–386.

94. Njagi, E. C., Huang, H., Stafford, L., Genuino, H., Galindo, H. M., Collins, J. B., Hoag, G. E., and Suib, S. L. (2011). Biosynthesis of iron and silver nanoparticles at room temperature using aqueous sorghum bran extracts. *Langmuir*, **27**, pp. 264–271.

95. Wang, Z., Fang, C., and Megharaj, M. (2014). Characterization of iron-polyphenol nanoparticles synthesized by three plant extracts and their fenton oxidation of azo dye. *ACS Sust. Chem. Eng.*, **2**, pp. 1022–1025.

96. Hoag, G. E., Collins, J. B., Holcomb, J. L., Hoag, J. R., Nadagouda, M. N., and Varma, R. S. (2009). Degradation of bromothymol blue by 'greener' nano-scale zero-valent iron synthesized using tea polyphenols. *J. Mater. Chem.*, **19**, pp. 8671–8677.

97. Markova, Z., Novak, P., Kaslik, J., Plachtova, P., Brazdova, M., Jancula, D., Siskova, K. M., Machala, L., Marsalek, B., Zboril, R., and Varma, R. (2014). Iron(II,III)-polyphenol complex nanoparticles derived from green tea with remarkable ecotoxicological impact. *ACS Sust. Chem. Eng.*, **2**, pp. 1674–1680.

98. Wang, Z. (2013). Iron complex nanoparticles synthesized by eucalyptus leaves. *ACS Sust. Chem. Eng.*, **1**, pp. 1551–1554.

99. Wang, Y., Fang, Z., Liang, B., and Tsang, E. P. (2014). Remediation of hexavalent chromium contaminated soil by stabilized nanoscale zero-valent iron prepared from steel pickling waste liquor. *Chem. Eng. J.*, **247**, pp. 283–290.

100. Zhao, X., Liu, W., Cai, Z., Han, B., Qian, T., and Zhao, D. (2016). An overview of preparation and applications of stabilized zero-valent iron nanoparticles for soil and groundwater remediation. *Water Res.*, **100**, pp. 245–266.

101. Saleh, N., Sirk, K., Liu, Y., Phenrat, T., Dufour, B., Matyjaszewski, K., Tilton, R. D., and Lowry, G. V. (2007). Surface modifications enhance nanoiron transport and NAPL targeting in saturated porous media. *Environ. Eng. Sci.*, **24**, pp. 45–57.

102. Kanel, S. R., Nepal, D., Manning, B., and Choi, H. (2007). Transport of surface-modified iron nanoparticle in porous media and application to arsenic(III) remediation. *J. Nanoparticle Res.*, **9**, pp. 725–735.

103. Comba, S. and Sethi, R. (2009). Stabilization of highly concentrated suspensions of iron nanoparticles using shear-thinning gels of xanthan gum. *Water Res.*, **43**(15), pp. 3717–3726.

104. Tiraferri, K., Chen, L., Sethi, R., and Elimelech, M. (2008). Reduced aggregation and sedimentation of zero-valent iron nanoparticles in the presence of guar gum. *J. Colloid Interface Sci.*, **324**, pp. 71–79.

105. He, F. and Zhao, D. (2007). Manipulating the size and dispersibility of zerovalent iron nanoparticles by use of carboxymethyl cellulose stabilizers. *Environ. Sci. Technol.*, **41**, pp. 6216–6221.

106. Schrick, B., Hydutsky, B. W., Blough, J. L., and Mallouk, T. E. (2004). Delivery vehicles for zerovalent metal nanoparticles in soil and groundwater. *Chem. Mater.*, **16**, pp. 2187–2193.

107. Li, Y., Jin, Z., Li, T., and Li, S. (2011). Removal of hexavalent chromium in soil and groundwater by supported nano zero-valent iron on silica fume. *Water Sci. Technol.*, **63**, pp. 2781–2787.

108. Xiao, S., Shen, M., Guo, R., Wang, S., and Shi, X. (2009). Immobilization of zerovalent iron nanoparticles into electrospun polymer nanofibers: Synthesis, characterization, and potential environmental applications. *J. Phys. Chem. C*, **113**, pp. 18062–18068.

109. Gu, C., Jia, H., Li, H., Teppen, B. J., and Boyd, S. A. (2010). Synthesis of highly reactive subnano-sized zero-valent iron using smectite clay templates. *Environ. Sci. Technol.*, **44**, pp. 4258–4263.

110. Liu, W., Tian, S., Zhao, X., Xie, W., Gong, Y., and Zhao, D. (2015). Application of stabilized nanoparticles for in situ remediation of metal-contaminated soil and groundwater: A critical review. *Curr. Pollut. Rep.*, **1**, pp. 280–291.

111. Gimbert, F., Morin-Crini, N., Renault, F., and Badot, P. M. (2008). Adsorption isotherm models for dye removal by cationized starch-based material in a single component system: Error analysis. *J. Hazard. Mater.*, **157**(1), pp. 34–46.

112. Hu, J., Chen, G., and Lo, I. (2006). Selective removal of heavy metals from industrial wastewater using maghemite nanoparticle: Performance and mechanisms. *J. Environ. Eng.*, **132**(7), pp. 709–715.

113. Badruddoza, A. Z. M., Shawon, Z. B. Z., Rahman, M. T., Hao, K. W., Hidajata, K., and Uddin, M. S. (2013). Ionically modified magnetic nanomaterials for arsenic and chromium removal from water. *Chem. Eng. J.*, **225**, pp. 607–615.

114. Pang, Y., Zeng, G., Tang, L., Zhang, Y., Liu, Y., Lei, X., Li, Z., Zhang, J., Liu, Z., and Xiong, Y. (2011). Preparation and application of stability enhanced magnetic nanoparticles for rapid removal of Cr(VI). *Chem. Eng. J.*, **175**, pp. 222–227.

115. Mohan, D. and Pittman C. U. Jr. (2006). Activated carbons and low cost adsorbents for remediation of tri- and hexavalent chromium from water. *J. Hazard. Mater.*, **137**(2), pp. 762–811.

116. Langmuir, I. (1916). The constitution and fundamental properties of solids and liquids. *J. Am. Chem. Soc.*, **38**, pp. 2221–2295.

117. Work, E. (2012). *Adsorption Technology in Water Treatment.* Berlin/Boston, Walter de Gruyter GmbH & Co. KG.

118. Freundlich, H. M. F. (1906). Over the adsorption in solution. *J. Phys. Chem.*, **57**, pp. 385–471.

119. Dubinin, M. M. and Radushkevich, L. V. (1947). The equation of the characteristic curve of the activated charcoal. *Proc. Acad. Sci. USSR Phys. Chem. Sect.*, **55**, pp. 331–337.

120. Wüst, W. F., Kober, R., Schlicker, O., and Dahmke, A. (1999). Combined zero- and first-order kinetic models of the degradation of TCE and cis-DCE with commercial iron. *Environ. Sci. Technol.*, **33**, pp. 4304–4309.

121. Johnson, T., Scherer, M. M., and Tratnyek, P. G. (1996). Kinetics of halogenated organic compound degradation by iron metal. *Environ. Sci. Technol.*, **30**, pp. 2634–2640.

## Chapter 10

# Nitrate Removal by Bimetallic Catalysts Supported by Iron Nanomaterials

**Shanawar Hamid,[a] Yoonseok Chang,[b] and Woojin Lee[c,*]**
[a]*Department of Structures and Environmental Engineering, University of Agriculture, Faisalabad 38000, Pakistan*
[b]*School of Environmental Science and Technology, Pohang University of Science and Technology (POSTECH), 77 Cheongam-ro Nam-gu, Pohang 37673, South Korea*
[c]*Department of Civil Engineering, Nazarbayev University, 53 Kabanbay Batyr Ave. Astana 010000, Republic of Kazakhstan*
shanawar@kaist.ac.kr

## 10.1 Introduction

### 10.1.1 Contamination of Waterbodies by Nitrate

High concentration of nitrate, toxic to human health and aquatic life, has been mostly attributed to anthropogenic activities and found in waterbodies over the globe. The current scenario of global water crisis demands effective treatment of nitrate for adequate water supply for domestic, industrial, and agricultural consumption. Recently, catalytic nitrate reduction has attracted attention due to its high removal capacity and selectivity.

*Iron Nanomaterials for Water and Soil Treatment*
Edited by Marta I. Litter, Natalia Quici, and Martín Meichtry

ISBN 978-981-4774-67-3 (Hardcover), 978-981-4669-49-8 (eBook)
www.panstanford.com

Nitrate is a highly soluble, stable, and leachable compound in environmental conditions. These unique properties of nitrate enhance its transportation from a source to major waterbodies, for example, rivers, lakes, aquifers, and sea. In recent decades, the contamination of surface water and groundwater by nitrate has accelerated by anthropogenic and nonpoint contaminant origins, for example, enhanced agricultural fertilization and untreated or poorly treated effluents from domestic and industrial sectors [1, 2]. The nitrate contamination poses a wide range of multidirectional and detrimental effects on both natural and engineered environments. It enhances algal growth in waterbodies, which consume dissolved oxygen and ultimately causing hypoxia [3]. Hypoxia is a condition in waterbodies where concentration of dissolved oxygen drops to extremely low levels and does not support sustenance of aquatic living organisms [1, 4]. Consumption of nitrate-contaminated drinking water could cause methemoglobinemia (blue baby syndrome) in infants [5]. Furthermore, it could be a precursor to carcinogenic *N*-nitroso compounds during its process in human intestines [5–8]. Due to the multidirectional effects, stringent regulations for the control of nitrate have been imposed by the European Union (EU) and United States Environmental Protection Agency (USEPA). The regulated maximum allowable concentrations (MCL) by EU and USEPA are 50 mg/L (11.3 mg/L nitrate – nitrogen) and 44.3 mg/L (10 mg/L nitrate – nitrogen), respectively.

This chapter describes the reduction of nitrate by iron nanomaterials supported Cu–Pd bimetallic catalysts. It provides fundamental understanding for the development of novel nitrate removal technology using iron nanomaterials and proper application to enormous nitrate pollution challenges in a safe and eco-friendly manner.

#### 10.1.1.1 Technologies for aqueous nitrate removal

The removal of aqueous nitrate without full understanding of its chemistry is critical due to the potential adverse effect on ecosystems, global food chain, and human health. Ideally, a complete nitrate removal into harmless end products is desired. Hence, numerous techniques have been utilized for the proper removal of nitrate, including adsorption, ion exchange, reverse osmosis, and biological denitrification. However, these techniques have their own advantages and disadvantages (Table 10.1). For example, physical and physicochemical processes (e.g., adsorption, reverse osmosis, ion exchange) produce highly concentrated solid streams of nitrate and additional by-products, which require further disposal [5, 6, 8–11]. The biological denitrification is relatively slow in kinetics, complex in operation, and produces excessive waste sludge [8, 10, 11]. Due to these reasons, environmental scientists and engineers are striving for new denitrification techniques that could achieve high removal with minimum or no by-products. Recent advancements in the development of material sciences and engineering from multidisciplinary research fields have enabled the environmental professionals to effectively treat critical pollutants such as nitrate. In this regard, the reduction of nitrate by monometallic and bimetallic catalysts has emerged as a feasible alternative due to its unique features such as fast kinetics, high reductive capacity and selectivity to harmless and eco-friendly nitrogen [5, 12–18]. A comprehensive comparison between catalytic nitrate reduction and other techniques is presented in Table 10.1 [19]. The comparison shows that catalytic nitrate reduction is superior to other techniques due to several advantages, i.e., harmless end product, zero waste production, complete nitrate removal, high operational flexibility and mobility, simple control, low space requirements, zero production of odor and noise, and immediate start up.

#### 10.1.1.2 Catalytic nitrate reduction

In general, bimetallic catalyst consists of a promoter metal (e.g., Cu, Sn, In, Ni, Ag), a noble metal (e.g., Pd, Pt, Au, Ru), and a support surface (e.g., zerovalent iron, iron oxides, titania, alumina) (Fig. 10.1). The reduction of nitrate on the surface of bimetallic catalysts is a multistep redox reaction as depicted in Fig. 10.1. Hydrogen is supplied as a reducing agent from the start of the reaction, which is activated by a noble metal ($M_N$, e.g., Pd) (Eq. 10.1). In the first step, zerovalent promoter metal ($M_P$) adsorbs and reduces nitrate to nitrite by abstraction of oxygen (Eq. 10.2). In the next step, nitrite is adsorbed by the noble metal sites and reduced to nitrogen or ammonium with the help of activated hydrogen (Eq. 10.3) and (Eq. 10.4). In the final step, oxidized promoter metal is reduced to its zerovalent state by the spillover of activated hydrogen from noble metal (Eq. 10.5) [5, 13, 20, 21].

$$H_2 + M_N{}^0 \rightarrow 2H_{ads}\text{-}M_N{}^0 \quad (10.1)$$

$$NO_3{}^- + M_P{}^0 \rightarrow NO_2{}^- + M_PO \quad (10.2)$$

$$2NO_2{}^- + 6H_{ads}\text{-}M_N{}^0 \rightarrow M_N{}^0 + N_2 + +2OH^- + 2H_2O \quad (10.3)$$

$$NO_2{}^- + 6H_{ads}\text{-}M_N{}^0 \rightarrow M_N{}^0 + NH_4{}^+ + 2OH^- \quad (10.4)$$

$$MO + H_{ads} \rightarrow M^0 + OH^- \quad (10.5)$$

**Table 10.1** Comparison of the characteristics of technologies for water remediation after nitrate contamination

| | Technology | | | |
|---|---|---|---|---|
| **Property** | **Ion exchange** | **Reverse osmosis** | **Biological denitrification** | **Catalytic Reduction** |
| Fate of nitrate | Adsorbed and concentrated | Concen-trated in a waste stream | Transformed to $N_2$ | Trans-formed to $N_2$ |
| Waste | Waste brine | Waste brine | Bacteria sludge | None |
| Chemical additives | Sodium chloride | Sulfuric acid and base | Ethanol and phosphoric acid | $H_2$ |
| Percentage of efficiency in water purification | 85–98% | 75–80% | 98% | 98–100% |
| Flexibility in variable operations | Medium | Medium | Low | High |
| Energy use | Low | High | Medium | Low |
| Space requirements | Limited | Limited | High | Low |
| Movable | Yes | Yes | No | Yes |
| Manageability | Good | Good | Poor | Good |
| Type of operations | Periodic regeneration | Continuous | Continuous | Continuous |
| Sensitivity to deactivation | Medium | High | High | Medium |
| Automatic control | Simple | Simple | Complex | Simple |

| Property | Technology<br>Ion exchange | Reverse osmosis | Biological denitrification | Catalytic Reduction |
|---|---|---|---|---|
| Startup time | Immediate | Immediate | Up to 1 month | Immediate |
| Monitoring required | Little | Little | Intensive | Intensive |
| Selectivity of the process | Low | Low | High | High |
| Odors | No | No | Yes | No |
| Noise | Some | High | None | None |
| Indicative cost (€/m$^3$) | 0.15–0.25 | 0.4–0.6 | 0.2–0.3 | 0.25–0.55 |
| Sensitivity of costs to scale-down | Medium | High | High | Low |
| Multipurpose use | None | Depending on molecular size | Some | Highly effective |

*Source*: Adapted from Ref. [19].

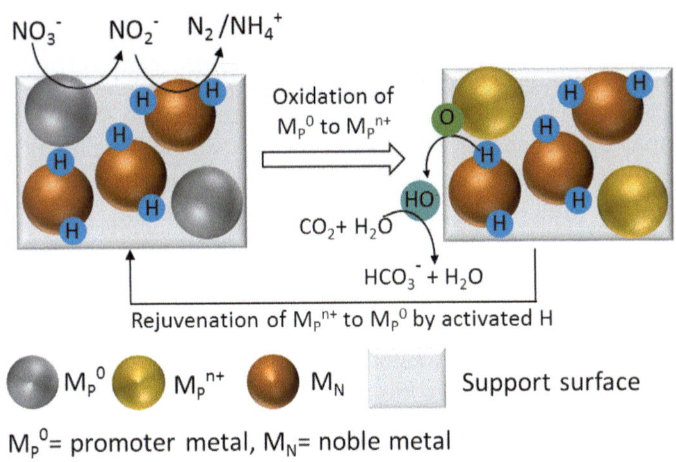


**Figure 10.1** Nitrate reduction mechanism by supported bimetallic catalysts. Reprinted from Ref. [20], Copyright 2016, with permission from Elsevier.

The aforementioned steps show that hydroxyl ions are constantly generated during catalytic reduction of nitrate, which may increase the system pH if proper buffering is not used. An increase in pH could have strong influence on the reactivity and selectivity of the bimetallic catalyst because hydroxyl ions adsorb on the catalyst surface and affect nitrate adsorption due to coulombic repulsion [13]. An increase in the system pH during nitrate reduction could also enhance the formation of ammonium and ultimately suppress nitrogen selectivity [14, 22, 23]. However, the use of a proper buffering agent to neutralize the increase in pH enhances the nitrate reduction kinetics and nitrogen selectivity. For example, use of carbon dioxide gas has been reported as an ideal buffer system, which could suppress ammonium selectivity [5, 13, 20, 21].

As discussed above and shown in Fig. 10.1, bimetallic catalysts consists of three major parts, i.e. promoter metal, noble metal, and support surface. The selection of optimal promoter–noble metal combination and their content ratio is achieved through optimization tests. In this regard, combination of Cu and Pd has been observed as most reactive and selective for nitrate reduction [5, 13, 21]. In general, bimetallic catalyst are composed 1–5% of promoter and noble metals and 95~99% support material.

The support material provides a suitable surface for anchoring the equally distributed bimetallic ensembles and separates the solid bimetallic catalysts from aqueous phase. It is obvious that the reaction kinetics and selectivity of catalytic nitrate reduction could be strongly influenced by the inherent properties of the support material because it is a major part of the catalyst. This is the reason why many bimetallic catalysts supported materials have been reported to date, for example, alumina [17, 24–26], zirconia [27], ceria, [17, 28], silica [29], zeolites [30], titania [31], activated carbon [32, 33], niobia [34], hydrotalcite [35], polymers [36–38], zerovalent iron [21, 22], hematite [5], and maghemite [13]. However, many important factors can influence the selection of a suitable support material, i.e., stability, ecotoxicity, by-products, leaching of bimetals, economy, and suitability for actual applications. Preferably, the use of an environmental friendly, economical, and stable support material is desired due to strict environmental regulations and overall cost of the catalytic process.

Iron is one of the most abundant elements on earth. It has unique properties because it is found in many oxidation states ranging from −2 to +6. However, its most abundant form is +2 and +3. Certain species of iron (e.g., zerovalent ion and iron oxides) have been extensively studied for environmental applications due to their economic aspects [39–41]. In the case of catalytic nitrate reduction, iron-based materials are excellent support surfaces for the development of bimetallic catalysts because they are economical, eco-friendly, stable, easily separable due to their magnetic properties, and easy to handle as compared to other materials mentioned above. This chapter discusses the usage of three iron species, i.e., nano-scale zerovalent iron (nZVI), maghemite, and hematite as support materials for the bimetallic catalyst for the catalytic reduction of nitrate.

## 10.2 Nitrate Reduction by Iron-Materials-Supported Bimetallic Catalysts

### 10.2.1 Nitrate Reduction by nZVI and nZVI-Supported Monometallic Catalysts

nZVI is one of the most widely used iron materials for diverse environmental applications in environmental engineering and technology field. It has been used as a reducing agent for in situ and ex situ remediation and even bench-scale studies for the reductive treatment of halogenated organic compounds, dyes, heavy metals, pesticides, and nitrates [42–45]. The suitability of nZVI for the remediation of diverse contaminants comes from its high reduction capacity, production of $H_2$ during its anaerobic corrosion in water (Eq. 10.6), low cost, easy operation, and magnetic properties [46, 47]. nZVI has shown high reduction kinetics and efficiency for the enhanced removal of aqueous nitrates [48–52]. It is well established that reduction of nitrate by nZVI produces undesirable ammonium as the major by-product [53–55], while nitrite and nitrogen as the minor by-products as shown in the following Eqs. (10.7–10.9) [53].

$$Fe^0 + 2H_2O \rightarrow Fe^{2+} + H_2 + 2OH^- \tag{10.6}$$

$$5Fe^0 + 2NO_3^- + 6H_2O \rightarrow 5Fe^{2+} + N_2 + 12OH^- \quad (10.7)$$

$$4Fe^0 + NO_3^- + 7H_2O \rightarrow 4Fe^{2+} + NH_4^+ + 10OH^- \quad (10.8)$$

$$Fe^0 + NO_3^- + H_2O \rightarrow Fe^{2+} + NO_2^- + 2OH^- \quad (10.9)$$

The above equations show that nZVI oxidizes rapidly during nitrate reduction regardless of the final product of the reaction implying that it is an irreversible catalytic reaction. These critical disadvantages, i.e., sudden decrease in reactivity, production of ammonia, rapid oxidation of nZVI, and agglomeration, have weakened the chances of nZVI as an acceptable and feasible environmental catalyst [53]. Hence, many efforts were made to improve the selectivity to eco-friendly nitrogen by suppressing production of ammonium using the addition of less precious secondary metals (i.e., nZVI–M, M = Cu, Ni, Ag, etc.) [46, 54, 55]. In particular, Cu addition has shown more promising results as compared to those by Ni, Ag, etc. [45, 53, 56]. In nZVI–Cu combination, the nZVI acts as a reducing agent for Cu by direct electron transfer from nZVI core (Eq. 10.10).

$$Fe^0 + Cu^{2+} \rightarrow Fe^{2+} + Cu^0 \quad (10.10)$$

In this mechanism, the zerovalent Cu acts as a catalyst for enhanced nitrate reduction kinetics and improved nitrogen selectivity [46, 55]. Additionally, the nZVI–Cu combination enhances the nitrate reduction because the agglomeration of nZVI particles is inhibited by secondary metal; thus more active surface area is exposed to nitrate, and accelerated adsorption and reduction of nitrate on the secondary metal [20, 53, 56]. However, in this case, the longevity of nZVI–Cu is also affected due to its accelerated oxidation by enhanced electron transfer. In nZVI–M combination, the nitrate reduction is driven only by redox reactions of two metals (i.e., $Fe^0 \rightarrow Fe^{2+}$ and $Cu^{2+} \rightarrow Cu^0 \rightarrow Cu^{2+}$). Other disadvantage of this reaction mechanism is that less precious metals such as Cu cannot benefit from $H_2$ produced during nZVI reaction with water, because they lack in $H_2$ activation capacity. Hence, addition of trace amounts of noble metals (e.g., Pd, Pt) has been also investigated for the contaminant hydrogenation due to their profound ability of adsorption and activation of $H_2$ [21, 46, 49, 53]. The stability of nZVI improved when Pd was added to nZVI; however, the nitrate reduction efficiencies by nZVI–Pd and nZVI–Pt were lower than that of nZVI–Cu, perhaps due to higher affinity of nitrate toward Cu as compared to that of Pd and

Pt [21, 46, 49]. In addition, H activated by Pd cannot reduce nitrate directly [46].

The other important effect of second metal addition to nZVI is the variation in end products, i.e., ammonium, nitrite, and nitrogen. As mentioned above, ammonium is the main by-product (~80–100%) in nitrate reduction by nZVI; however, addition of Cu to nZVI suppresses the ammonium production but enhances nitrite production (Eq. 10.11) [21, 46, 53]. nZVI and Cu surface lacks in adsorption and/or reduction affinity toward nitrite; therefore, nitrite accumulates in the aqueous phase [21, 53].

$$Cu^0 + NO_3^- \rightarrow CuO + NO_2^- \qquad (10.11)$$

### 10.2.2 Nitrate Reduction by nZVI–Cu–Pd Catalysts in Batch Test

The reactivity, selectivity, and stability of nZVI and monometallic nZVI–Cu, nZVI–Pd are not satisfactory as discussed in the previous section. Because of the remarkable catalytic properties of respective materials, i.e., Cu (higher nitrate affinity) and Pd (enhanced $H_2$ activation capacity), nZVI (strong reduction capacity and $H_2$ production) could be utilized as environmental catalyst by coating it with Cu and Pd to enhance the reactivity and stability for the nitrate reduction. Researchers have investigated nitrate reduction by nZVI-supported bimetallic Cu–Pd [21, 46, 53]. It has been reported that nZVI-supported Cu–Pd showed elevated nitrogen selectivity. Interestingly, nitrate removal was less than that by nZVI–Cu and nZVI, respectively. It was reported that addition of Pd and Cu together occupy the active surface area of nZVI; thus nitrate reduction and $H_2$ production are suppressed. Second, there is an electron loss from nZVI to Pd due to the formation of zerovalent Pd. Zerovalent Pd does not support nitrate reduction directly but only activates $H_2$. However, the addition of Pd enhances nitrogen selectivity [46, 53]. The nitrogen production comes from nitrite reduction route in the presence of Pd. In the first step, $H_2$ is activated by Pd (Eq. 10.12) [20]. $H_{ads}$-Pd possesses strong affinity and reduction tendency toward nitrite and quickly reduces it to nitrogen (Eq. 10.13) or ammonium (Eq. 10.14).

$$H_2 + Pd \rightarrow 2H_{ads}\text{-}Pd \qquad (10.12)$$

$$2NO_2^- + 6H_{ads}\text{-}Pd \rightarrow Pd + N_2 + 2OH^- + 2H_2O \qquad (10.13)$$

$$NO_2^- + 6H_{ads}\text{-}Pd \rightarrow Pd + NH_4^+ + 2OH^- \qquad (10.14)$$

Nitrogen selectivity during the nitrate reduction is enhanced to some extent when Cu–Pd is added to nZVI surface [21]. However, this could be further improved by augmenting $H_2$ supply from external source. It was reported that provision of additional $H_2$ source enhances nitrate removal, kinetics, and nitrogen selectivity by twofold [46]. An enhanced stability of nZVI–Cu–Pd was also reported during the reduction of nitrate over five cycles by Bae et al. [46].

### 10.2.3 Nitrate Reduction by nZVI–Cu–Pd in Continuous Flow

nZVI and nZVI–Cu–Pd could achieve complete nitrate removal in batch experiments. However, batch tests cannot give much information about the longevity of nZVI and its stability due to high reactivity for a limited amount of batch concentration of nitrate. Hamid et a. [21] investigated the stability, nitrate removal, and by-product selectivity of nZVI and nZVI-supported Cu–Pd bimetallic catalysts and reported that catalytic behavior of nZVI and nZVI–Cu–Pd catalysts in continuous mode is different from that of nZVI and nZVI–Cu–Pd in batch system. The key findings are given in Fig. 10.2. nZVI achieved about 80% removal in one bed volume (1 h continuous flow reaction), but gradually lost its reactivity. TEM images (Fig. 10.3) show that round-shaped nZVI particles changed to squared-shaped iron oxide particles during nitrate reduction due to oxidation, although anaerobic conditions were well maintained. Similar results were observed for the individual addition of $H_2$, Cu, Pd, and Cu–Pd, i.e., nZVI+$H_2$, nZVI–Cu, nZVI–Pd, and nZVI–Cu–Pd. These results showed that addition of $H_2$ to nZVI alone does not affect general reaction mechanism of nitrate reduction by nZVI. These results also indicate that addition of Cu and Pd could not enhance the stability of nZVI in continuous system. However, there was a significant improvement in stability when nZVI–Pd was supplied with $H_2$ (Fig. 10.2). nZVI–Pd+$H_2$ successfully reduced ~90% influent nitrate for 3-bed volumes while nZVI–Pd without $H_2$ could not. This enhanced stability could be due to $H_2$ activation by Pd, which could also reduce iron oxide to $Fe^0$ and thus inhibits the rapid passivation of nZVI [21]. The nitrate reduction stability was further enhanced by the nZVI–Cu–Pd+$H_2$ system, which showed a complete nitrate removal for more than 9-bed volumes (9 h of continuous operation). This was attributed to the augmented effect of Cu, $H_2$, and Pd, which only works when all three are supplied together. The detailed mechanism of nitrate reduction is depicted in Fig. 10.4, which shows that nitrate is reduced in two pathways, i.e., directly on nZVI surface, which is mostly converted to ammonium as described in Eq. 10.8. Second, nitrate is adsorbed on $Cu^0$ surface and reduced to nitrite (Eq. 10.11). During these processes, $Cu^0$ is oxidized to CuO. Nitrite is desorbed from Cu to bulk aqueous phase and then adsorbed on the surface of Pd and reduced by Pd-$H_{ads}$ to nitrogen (Eq. 10.13) or ammonium (Eq. 10.14).

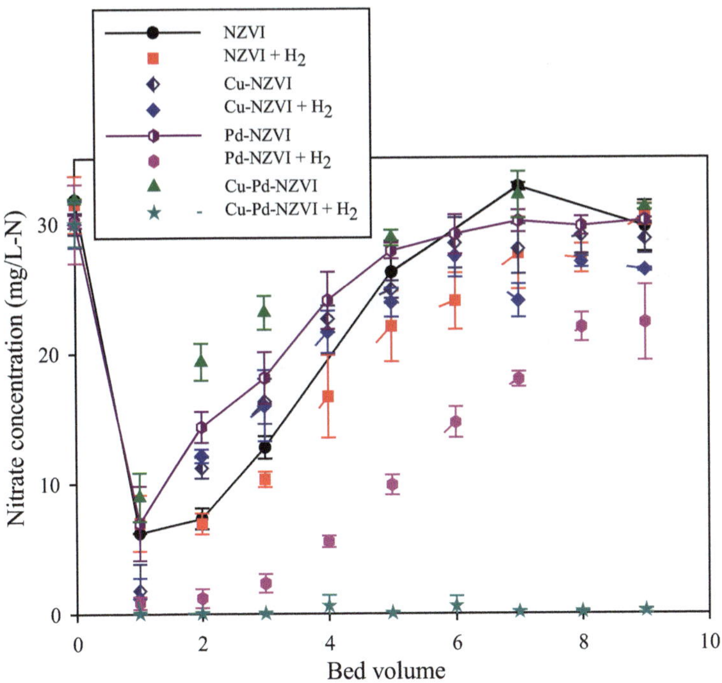


**Figure 10.2** Effect of Cu, Pd, and $H_2$ addition on nitrate reduction and process stability. Reprinted with permission from Ref. [21]. Copyright 2015 American Chemical Society.

The nitrogen selectivity in this study was ~48%, which shows that a considerable amount of the nitrite is reduced to nitrogen on Cu–Pd ensembles. Pd continuously activates the $H_2$, which not only rejuvenates CuO to $Cu^0$ but also inhibits the passivation of nZVI. Thus, a sustainable and selective nitrate reduction could be achieved by nZVI–Cu–Pd in the presence of $H_2$ supply.

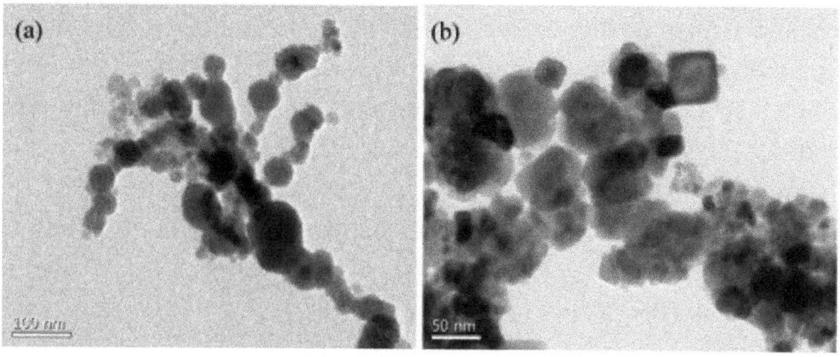


**Figure 10.3** TEM images of nZVI (a) before and (b) after nitrate reduction in a continuous flow system.

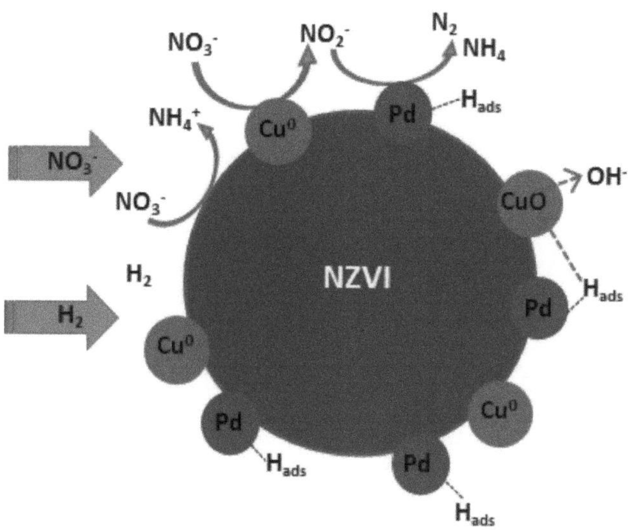


**Figure 10.4** Nitrate reduction mechanism by nZVI–Cu–Pd in the presence of $H_2$.

## 10.3 Nitrate Reduction by Iron-Oxides-Supported Cu–Pd Bimetallic Catalysts

Maghemite has been reported as an eco-friendly nano-magnet and environmentally benign material [13, 39]. It is an abundant and stable material with super-magnetic properties [41]. Jung et al. [13] developed an eco-friendly and selective Cu–Pd bimetallic catalyst supported by maghemite. They investigated the reaction mechanism with the help of several verification tools (i.e., TEM, XRD, XPS, TPR) and found that maghemite surface provides excellent environment for a uniform and closely packed distribution of Cu and Pd. It was reported that maghemite is not reduced by $NaBH_4$ or $H_2$ during the experimental process, indicating that it has a very stable surface [13].

Explaining the reaction mechanism, it has been also reported that unlike nZVI, maghemite does not support nitrate removal by adsorption and reduction, i.e., the bimetallic catalysis is the main removal mechanism as depicted in Fig. 10.1. Maghemite-supported Cu–Pd bimetallic catalyst achieved complete nitrate removal with faster reduction kinetics than that by Cu–Pd bimetallic catalyst supported by alumina and titania [13]. This was attributed to lower pore volume of the maghemite as compared to that of alumina and titania. The Cu–Pd maghemite also achieved about 47% nitrogen selectivity at optimal conditions. Maghemite-supported Cu–Pd also achieved a sustainable nitrate removal during three repeated removal cycles [13].

Hematite is another eco-friendly and abundant iron oxide used in several environmental applications [57, 58]. Jung et al. [5] reported a successful use of hematite as a support material for the development of a reactive and selective Cu–Pd bimetallic catalyst. Hematite-supported Cu–Pd achieved a complete and accelerated nitrate removal in 90 min with ~71% nitrogen selectivity. The hematite Cu–Pd has a higher turnover frequency (TOF) as compared to Cu–Pd bimetallic catalysts supported by silica, cerium oxide, and alumina, showing that it had highest active sites for nitrate reduction [5]. The effect of surface charge on nitrate reduction was also reported by measuring point of zero charge. It was reported that surface of hematite-supported Cu–Pd could be positively charged in contrast to that of alumina-, silica-, and cerium-oxide-supported bimetallic catalysts [5]. Thus, it can provide more favorable environment for nitrate adsorption. However, hematite support was inert itself during nitrate reduction and it did not remove any nitrate via adsorption and reduction [5].

## 10.4 Effect of Cu, Pd, and $H_2$ Variation on By-product Selectivity in Iron-Materials-Supported Bimetallic Catalyst Systems

Nitrate reduction and stability of bimetallic catalysts supported by iron materials (i.e., hematite, maghemite, and nZVI) are considerably enhanced when Cu, Pd, and $H_2$ are provided together [5, 13, 21]. However, the selectivity of by-products, i.e., nitrite, nitrogen, and ammonium is also dependent on their relative concentrations. A systematic comparison of experimental factors affecting the product selectivity and their expected trends are shown in Table 10.2. Nitrite selectivity in the presence of $Cu^0$ promotes nitrate reduction in bimetallic catalyst. An increase in Cu content to a certain maximum level enhances nitrate reduction and nitrite production kinetics [21, 46, 53]. However, beyond a certain maximum limit, nitrate reduction kinetics decreases due to overlapping and deposition of Cu on Pd-active sites [13].

On the other hand, an increase in Pd loading enhances nitrate reduction kinetics, while it lowers nitrite selectivity. This is because an increase in Pd enhances activated $H_{ads}$-Pd by consuming more $H_2$. An enhanced supply of $H_{ads}$-Pd in the system enhances nitrite reduction kinetics (Eq. 10.13). In addition to this, it also results in accelerated spillover of $H_{ads}$ from Pd-$H_{ads}$ to CuO surface, which ultimately, accelerates rejuvenation of CuO to $Cu^0$. An increase in $H_2$ supply enhances both nitrate reduction kinetics and ammonium selectivity, and it suppresses the selectivities of nitrite and nitrogen. The enhancement of nitrate reduction kinetics and suppression of nitrite selectivity is due to the excessive activation of $H_{ads}$ as explained above. An increase in $H_{ads}$ enhances the chances of recombination of N–H to form more ammonium as compared to recombination of

N–N to form nitrogen. This is the reason why nitrogen selectivity is suppressed with the increase in $H_2$ supply. A brief overview of the effects of Cu, Pd, and $H_2$ on by-products selectivity is given in Table 10.2.

**Table 10.2** Effect of increase in Cu, Pd, and $H_2$ on the formation of various by-products and $Cu^0$ in iron-materials-supported Cu–Pd bimetallic systems

| | | Trends | | | | |
|---|---|---|---|---|---|---|
| **Factor** | **Condition** | **Nitrate reduction kinetics** | **Nitrite selectivity** | **Ammonium selectivity** | **Nitrogen selectivity** | **$Cu^0$** |
| Cu | Increase | *Volcanic shape trend | Increase | Decrease | *Volcanic shape trend | Decrease |
| Pd | Increase | Increase | Decrease | Increase | *Volcanic shape trend | Increase |
| $H_2$ | Increase | Increase | Decrease | Increase | *Volcanic shape trend | Increase |

*The values increase to a certain maximum level with increase in the assigned factor but then decrease with further increase in the assigned factor.

It should be noted that there is no absolute optimal ratio for Cu/Pd and $H_2$ supply. The optimal ratio is different for different support surfaces and experimental conditions, which can be adjusted based on individual experimental design as shown in Table 10.3.

## Acknowledgments

The authors are sincerely grateful to all the members of Environmental Geobiochemical Research Group in KAIST. Special thanks should be given to Dr. Sungjun Bae, Ms. Sungyoon Jung, and Mr. Junyoung Jung, the core environmental catalyst scientists of EGRL, for their stunning experimental and analytical contributions. This research work was supported by the Korean Ministry of Environment (project no. RE201402059).

**Table 10.3** Reactivities and selectivities of Cu–Pd catalysts supported by different materials

| Initial $NO_3^-$ (mg/L) | Catalyst loading (g/L) | Catalyst | $H_2$ supply (mL/min) | $NO_3^-$ removal (%) | $N_2$ selectivity (%) | $NH_4^+$ selectivity (%) | Ref. |
|---|---|---|---|---|---|---|---|
| 132.8 | 1.25 | α-$Fe_2O_3$ 1.6%Cu-2.8%Pd | 30 | 96.4 | 70 | 30 | [5] |
| 100 | 1.00 | $TiO_2$ 1%Cu-2%Pd | 60 | 90 | 76 | 19 | [6] |
| 100 | 1.62 | $Al_2O_3$ 1.25%Cu-5%Pd | 500 | 100 | 82 | 5.10 mg/L | [8] |
| 132.8 | 0.16 | γ-$Fe_2O_3$ 0.5%Cu-0.5%Pd | 100 | 99 | 55 | 43 | [13] |
| 100 | 0.51 | ACo 1%Cu-2%Pd | 200 | 99 | 43 | 57 | [15] |
| 132.8 | 2 | nZVI 1.5% Cu-0.5%Pd | 30 | 100 (for 24 h) | 68–48 | 25–46 | [21] |
| 177.14 | 7.69 | Iron 0.5%Cu -0.3%Pd | — | 91.5 | 29.60 | 65.60 | [22] |
| 100 | 0.6 | $Al_2O_3$ 1.7%Cu-2%Pd | 90 | 100 | 74 | 7.60 mg/L | [25] |
| 100 | 0.88 | $Nb_2O_5$ 0.5%Cu-2%Pd | 50 | 100 | 89 | 3.1 mg/L | [34] |
| 100 | 0.88 | $Al_2O_3$ 0.5%Cu-2%Pd | 50 | 100 | 87 | 3.9 mg/L | [34] |
| 100 | 0.6 | Dowex 1*4 1%Cu-4%Pd | $H_2:N_2 = 64$ | 52 | 82 | 4 | [36] |
| 100 | 0.6 | nZVI 0.5%Cu-0.5%Pd | — | 97 | 18 | 80 | [46] |
| 100 | 0.80 | ASA 1.25%Cu -5%Pd | 200 | 100 | 90.20 | 9.80 | [59] |

| Initial $NO_3^-$ (mg/L) | Catalyst loading (g/L) | Catalyst | $H_2$ supply (mL/min) | $NO_3^-$ removal (%) | $N_2$ selectivity (%) | $NH_4^+$ selectivity (%) | Ref. |
|---|---|---|---|---|---|---|---|
| 100 | 2.22 | $Al_2O_3$ 1%Cu -4%Pd | 0.15 atm | 100 | 82.30 | 17.70 | [60] |
| 100 | 0.51 | CNT-$TiO_2$ 1%Cu -1%Pd | 200 | 100 | 66 | 34 | [61] |
| 100 | 0.51 | $TiO_2$ 1%Cu-1%Pd | 200 | 95 | 17 | 83 | [61] |
| 100 | 0.51 | Ac-Mn 1%Cu -1%Pd | 200 | 59 | 60 | 40 | [61] |
| 221.43 | 1.00 | $TiO_2$ 1%Cu-3%Pd | 100 | 100 | 50.20 | 49.70 | [62] |
| 100 | 1.00 | AC 0.4%Cu -2%Pd | 333 | 74.4 | 51.30 | 48.30 | [63] |
| 200 | 1 | $SiO_2$ 0.6%Cu -5%Pd | 84 | 95.5 | 11.6 | 77.2 | [64] |
| 200 | 1 | $Al_2O_3$ 0.6%Cu 5%Pd | 84 | 45.7 | 56.9 | 4.7 | [64] |
| 200 | 1 | $ZrO_2$ 0.6%Cu -5%Pd | 84 | 84.5 | 40.4 | 45.0 | [64] |
| 200 | 1 | AC 0.6%Cu -5%Pd | 84 | 97.1 | 78.3 | 21.4 | [64] |
| 200 | 1 | AC 3%Cu -5%Pd | 84 | 100 | 18.3 | 81.7 | [64] |
| 100 | 0.57 | $ZrO_2$ 1.25%Cu -5%Pd | 14 | 100 | — | 8.1 | [65] |
| 100 | 0.57 | $SnO_2$ 1.25%Cu -5%Pd | 4 | 100 | — | 6 | [65] |

## References


1. World Health Organization. (2008). *Guidelines for Drinking-Water Quality*, 3rd Ed. (Vol.1), Geneva.
2. Rabalais, N. N. (2002). Nitrogen in aquatic ecosystems, *Ambio*, **31**, pp. 102–112.
3. Diaz, R. and Rosenberg, R. (1995). Marine benthic hypoxia: A review of its ecological effects and the behavioural response of benthic macrofauna, *Oceanogr. Mar. Boil.*, **33**, pp. 245–303.
4. Lohrenz, S. E., Fahnenstiel, G. L., and Redalje, D. G. (1994). Spatial and temporal variations in photosynthesis parameters in relation to environmental conditions in coastal waters of the northern Gulf of Mexico, *Estuaries*, **17**, pp. 779–795.
5. Jung, S., Bae, S., and Lee, W. (2014). Development of Pd-Cu/hematite catalyst for selective nitrate reduction, *Environ. Sci. Technol.*, **48**, pp. 9651–9658.
6. Gao, W., Guan, N., Chen, J., Guan, X., Jin, R., Zeng, H., Liu, Z., and Zhang, F. (2003). Titania supported Pd-Cu bimetallic catalyst for the reduction of nitrate in drinking water, *Appl. Catal. B: Environ.*, **46**, pp. 341–351.
7. Gulis, G., Czompolyova, M., and Cerhan, J. R. (2002). An ecologic study of nitrate in municipal drinking water and cancer incidence in Trnava District, Slovakia, *Environ. Res. Sect. A.*, **88**, pp. 182–187.
8. Horold, S., Vorlop, K. D., Tacke, T., and Sell, M. (1993). Development of catalysts for a selective nitrate and nitrite removal from drinking water, *Catal. Today*, **17**, pp. 21–30.
9. Wada, K., Hirata, T., Hosokawa, S., Iwamoto, S., and Inoue, M. (2012). Effect of supports on Pd-Cu bimetallic catalysts for nitrate and nitrite reduction in water, *Catal. Today*, **185**, pp. 81–87.
10. Pintar, A. (2003). Catalytic processes for the purification of drinking water and industrial effluents, *Catal. Today*, **77**, pp. 451–465.
11. Kapoor, A. and Viraraghavan, T. (1997). Nitrate removal from drinking water: Review, *J. Environ. Eng. Asce.*, **23**, pp. 371–380.
12. Zhang, R., Shuai, D., Guy, K. A., Shapley, J. R., Strathmann, T. J., and Werth, C. J. (2013). Elucidation of nitrate reduction mechanisms on a Pd-In bimetallic catalyst using isotope labeled nitrogen species, *Chem. Cat. Chem.*, **5**, pp. 313–321.
13. Jung, J., Bae, S., and Lee, W. (2012). Nitrate reduction by maghemite supported Cu-Pd bimetallic catalyst, *Appl. Catal. B: Environ.*, **127**, pp. 148–158.

14. Chaplin, B. P., Reinhard, M., Schneider, W. F., Schuth, C., Shpley, J. R., Strathmann, T. J., and Werth, C. J. (2012). Critical review of Pd based catalytic treatment of priority contaminants in water, *Environ. Sci. Technol.*, **46**, pp. 3655–3670.

15. Soares, O. S. G. P., Orfao, J. J. M., and Pereira, M. F. R. (2009). Bimetallic catalysts supported on activated carbon for the nitrate reduction in water: Optimization of catalysts composition, *Appl. Catal., B: Environ.*, **91**, pp. 441–448.

16. Zhang, F., Miao, S., Yang, Y., Zhang, X., Chen, J., and Guan, N. (2008). Size dependent hydrogenation selectivity of nitrate on Pd-Cu/$TiO_2$ catalysts, *J. Phys. Chem. C.*, **112**, pp. 7665–7671.

17. Epron, F., Gauthard, F., and Barbier, J. (2002). Influence of oxidizing and reducing treatments on the metal-metal interactions and on the activity for nitrate reduction of a Pt-Cu bimetallic catalyst, *Appl. Catal. A: Gen.*, **237**, pp. 253–261.

18. Vorlop, K. D. and Tacke, T. (1989). Erste schritte auf dem weg zer edelmetallkatalysierten nitrat- und nitrit- entfernung aus trinkwasser, *Chem. Ing. Technol.*, **61**, pp. 836-837.

19. Centi, G. and Perathoner, S. (2003). Remediation of water contamination using catalytic technologies, *Appl. Catal. B: Environ.*, **41**, pp. 15–29.

20. Hamid, S., Macharla, A. K., and Lee, W. (2016). Highly reactive and selective Sn-Pd bimetallic catalyst supported by nanocrystalline ZSM-5 for aqueous nitrate reduction, *Appl. Catal. B: Environ.*, **187**, pp. 37–46.

21. Hamid, S., Bae, S., Lee, W., Amin, M. T., and Alazba, A. A. (2015). Catalytic nitrate removal in continuous bimetallic Cu-Pd/NZVI system, *Ind. Eng. Chem. Res.*, **54**, pp. 6247–6257.

22. Liou, Y. H., Lin, C. J., Weng, S. C., Ou, H. H., and Lo, S. L. (2009). Selective decomposition of aqueous nitrate into nitrogen using iron deposited bimetals, *Environ. Sci. Tech.*, **43**, pp. 2482–2488.

23. Palomares, A. E., Prato, J. G., Marquez, F., and Corma, A. (2003). Denitrification of natural water on supported Pd/Cu catalysts, *Appl. Catal. B: Environ.*, **41**, pp. 3–13.

24. Prusse, U. and Vorlop, K. D. (2001). Supported bimetallic palladium catalysts for water phase nitrate reduction, *J. Mol. Catal. A: Chem.*, **173**, pp. 313–328.

25. Sa, J., Gross, S., and Vinek, H. (2005). Effect of the reducing step on the properties of Pd-Cu bimetallic catalysts used for denitration, *Appl. Catal. A: Gen.*, **294**, pp. 226–234.

26. Witonska, I., Karski, S., and Goluchowska, J. (2007). Kinetic studies on the hydrogenation of nitrate in water using $Rh/Al_2O_3$ and $Rh\text{-}Cu/Al_2O_3$ catalysts, *Kinet. Catal.*, **48**, pp. 823–828.

27. Strukul, G., Gavagnin, R., Pinna, F., Modaferri, E., Perathoner, S., Centi, G., Marella, M., and Tomaselli, M. (2000). Use of palladium based catalysts in the hydrogenation of nitrates in drinking water: from powders to membranes, *Catal. Today*, **55**, pp. 139–149.

28. Barrabes, N., Dafinov, A., Medina, F., and Sueiras, J. E. (2010). Catalytic reduction of nitrates using $Pt/CeO_2$ catalysts in a continuous reactor, *Catal. Today*, **149**, pp. 341–347.

29. Garron, A., Lazar, K., and Epron, F. (2005). Effect of the support on tin distribution in $Pd\text{-}Sn/Al_2O_3$ and $Pd\text{-}Sn/SiO_2$ catalysts for application in water denitration, *Appl. Catal. B: Environ.*, **59**, pp. 57–69.

30. Soares, O. S. G. P., Marques, L., Freitas, C. M. A. S., Fonseca, A. M., Parpot, P., Órfão, J. J. M., Pereira, M. F. R., and Neves, I. C. (2015). Mono and bimetallic NaY catalysts with high performance in nitrate reduction in water, *Chem. Eng. J.*, **281**, pp. 411–417.

31. Sa, J., Berger, T., Fottinger, K., Riss, A., Anderson, J. A., and Vinek, H. (2005). Can $TiO_2$ promote the reduction of nitrates in water? *J. Catal.*, **234**, pp. 282–291.

32. Barrabes, N., Just, J., Dafinov, A., Medina, F., Fierro, J. L. G., Sueiras, J. E., Salagre, P., and Cesteros, Y. (2006). Catalytic reduction of nitrate on Pt-Cu and Pd-Cu on active carbon using continuous reactor: The effect of copper nanoparticles, *Appl. Catal. B: Environ.*, **62**, pp. 77–85.

33. Matatov-Meytal, U. and Sheintuch, M. (2009). The relation between surface composition of Pd-Cu/ACC catalysts prepared by selective deposition and their denitrification behavior, *Catal. Commun.*, **10**, pp. 1137–1141.

34. Maia, M. P., Rodrigues, M. A., and Passos, F. B. (2007). Nitrate catalytic reduction in water using niobia supported palladium-copper catalysts, *Catal. Today*, **123**, pp. 71–176.

35. Palomares, A. E., Franch, C., and Corma, A. (2010). Nitrates removal from polluted aquifers using (Sn or Cu)/Pd catalysts in a continuous reactor, *Catal. Today.*, **149**, pp. 348–351.

36. Gasparovicova, D., Kralik, M., Hronec, M., Vallusova, Z., Vinek, H., and Corain, B. (2007). Supported Pd-Cu catalysts in the water phase reduction of nitrates: Functional resin versus alumina, *J. Mol. Catal. A: Chem.*, **264**, pp. 93–102.

37. Neyertz, C., Marchesini, F. A., Boix, A., Miró, E., and Querini, C. A. (2010). Catalytic reduction of nitrate in water: Promoted palladium catalysts supported in resin, *Appl. Catal. A: Gen.*, **372**, pp. 40–47.

38. Dodouche, I., Barbosa, D. P., Rangel, M. D., and Epron, F. (2009). Palladium-tin catalysts on conducting polymers for nitrate removal, *Appl. Catal. B: Environ.*, **93**, pp. 50–55.

39. Girginova, P. I., Daniel-da-Silva, A. L., Lopes, C. B., Figueira, P., Otero, M., Amaral, V. S., Pereira, E., and Trindade, T. (2010). Silica coated magnetite particles for magnetic removal of $Hg^{2+}$ from water, *J. Colloid. Interface Sci.*, **345**, pp. 234–240.

40. Chomoucka, J., Drbohlavova, J., Huska, D., Adam, V., Kizek, R., and Hubalek, J. (2010). Magnetic nanoparticles and targeted drug delivering, *Pharmacol. Res.*, **62**, pp. 144–149.

41. Gyergyek, S., Makovec, D., Mertelj, A., Huskic, M., and Drofenik, M. (2010). Superparamagnetic nanocomposite particles synthesized using the mini-emulsion technique, *Colloids Surf. A.*, **366**, pp. 113–119.

42. Crane, R. A. and Scott, T. B. (2012). Nanoscale zero-valent iron: Future prospects for an emerging water treatment technology, *J. Hazard. Mater.*, **211–212**, pp. 112–125.

43. Li, X. and Elliott, D. W. (2006). Zero-valent iron nanoparticles for abatement of environmental pollutants: Materials and engineering aspects, *Crit. Rev. Solid State Mater. Sci.*, **31**, pp. 111–122.

44. Sihn, Y., Bae, S., and Lee, W. (2013). Formation of surface mediated iron colloids during U(VI) and NZVI interaction, *Adv. Environ. Res.*, **2**, pp. 167–177.

45. Bigg, T. and Judd, S. J. (2000). Zero-valent iron for water treatment, *Environ Tech.*, **21**, pp. 661–670.

46. Bae, S., Hamid, S., Jung, J., Sihn, Y., and Lee, W. (2016). Effect of promoter and noble metals and suspension pH on catalytic nitrate reduction by bimetallic nanoscale $Fe^0$ catalysts, *Environ. Tech.*, **37**, pp. 1077–1087.

47. Bae, S. and Hanna, K. (2015). Reactivity of nanoscale zero-valent iron (NZVI) in unbuffered system: Effect of pH and Fe(II) dissolution, *Environ. Sci. Tech.*, **49**, pp. 10536–10543.

48. Cheng, I. F., Muftikian, R., Fernando, Q., and Korte, N. (1997). Reduction of nitrate to ammonia by zero-valent iron, *Chemosphere*, **35**, pp. 2689–2695.

49. Liou, Y. H., Lo, S. L., Lin, C. J., Kuan, W. H., and Weng, S. C. (2005). Chemical reduction of an unbuffered nitrate solution using catalyzed

and uncatalyzed nanoscale iron particles, *J. Hazard. Mater.*, **127**, pp. 102–110.

50. Yang, G. C. and Lee, H. L. (2005). Chemical reduction of nitrate by nanosized iron: Kinetics and pathways, *Water Res.*, **39**, pp. 884–894.
51. Hwang, Y. H., Kim, D. G., and Shin, H. S. (2011). Mechanism study of nitrate reduction by nano zero valent iron, *J. Hazard. Mater.*, **185**, pp. 1513–1521.
52. Su, Y., Adeleye, A. S., Huang, Y., Sun, X., Dai, C., Zhou, X., Zhang, Y., and Keller, A. A. (2014). Simultaneous removal of cadmium and nitrate in aqueous media by nanoscale zerovalent iron (nZVI) and Au doped nZVI particles, *Water Res.*, **63**, pp. 102–111.
53. Liu, H., Guo, M., and Zhan, Y. (2014). Nitrate removal by Fe0/Pd/Cu nano-composite in groundwater, *Environ. Technol.*, **35**, pp. 917–924.
54. Khalil, A. M. E., Eljamal, O., Jribi, S., and Matsunaga, N. (2016). Promoting nitrate reduction kinetics by nanoscale zero valent iron in water via copper salt addition, *Chem. Eng. J.*, **287**, pp. 367–380.
55. Kang, H., Xiu, Z., Chen, J., Cao, W., Guo, Y., Li, Y., and Jin, Z. (2012). Reduction of nitrate by bimetallic Fe/Ni nanoparticles, *Environ Tech.*, **33**, pp. 2185–2192.
56. Hosseini, S. M., Ashtiani, B. A., and Kholghi, M. (2011). Nitrate reduction by nano-Fe/Cu particles in packed column, *Desalination*, **276**, pp. 214–221.
57. Zhong, D. K., Cornuz, M., Sivula, K., Gratzel, M., and Gamelin, D. R. (2011). Photo-assisted electrodeposition of cobalt-phosphate (Co-Pi) catalyst on hematite photoanodes for solar water oxidation, *Energy Environ. Sci.*, **4**, pp. 1759–1764.
58. Yang, X., Du, C., Liu, R., Xie, J., and Wang, D. (2013). Balancing photovoltage generation and charge-transfer enhancement for catalyst decorated photoelectrochemical water splitting: A case study of the hematite/MnOx combination, *J. Catal.*, **304**, pp. 86–91.
59. Xie, Y., Cao, H., Li, Y., Zhang, Y., and Crittenden, J. C. (2011). Highly selective PdCu/amorphous silica-alumina (ASA) catalysts for groundwater denitration, *Environ. Sci. Technol.*, **45**, pp. 4066–4072.
60. Chen, Y. X., Zhang, Y., and Chen, G. H. (2003). Appropriate conditions or maximizing catalytic reduction efficiency of nitrate into nitrogen gas in groundwater, *Water Res.*, **37**, pp. 2489–2495.
61. Soares, O. S. G. P., Orfão, J. J. M., and Pereira, M. F. R. (2011). Nitrate reduction in water catalysed by Pd-Cu on different supports, *Desalination*, **279**, pp. 367–374.

62. Bae, S., Jung, J., and Lee, W. (2013). The effect of pH and zwitterionic buffers on catalytic nitrate reduction by $TiO_2$ bimetallic catalyst, *Chem. Eng. J.*, **232**, pp. 327–337.

63. Trawczynski, J., Gheek, P., Okal, J., Zawzdzki, M., and Gomez, M. J. I. (2011). Reduction of nitrate on active carbon supported Pd-Cu catalysts, *Appl. Catal. A: Gen.*, **409–410**, pp. 39–47.

64. Yoshinaga, Y., Akita, T., Mikami, I., and Okuhara, T. (2002). Hydrogenation of nitrate in water to nitrogen over Pd-Cu supported on active carbon, *J. Catal.*, **207**, pp. 37–45.

65. Gavagnin, R., Biasetto, L., Pinna, F., and Strukul, G. (2002). Nitrate removal in drinking waters: The effect of tin oxides in the catalytic hydrogenation of nitrate by $Pd/SnO_2$ catalysts, *Appl. Catal. B: Environ.*, **38**, pp. 91–99.

# Chapter 11

# Iron or Iron-Based Bimetallic Nanoparticle-Immobilized Electrospun Polymer Nanofibers for Environmental Remediation Applications

**Shili Xiao**[a] **and Xiangyang Shi**[b]
[a]*School of Textile Science and Engineering, Wuhan Textile University, Wuhan, Hubei, 430073, People's Republic of China*
[b]*State Key Laboratory for Modification of Chemical Fibers and Polymer Materials, College of Chemistry, Chemical Engineering and Biotechnology, Donghua University, Shanghai, 201620, People's Republic of China*
xshi@dhu.edu.cn

## 11.1 Introduction

Environmental pollution shows high impact on human health by disrupting the respiratory, cardiovascular, and neurological systems and destroys the nature on which we are dependent for survival [1]. Therefore, environmental remediation is imperative. Among many contemporary remediation technologies that have been developed, nanotechnology has emerged to be one of the most powerful

*Iron Nanomaterials for Water and Soil Treatment*
Edited by Marta I. Litter, Natalia Quici, and Martín Meichtry

ISBN 978-981-4774-67-3 (Hardcover), 978-981-4669-49-8 (eBook)
www.panstanford.com

techniques for groundwater and contaminated soil treatment, which has brought a new aspect into scientific discipline and technology divisions due to the super-functional properties of these materials [1, 2]. In particular, zerovalent iron nanoparticles (nZVI) have received tremendous scientific and technological interest as one of the most effective nanomaterials used for environmental remediation purposes.

Due to its large specific area and strong reducing potential, nZVI have been widely used in the dechlorination of chlorinated organic contaminants such as trichloroethylene (TCE) and polychlorinated biphenyls (PCBs) [3], in the sequestration of toxic metalloid and metal ions such as As(V) and Cr(VI) [4, 5], in the degradation of nuclear wastes, explosives, and herbicides, and in the decoloration of dye effluents from the textile industry [6–8]. However, nZVI developed using various methods [9–11] are prone to agglomeration during the process of contaminant degradation and their transport process in the subsurface environment, which often leads to a reduced reactivity for contaminant removal, thus limiting their environmental applications. To mitigate the agglomeration of nZVI and thus improve their transport and delivery capacity through porous media, some polymeric surfactants [e.g., poly(vinyl alcohol-*co*-vinyl) acetate-*co*-itaconic acid] and stabilizers such as poly(acrylic acid), starch, and polyglycol, etc., have been used to immobilize the nanoparticles (NPs) onto microsized particles without compromising the surface reactivity of nZVI [12–14]. In this case, the produced nZVI was more dispersible and stable. Nevertheless, the stabilized nZVI are difficult to be recycled once they are used to treat contaminants in aqueous environment. Although immobilizing nZVI onto solid supports such as polymeric membranes and activated carbon to avoid secondary water contamination could be an additional option, these solid supports have low specific surface area, which could significantly affect the reactivity of nZVI. Therefore, immobilizing nZVI onto a continuous medium with a high surface-area-to-volume ratio and good porosity is anticipated to meet the requirements for environmental remediation applications. Most recently, polymer nanofibrous mats for environmental application have witnessed a growing interest [15]. These nanofibrous mats have low weight and high porosity, thus offering high permeability, as well as small pore size, making them suitable for microfiltration applications. Moreover, their very high aspect ratio and flexible surface functionality of polymer nanofibers make them an excellent absorption medium for the removal of contaminants [16, 17]. In particular, the polymer nanofibrous mats are easy to be separated from the aqueous environment without introducing secondary contamination. To fabricate polymeric nanofibers, a series of processing techniques such as template synthesis [18], self-assembly [19], centrifugal spinning [20], and electrospinning have been developed. Among these methods, electrospinning has drawn special attention as a promising and straightforward technology that produces continuous fibers with diameters in the range from tens of nanometers to several micrometers.

Through electrospinning, various synthetic and natural polymeric nanofibers with a high aspect ratio and a high specific surface area have been produced in the past two decades. In addition, these nanofibers could be functionalized by surface modification or by incorporating with functional additives such as bioactive molecules, semiconductors, catalyst NPs or sensing agents, which made them find a wide potential applications in tissue engineering [21], electrochemical energy storage [22], antibacterial membranes [23, 24], and environmental remediation [25]. In this chapter, we review the recent progress toward the development of electrospun polymer/iron or iron-based bimetallic NP composites for environmental remediation applications.

## 11.2 Iron and Iron-Based Bimetallic Nanoparticles for Environmental Remediation

The synthesis and application of nZVI was firstly reported by W. Zhang in 1997 [26]. Since then, much effort has been devoted to the synthesis of nanoscale iron, modification of the NP surface property, and enhancement of the efficiency for field delivery and reactions, which has been summarized in some reviews [1, 2, 11, 27–29].

### 11.2.1 Characterization of Iron and Iron-Based Bimetallic NPs

As the most abundant transition metal, iron typically exists in the environment as Fe(II) and Fe(III) oxides. Therefore, ZVI is a manufactured material. Commonly, nZVI are prepared by reducing Fe(II) or Fe(III) ions in aqueous solution using sodium borohydride (Eq. 11.1) due to its relative simplicity, which can be done in almost any wet chemistry lab with the need of only two common reagents. The synthesized nZVI exhibit a typical core–shell structure, shown in Fig. 11.1a. However, most particles formed chain-like aggregates when nZVI were synthesized without any surfactants or stabilizers (Fig. 1.1b). The core consists primarily of zerovalent or metallic iron, while the mixed valent (i.e., Fe(II) and Fe(III)) oxide shell is formed as a result of oxidation of the metallic iron [27]. The composition of nZVI was confirmed by X-ray diffraction (Fig. 11.1c) [28]. Thus, during the contamination treatment process, the ZVI core provides the reducing power for reactions with environmental contaminants. And the shell provides sites for chemical complex formation (e.g., chemosorption). In order to increase the degradation rates of nZVI to certain environmental pollutants that are retarded by the oxidation of nZVI surface, bimetallic iron NPs, in which a second and less reactive metal such as Pb, Ni, Pt, Zn, Cu, or Ag is introduced. In this system, iron-based bimetallic NPs are commonly synthesized by soaking the freshly prepared nZVI in a solution of the second metal salt or simultaneous reduction (co-reduction) of iron and additive metal precursors by sodium borohydride in one reaction pot [30]. It is reported that the second noble metal promotes iron oxidation and may act as a catalyst for electron transfer and hydrogenation. A series of studies have demonstrated that bimetallic iron NPs such as Pb/Fe, Zn/Fe, Cu/Fe, Ni/Fe, Ag/Fe, etc. can achieve significant high degradation rates and reduce the formation of toxic by-products [30–34]. For instance, Yu et al. [34] uniquely used pulsed laser ablation method in liquid to prepare iron@graphite/palladium bimetallic NPs (Fe@C/Pd) for enhanced dechlorination of m-dichlorobenzene (m-DCB). They reported that the Fe@C/Pd showed excellent dechlorination efficiency for m-DCB with 100% degradation within 75 min. The graphitic layer on the Fe NPs played roles not only as an oxidation resistant for the Fe NPs to surroundings, but also as a supporter of the Pb NPs. Gautam [33] synthesized Fe/Zn bimetallic NPs via a coprecipitation method for the adsorptive removal of carcinogenic dye malachite green and Congo red. In another study, Gao [31] systematically compared the degradation mechanism of microcystin-LR using ZVI and bimetallic Fe/Ni and Fe/Pd NPs. They demonstrated that bimetallic Fe/Ni and Fe/Pb NPs removed much more microcystin-LR than iron NPs due to the catalytic effect of Ni and Pb. Moreover, Fe/Pb displayed a faster degradation rate because of the better catalytic property of Pb than Ni. Moreover, other synthesis methods such as solution deposition process (SDP), hydrogen reduction process (HRP), hydrogen reduction of ferrihydrite co-precipitated with the metal cations (HRCO), and electroless plating and conventional displacement placing are also introduced to prepare the Fe-based bimetallic NPs [35, 36].

$$4Fe^{3+} + 3BH_4^- + 9H_2O \rightarrow 4Fe^0 \downarrow + 3H_2BO_3^- + 12H^+ + 6H_2 \quad (11.1)$$

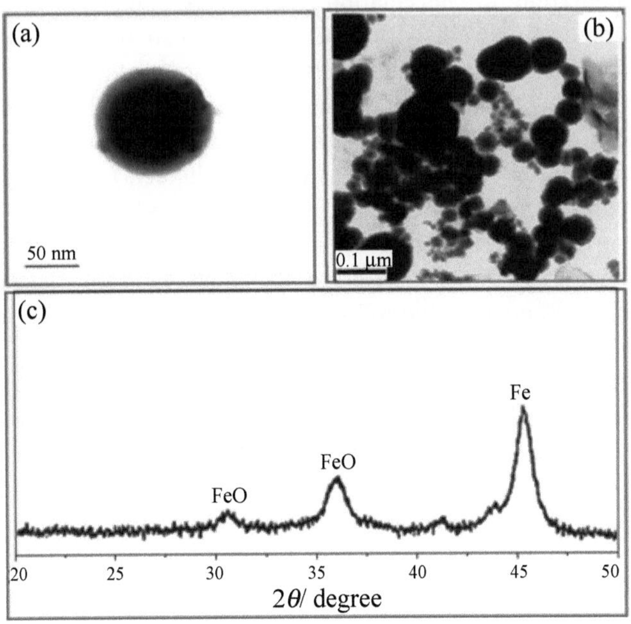


**Figure 11.1** Characterization of ZVI nanoparticles. (a) TEM image of core–shell structure of ZVI NP; (b) TEM image of chain-like aggregates of ZVI NPs; and (c) X-ray diffractogram of ZVI NPs. Reprinted from Ref. [28], Copyright 2006, with permission from Elsevier.

### 11.2.2 Surface Modification of Iron-Based NPs

Iron-based NPs are colloids in nature and exhibit a strong tendency to aggregate as well as adhere to the surfaces of natural materials such as soil and sediment due to their high surface energy, van der Waals and magnetic forces. In order to minimize the agglomeration and thus produce more dispersible and stable nZVI, different methods have been developed, which can be categorized into the following two aspects: (1) Polymers have been commonly used to control the size and prevent the aggregation of metal NPs. In a study, a nontoxic, biodegradable polymeric surfactant poly(vinyl alcohol-co-vinyl acetate-co-itaconic acid) (PV3A) was employed as a dispersant to produce dispersible nZVI. The PV3A-stabilized nZVI had a relatively smaller mean size of 15.5 nm, whereas in the absence of PV3A, the formed nZVI had a mean size of 105.7 nm. Batch experiment demonstrated that there was no noticeable sedimentation or flocculation with PV3A-stabilized nZVI over 6 months [37]. nZVI with different sizes can be synthesized by using carboxymethyl cellulose (CMC) as stabilizer to prevent the NPs from agglomeration through electrostatic stabilization [38]. Besides, several other polymers have also been employed to successfully stabilize nZVI, such as poly(acrylic acid), starch, polyglycol, polystyrene sulfonates, and polyvinyl pyrrolidone [10, 39–41]. Among these, anionic polyelectrolytes are most effective because aquifer materials at neutral pH are normally negatively charged [41, 42]. (2) Although stable nZVI can be obtained after coating polymers and surfactants, these dispersible nZVI are difficult to recycle once they are used to treat contaminants. To avoid secondary water contamination, immobilizing iron-based NPs onto solid supporters that can be easily separated from the contaminated water solution is another option. For instance, a tunable synthesis and immobilization of nZVI into polyelectrolyte multilayers was reported [13]. In this study, poly(acrylic acid) (PAA)/polyallylamine hydrochloride (PAH) multilayers were assembled onto micrometer-sized particle surfaces through layer-by-layer self-assembly approach firstly. Then the polyelectrolyte multilayer-assembled microparticles were used to complex Fe(II) ions in aqueous solution through an electrostatic interaction with the carboxylic acid groups of PAA. Followed by an in situ chemical reduction, nZVI can be formed and immobilized onto the microparticles. The synthesized nZVI displayed an excellent reactivity for the dechlorination of trichloroethylene (TCE). Further, nanoporous polyacrylonitrile-based oxidized membrane was employed to immobilize nZVI via firm chelation and reduction reaction, which effectively prevent the aggregation and release of nZVI [43]. The developed nZVI-immobilized membrane presented effective decoloration efficiency to both anionic methyl blue and cationic methylene blue. Recently, Yuan et al. [44] synthesized Fe/Ni bimetallic NPs supported by commercial polystyrene cation exchange resin D072. Results showed that Fe/Ni-D072 with 4%wt of Ni/Fe loading possesses superior TCE dechlorination performance compared to Fe-D072 and Ni-D072. The leaching of Fe and Ni ions was far below the maximum level established by the US Environment Protection Agency and the Fe/Ni-D072 composite can be partially recovered [44]. Some other carriers or membranes such as mesoporous carbon [32], PAA/polyvinylidene fluoride (PVDF) membrane, and PEG/nylon membrane [45–47] have been also introduced to immobilize Fe-based NPs. However, the solid supports reported in literature have low specific surface area, which could significantly affect the reactivity of nZVI. Therefore, developing continuous medium with a high surface-area-to-volume ratio and good porosity is anticipated to meet the requirements for environmental remediation applications.

## 11.3 Electrospinning Technology

The electrospinning technique was patented by Formhals in 1934 [48], wherein an experimental setup was outlined for the production of polymer filaments with the diameters ranging from 0.01 to 10 μm. The electrospinning technique can be considered as a variant of the electrospraying process, in which electrostatic force is used to supplement or replace conventional mechanical forces. However, electrospinning did not receive considerable attention until the early 1990s. Since Reneker et al. [49] used the electrospinning technology to successfully produce small-diameter, continuous filaments, the technology gained more attention and increasing efforts have been devoted to the fabrication of various nanofibers, the mechanism of nanofiber formation, the factors affecting the nanofiber morphology, and the functionalization and application of electrospun nanofibers [15, 49–51].

A typical electrospinning setup mainly consists of three components: a high-voltage supplier, a capillary tube with a spinneret or needle of small diameter, and a grounded metal collector, as shown in Fig. 11.2a [50]. A high voltage is used to create an electrically charged jet of polymer solution or melt out of the spinneret and form an electrical field between the polymer solution and the collector. Under a sufficiently high electrical field, electrical force would overcome the surface tension of the polymer solution, deforming the pendent drop at the tip of the spinneret into a "Taylor cone" and resulting in the ejection of a thin jet. The charged jet undergoes a stable stretching first, then it starts bending (as shown in Fig. 11.2b [52]) and whipping randomly, leading to further stretching due to the combined effects of solvent evaporation and charge repulsion. This unstable jet is ultimately solidified and deposited onto the collector in a randomly oriented non-woven structure.

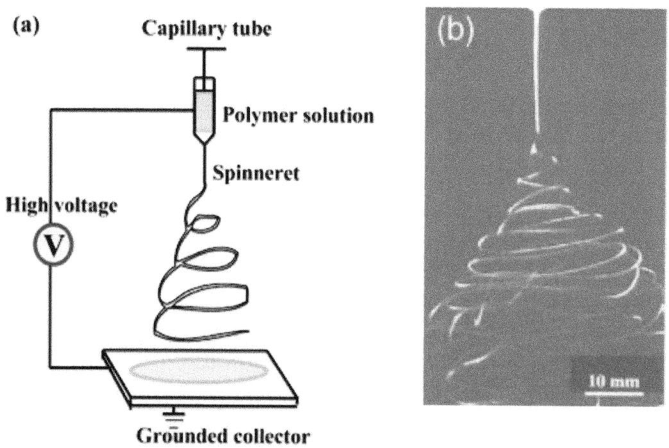


**Figure 11.2** (a) Schematic diagram showing a laboratory setup of electrospinning; (b) photograph of typical electrospinning jets captured by a high-speed video showing the bending instability of the jet. Reprinted from Refs. [50 and 52], Copyright 2007, with permission from Elsevier.

The formation and physical properties (such as diameter, shape, and surface morphology) of nanofibers can be influenced by a number of parameters, which can be grouped as follows:

(1) the intrinsic properties of the polymer solution, such as molecular weight and molecular weight distribution of the polymer, the conformation of polymer chains; (2) solution property, for example, types of solvents, viscosity (depending on the polymer concentration and intermolecular force), surface tension, and electric conductivity; (3) the operational conditions, such as the intensity of the electrical field, the feeding rate of the polymer solution, and the distance between the spinneret and collector; (4) the environmental conditions, such as the humidity and temperature. By appropriately varying one or more of the above parameters, nanofibers can be successfully electrospun from a rich variety of materials, including natural and synthetic polymers, biopolymers (e.g., protein and DNA), ceramics, and some small molecules such as phospholipids. The electrospun nanofibers or nanostructured materials have been widely used as tissue engineering scaffolds, optical sensors, catalysts, protective clothing, drug-delivery systems, and wound dressing, because of their superior properties such as high aspect ratio, large specific surface area, and high porosity and stability in liquid media [25, 53–55]. Especially, the fully interconnected open pore structures and controllable pore size distribution of nanofibrous mats make them good alternatives in water filtration. Currently, electrospun nanofibrous mats for wastewater purification are under extensive investigations.

## 11.4 Synthesis of Electrospun Iron or Iron-Based Bimetallic NP/Polymer Composite

Considering the limitations of iron or iron-based NPs during the treatment of contaminants and the advantages of electrospun nanofibers, Xiao et al. [56] creatively immobilized iron NPs into electrospun nanofibers to fabricate a "nano in nano" structural composite. In this study, PAA was introduced to complex ferric ions for the in situ formation of iron NPs into nanofibers. Briefly, aqueous solutions of PVA and PAA were mixed to have a mass ratio of 1:1 at a constant total polymer concentration (12%), where the 1:1 mass ratio was proven to achieve the highest viscosity of the mixture solution. Then PAA/PVA nanofibers were obtained by electrospinning PAA/PVA mixture solution. The freshly prepared PAA/PVA nanofibrous mats were crosslinked upon heat treatment at 145°C for 30 min. Subsequently, the nanofibrous mats were immersed into an aqueous solution of ferric trichloride for 3 h to allow ferric cations to complex with available free carboxyl groups on PAA through ion exchange, followed by rinsing with water for three times. Sodium borohydride solution was dropped onto the Fe(III)-containing fibrous mat to form nZVI-immobilized nanofibrous mats (Fig. 11.3). It showed that the iron NP-immobilized PAA/PVA nanofibrous mats possessed a three-dimensional porous structure, where the iron NPs were well distributed in the PAA/PVA nanofibers with a mean diameter of 1.6 nm and the immobilized iron NPs did not significantly change the morphology of nanofibers (Fig. 11.4). Importantly, the iron NP-immobilized nanofibers displayed improved contaminants removal property compared to the nZVI, and the degradation efficiency of iron NP-immobilized nanofibers is dependent on the size and loading percentage of iron NPs within nanofibers [57]. Moreover, the iron NP-immobilized nanofibrous mats can be easily separated from contaminated water after the remediation process. By incorporating multiwalled carbon nanotubes, the iron NP-immobilized nanofibrous mats showed enhanced mechanical property [58].

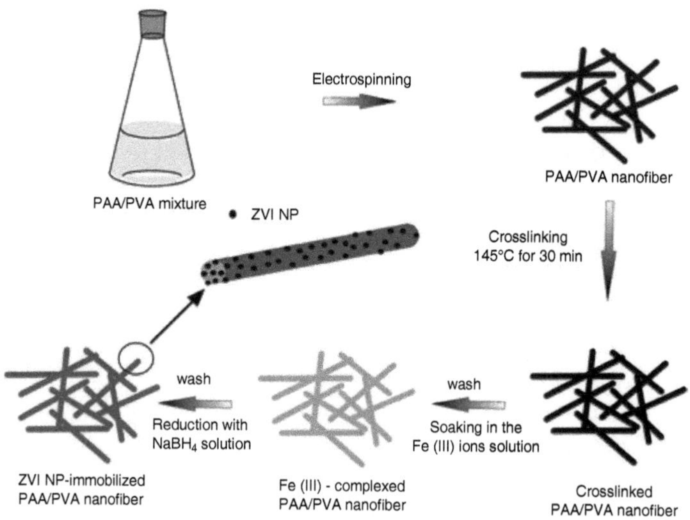


**Figure 11.3** Schematic illustration of immobilizing ZVI NPs into PAA/PVA nanofibers. Reprinted with permission from Ref. [56]. Copyright 2009 American Chemical Society.

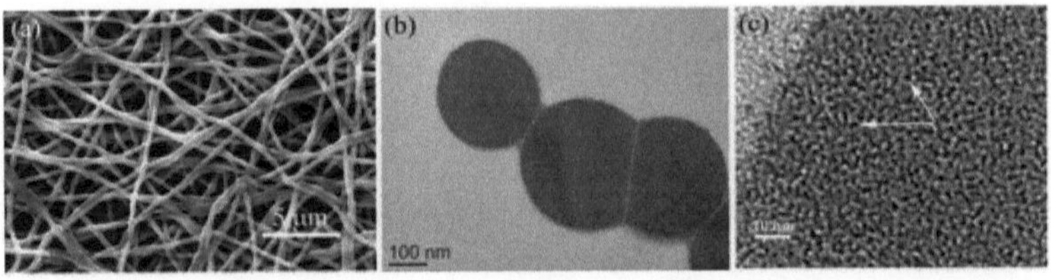


**Figure 11.4** (a) SEM and (b) TEM images of ZVI NP-immobilized PAA/PVA nanofibers. (c) Magnified TEM image of ZVI NPs. Reprinted with permission from Ref. [56]. Copyright 2009 American Chemical Society.

Further, combining the layer-by-layer self-assembly approach and electrospinning technology, a facile approach to synthesizing and immobilizing iron nanoparticles onto nanofibers was developed by Xiao et al. [59]. In this approach, negatively charged cellulose acetate (CA) nanofibers fabricated by electrospinning were assembled with multilayers of poly(diallyldimethylamonium chloride) (PDADMAC) and PAA through electrostatic layer-by-layer assembly. The formed PAA/PDADMAC multilayers onto CA nanofibers were then used as a nanoreactor to complex Fe(II) ions through the binding with free carboxyl groups of PAA for subsequent reductive formation of nZVI. The cross-sectional TEM images indicated that iron NPs were dispersed in the PAA/PDADMAC multilayer film deposited onto the CA nanofibers and had a relatively dense and uniform distribution. They showed that the produced hybrid nanofibrous mats containing iron NPs exhibited superior capability compared to nZVI to decolorize an organic dye acid fuchsin and the loading capacity of iron NPs could be tuned by changing the number of PAA/PDADMAC layers and the cycles of binding/reduction process. Increasing the number of binding/reduction cycles led to a slight bigger size of iron NPs, which is not beneficial for improving the reactivity of iron NPs.

It is known that placing a trace amount of metal catalyst onto nZVI can significantly improve their performance of reductive dechlorination. Ma et al. have synthesized Fe/Pd bimetallic NP-immobilized nanofibrous mats for enhanced dechlorination of trichloroethylene [60]. Crosslinked PAA/PVA nanofibrous mats were first prepared and then dipped into an aqueous ferric trichloride solution for 3 h to allow ferric ions to complex with the available free PAA carboxyl groups through ionic exchange, followed by rinsing with water three times to remove non-complexed Fe(III) ions. nZVI immobilized in the PAA/PVA nanofibrous mats were formed by reducing Fe(III) using sodium borohydride solution for 30 min, followed by three times rinsing of water. The formed nZVI-containing nanofibrous mats were then immersed into an acetone solution of palladium chloride for 1 h. The Fe/Pd NPs were formed by the deposition of Pd on the nZVI surfaces through the following reaction:

$$Pd^{2+} + Fe^0 \rightarrow Pd^0 + Fe^{2+} \tag{11.2}$$

Cross-sectional TEM images of the Fe/Pd NP-immobilized nanofibrous mats showed that individual NPs with a relatively dense and uniform distribution along the cross section of nanofibers can be clearly observed. The mean diameter of the Fe/Pd NPs was estimated to be 2.8 ± 0.92 nm, bigger than the nZVI without deposition of metallic Pd as reported in literature [56]. This was presumably due to the replacement reaction that stimulates the Ostwald ripening of the particles. Nevertheless, the Fe/Pd NPs immobilized within PAA/PVA nanofibers are much smaller than those prepared using a wet chemistry approach reported in the literature [46, 61]. Similarly, Fe/Ni NP-immobilized nanofibers were developed by Ma et al. [62], wherein the mean diameter of formed Fe/Ni NPs was 1.5 ± 0.3 nm.

As a widely used environmental remediation material, iron NPs own strong reductive activity to contaminants. By immobilizing iron NPs within polymer nanofibers to form "nano in nano" composite nanofibers, one can design a desired environmental remediation material that possesses many advantages such as improved contaminant removal capability, good mechanical property, easiness to be separated from contaminated water, and recyclability.

## 11.5 Electrospun Iron or Iron-Based Bimetallic NP/Polymer Nanofibers for Environmental Remediation

### 11.5.1 Improved Contaminant Removal Capability

Due to the fact that iron NP powders aggregate easily during synthesis and environmental remediation application processes, iron NPs exhibit a decreased capability for contaminants removal. Immobilizing iron NPs into electrospun nanofibers by in situ chemical reduction provides an effective way to mediate the problems mentioned above. The iron NP-immobilized electrospun nanofibrous mats show an improved contaminants removal capability [56, 57, 59, 60, 63]. A model contaminant, acid fuchsin dye (a common organic dye containing benzene rings in textile industry, which cannot be easily decomposed using traditional chemical and biological methods) decoloration experiment showed that iron NP-immobilized polymer nanofibrous mats were able to rapidly decolorize about 85% of the acid fuchsin without any additives at room temperature within 5 min (the content of iron NP-immobilized nanofibrous mat was 0.67 g/L). The red color of the acid fuchsin solution was suddenly decolorized after 5 min and then gradually faded within a time frame of 40 min (Fig. 11.5). When the exposure time was extended to 40 min, only gradual decoloration was observed, and the decoloration efficiency could reach up to 95.8%. The decoloration effect of the iron NP-immobilized polymer nanofibers is solely related to the reactive nature of iron NPs immobilized in the polymer nanofibers. When PAA/PVA nanofibrous mats without iron NP were exposed to the same acid fuchsin solution, no decoloration was observed. In sharp contrast, iron NP powder showed a much lower decoloration capability to acid fuchsin than those immobilized into polymer nanofibers. No sudden decoloration effect after 5 min was observed [56].

Besides acid fuchsin, iron NP-immobilized polymer nanofibrous mats also show superior capability in decoloration of other dyes in the textile wastewater, such as methyl blue and acridine orange, in degradation of TCE, and in the removal of copper(II) ions [58, 64]. For instance, systematic copper ions removal experiments demonstrated that the uniform iron NPs immobilized into nanofibers offered high specific area of $Fe^0$ that enable effective, high capacity, and strong sorption of Cu(II) ions via chemical reduction and deposition. Moreover, the incorporation of multiwalled carbon nanotubes (MWCNTs) into iron NP-immobilized nanofibers did not compromise the Cu(II) removal ability of nanofibrous mats. For the iron NP-immobilized PAA/PVA nanofibrous mats containing MWCNTs, the Cu(II) removal capability sharply increased with the contact time within the first 30 min, and then gradually approached equilibrium (75.3 mg/g) in approximately 60 min. Nevertheless, MWCNT-incorporated PAA/PVA nanofibrous mats exhibited much lower Cu(II) ions removal capacity, with Cu(II) uptake rapidly reached 32.0 mg/g within the first 10 min and then leveled off as time elapsed. It is postulated that the sorption on the Fe-free mats was primarily driven by the complexation of Cu(II) ions with the free carboxyl groups in the PAA. This process involves the release of protons, and therefore it is pH-dependent. Whereas for the Fe-containing fibrous mats, besides the free carboxyl groups available for Cu(II) complexation, Cu(II) can also be chemically reduced by $Fe^0$ and deposited on the iron NP surfaces to form an alloy, which might be the mechanism dominating the sorption on Fe-containing mats. This process is not pH-dependent [64].

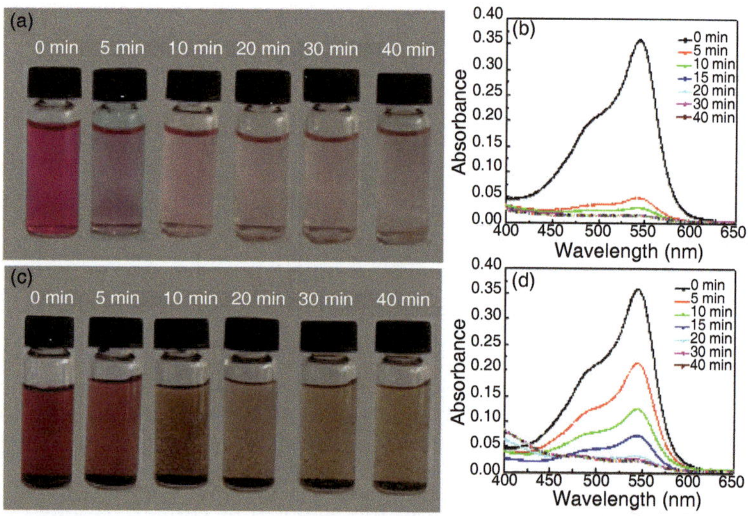


**Figure 11.5** Photographs of the acid fuchsin solution treated with ZVI NP-containing polymer nanofibrous mats (a) and ZVI NPs synthesized using literature method (c) at different time intervals. UV-Vis spectra of a solution of acid fuchsin (60 mg/L, 20 mL) in the presence of ZVI NP-containing polymer nanofibrous mats (b) and ZVI NPs synthesized using literature method (d) at a time interval of 0, 5, 10, 20, 30, and 40 min. Reprinted with permission from Ref. [56]. Copyright 2009 American Chemical Society.

In another study, Ma et al. prepared hybrid nanofibrous mats by immobilizing iron/palladium bimetallic NPs for dechlorination of trichloroethylene (TCE) [60]. Meanwhile, Fe/Pd colloidal NPs, nZVI (Fe)- immobilized nanofibrous mat and Pd-immobilized nanofibrous mat were synthesized for comparison. Upon exposure of the colloid

Fe/Pd NPs, nZVI-immobilized nanofibrous mats, and Fe/Pd NP-immobilized nanofibrous mats to an aqueous solution of TCE (10 mg/L), respectively, all the three systems were able to degrade TCE. During this experiment, the final Fe concentration was 0.7 g/L. The remaining fraction of TCE treated with colloid Fe/Pd NPs, nZVI- and Fe/Pd NP-immobilized nanofibrous mats was measured to be 0.41%, 0.88%, and 0.38%, respectively, after 3.5 h. However, no significant dechlorination was obtained when Pd NP-immobilized nanofibrous mat was used; the remaining fraction of TCE was up to 93.04% under similar conditions. At low TCE concentration (10 mg/L), no significant difference in the dechlorination efficiency was observed using the above three different systems. However, when the initial TCE concentration increased, both nZVI-contained nanofibrous mats and Fe/Pd colloid NPs exhibited decreased dechlorination efficiency, whereas Fe/Pd NP-containing nanofibrous mats kept the highest dechlorination efficiency for all cases. At a high TCE concentration (100 mg/L), the dechlorination efficiency of Fe/Pd NP-immobilized nanofibrous mats was as high as 95.9%. In contrast, the dechlorination efficiencies of nZVI-immobilized mats and the colloidal Fe/Pd NPs were only 77.3% and 38.2%, respectively (Fig. 11.6). The higher dechlorination efficiency of Fe/Pd NP-containing nanofibrous mats in comparison with Fe/Pd NPs proved that immobilizing Fe/Pd NPs into nanofibers could avoid the aggregation of colloidal Fe/Pd NPs during the remediation process, resulting in enhanced dechlorination efficiency [60]. Further, Ma et al. [62] developed Fe/Ni NP-immobilized nanofibrous mats through two different routes: (1) synthesis of Fe NPs followed by post-coating Ni on Fe NPs (referred to post-coated Fe/Ni NPs) and (2) co-reduction of nickel and iron precursors within the nanofibers (referred to co-reduced Fe/Ni NPs). Orange G and methyl blue decoloration experiments demonstrated that the co-reduced Fe/Ni-containing nanofibrous mats have much higher decoloration efficiency than the single iron NP-immobilized nanofibrous mats; whereas the mats containing post-coated Fe/Ni NPs have lower decoloration efficiency than the single iron NP-containing mats. In the presence of co-reduced Fe/Ni NP-containing nanofibrous mats, the orange color of orange G solution or the blue color of methyl blue solution faded quickly within 5–10 min and showed almost complete dye degradation after 40 min. In contrast, in the presence of the post-coated Fe/Ni-containing nanofibrous mats, the color of orange G or methyl blue solution faded slowly with the exposure time. The decoloration trend of nZVI-containing fiber mats is in-between the two types of the Fe/Ni NP-containing mats. The variation in decoloration of model dyes by the developed Fe/Ni NP-containing nanofibrous mats might be elucidated on the basis of combining modes of ZVI and Ni NPs. The close contact of Ni NPs with nZVI is able to afford an effective catalytic reaction at the nZVI surfaces [65], whereas the physical mixture of nZVI and Ni NPs can decrease the catalysis efficiency of Ni NPs [66]. Besides, insufficient or excessive coating of Ni may cause the formation of an Fe-rich surface or Ni-rich surface, which lowers the catalytic efficiency of Ni or lowers the reactivity of nZVI, respectively [67]. Therefore, the post-coated Fe/Ni NPs having a high amount of Ni coating on the surface of nZVI likely reduced the reactivity of ZVI core NPs, resulting in a lower decoloration efficiency than nZVI. Whereas, the co-reduced Fe/Ni NPs have a good balance of the nZVI reactivity and Ni NP catalytic activity, leading to a higher decoloration efficiency than nZVI.

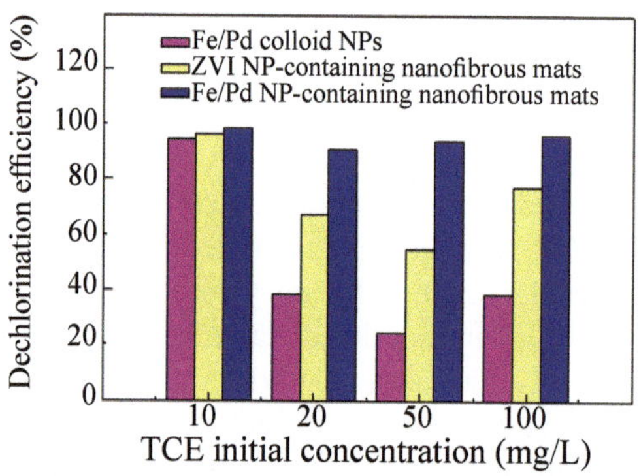


**Figure 11.6** Removal efficiency of TCE by Fe/Pd colloid NPs, ZVI NP- and Fe/Pd NP-containing nanofibrous mats with different initial TCE concentrations. The reaction time was 1.5 h. Reprinted from Ref. [60], Copyright 2012, with permission from Elsevier.

## 11.5.2 Enhanced Mechanical Properties

Mechanical properties of electrospun nanofibrous mats are of vital importance for their practical applications. Carbon nanotubes (CNTs) have received immense scientific and technological interest due to their unique hollow structure consisting of concentric graphene cylinders, which impart CNTs with excellent mechanical properties, high electrical conductivity, good chemical stability, and thermal conductivity [68, 69]. In particular, CNTs have been considered as the ideal nanofillers for reinforcing electrospun polymer materials due to their excellent mechanical properties [70, 71]. In a recent study, Xiao et al. [58] prepared mechanically enhanced nZVI-immobilized nanofibrous mats by incorporating MWCNTs. As can be seen from Fig. 11.7, the mechanical properties of nanofibrous mats with nZVI immobilization were significantly enhanced after incorporating 1 wt% MWCNTs. For the nanofibrous mats immobilized with nZVI, the mean tensile strength and Young's modulus were 4.18 MPa and 16.87 MPa, respectively. After incorporating MWCNTs, the mean tensile strength and Young's modulus increased up to 5.50 MPa and 23.3 MPa, respectively, demonstrating that the incorporation of MWCNTs into nanofibers can significantly enhance the mechanical properties of the corresponding nanofibers. The enhancement may be ascribed to the alignment of CNTs in the PAA/PVA nanofibers.

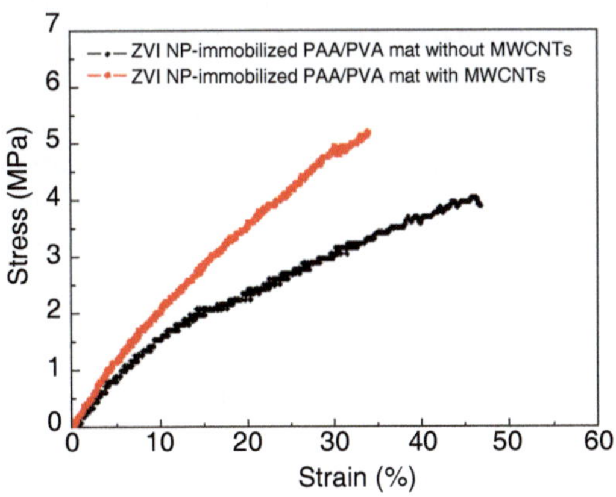


**Figure 11.7** Typical stress–strain curves of ZVI NP-immobilized nanofibrous mats with and without MWCNTs. Reprinted from Ref. [58], Copyright 2010, with permission from the Royal Society of Chemistry.

### 11.5.3 Recyclability and Easy Separation from Wastewater

For practical environmental remediation applications, the reusability and recyclability of the materials are of vital importance. The iron or iron-based bimetallic NP-immobilized nanofibrous mats developed are reusable and recyclable [56, 58, 60]. After exposure to sodium borohydride aqueous solution for 10 min, the nanofibrous mats could be regenerated for further remediation process. The regenerated iron-containing nanofibrous mats exhibited similar performance in the repeated contaminant removal experiments to that of the freshly prepared mats [56, 58, 60]. For instance, nZVI-immobilized PAA/PVA nanofibrous mats performed well in the decoloration of acid fuchsin and methyl blue; the decoloration efficiencies of acid fuchsin and methyl blue were 85% and 91%, respectively, after three cycles of remediation process [56, 58]. Moreover, the dechlorination efficiency of Fe/Pd NP-containing nanofibrous mats could reach up to 90% in the fourth dechlorination experiment after being regenerated with $NaBH_4$ [60]. In addition, the practical environmental remediation application of NPs often requires the NPs to be easily recyclable after the remediation process, instead of leaving the particles dispersed in water to generate secondary contamination. It has been reported that the nZVI- and Fe/Pd bimetallic NP-immobilized nanofibrous mats can be easily separated from the contaminated water by simply taking out of nanofibrous mats. The iron and Fe/Pd bimetallic NPs are quite stable and do not escape from the fibrous mats during the remediation process. Inductively coupled plasma atomic emission spectroscopy studies show that no iron or Pd is released from the nanofibrous mats even if the nanofibrous mats are exposed to water for a month [56, 60].

## 11.6 Conclusions and Outlook

In summary, this chapter reviews the electrospun iron or iron-based bimetallic NP-immobilized nanofibrous mats for environmental remediation applications. The immobilization of iron NPs within polymer nanofibers does not compromise the reductive activity of iron NPs to contaminants; it instead avoids their agglomeration during the contaminant treatment in aqueous environment and significantly improves their contaminants removal rate due to their uniform distribution in the nanofibers. The developed "nano in nano" composite remains a porous three-dimensional fibrous structure with a high surface area, which is beneficial for the capture of contaminants and subsequent degradation by iron NPs. In addition, the composite functionalized with iron NPs or iron-based bimetallic NPs has considerable mechanical properties and is easily separated from the contaminated water and reusable, without introducing secondary contamination. Therefore, the developed electrospun composite nanofibrous mats containing iron NPs may be used as a novel nanofiltration material.

The research area of electrospun iron/polymer composite nanofiber-based environmental remediation materials still remains open since serials of issues regarding their fabrication and applications remain to be addressed. For instance, other approaches for preparing iron-immobilized nanofibrous mats can be employed, such as coaxial electrospinning technology to improve the iron content in the nanofibers. In addition, different polymer systems having different properties may be used to form the composite nanofibers with different properties. And other iron-based bimetallic NPs (Fe/Cu, Fe/Ag, etc.) and iron-doped semiconductive NPs (Fe/$TiO_2$, Fe/ZnO, etc.) can also be immobilized into or onto polymer nanofibers to form electrospun inorganic/organic composite nanofibers with desired properties for environmental remediation and other applications. Furthermore, current efforts to use electrospun iron NP-immobilized nanofibrous mats for contaminants treatment are limited to several model dyes and organic compounds (e.g., TCE), other contaminants widely existing in the wastewater should be evaluated to test the contaminant removal efficacy of iron-based NP-immobilized nanofibrous mats.

## Acknowledgments

This research is financially supported by the Key Laboratory of Textile Science & Technology, Ministry of Education, "111 Project", B07024, the Program for Professor of Special Appointment (Eastern Scholar) at Shanghai Institution of Higher Learning, the National Natural Science Foundation of China (51403164), and the Education Commission of Hubei Province (No. 164004).

## References


1. Reddy, A. V. B., Yusop, Z., Jaafar, J., Reddy, Y. V. M., Aris, A. B., Majid, Z. A., Talib, J., and Madhavi, G. (2016). Recent progress on Fe-based nanoparticles: Synthesis, properties, characterization and environmental applications, *J. Environ. Chem. Eng.*, **4**, pp. 3537–3553.
2. Lefevre, E., Bossa, N., Wiesner, M. R., and Gunsch, C. K. (2016). A review of the environmental implications of in situ remediation by nanoscale zero valent iron (nZVI): Behavior, transport and impacts on microbial communities, *Sci. Total Environ.*, **565**, pp. 889–901.
3. Petersen, E. J., Pinto, R. A., Shi, X., and Huang, Q. (2012). Impact of size and sorption on degradation of trichloroethylene and polychlorinated biphenyls by nano-scale zerovalent iron, *J. Hazard. Mater.*, **243**, pp. 73–79.
4. Ponder, S. M., Darab, J. G., and Mallouk, T. E. (2000). Remediation of Cr(VI) and Pb(II) aqueous solution supported nanoscale zero-valent iron, *Environ. Sci. Technol.*, **34**, pp. 2564–2569.
5. Kanel, S. R., Greneche, J.-M., and Chol, H. (2006). Arsenic(V) removal from groundwater using nanoscale zero-valent iron as a colloidal reactive barrier material, *Environ. Sci. Technol.*, **40**, pp. 2045–2050.
6. Darab, J. G., Amonette, A. B., Burke, D. S. D., Orr, R. D., Ponder, S. M., Schrick, B., Mallouk, T. E., Lukens, W. W., Caulder, D. L., and Shuh, D. K. (2007). Removal of pertechnetate from simulated nuclear waste streams using supported zerovalent Iron, *Chem. Mater.*, **19**, pp. 5703–5713.
7. Epolito, W. J., Yang, H., Bottomley, L. A., and Pavlostathis, S. G. (2008). Kinetics of zero-valent iron reductive transformation of the anthraquinone dye Reactive Blue 4, *J. Hazard. Mater.*, **160**, pp. 594–600.
8. Naja, C., Halasz, A., Thiboutout, S., Ampleman, G., and Hawari, J. (2008). Degradation of hexahydro-1,3,5-trinitro-1,3,5-triazine(RDX) using zerovalent iron nanoparticles, *Environ. Sci. Technol.*, **42**, pp. 4364–4370.
9. Li, X.-Q., Brown, D. G., and Zhang, W.-X. (2007). Stabilization of biosolids with nanoscale zero-valent iron(nZVI), *J. Nanopart. Res.*, 9, pp. 233–243.
10. Yang, G. C. C., Tu, H.-C., and Hung, C.-H. (2007). Stability nanoiron slurries and their transport in the subsurface environment, *Sep. Purif. Technol.*, **58**, pp. 166–172.

11. Zhao, X., Liu, W., Cai, Z., Han, B., Qian, T., and Zhao, D. (2016). An overview of preparation and application of stabilized zero-valent iron nanoparticles for soil and groundwater remediation, *Water Res.*, **100**, pp. 245–266.

12. He, F. and Zhao, D. (2007). Manipulating the size and dispersibility of zerovalent iron nanoparticles by use of carboxymethyl cellulose stabilizers, *Environ. Sci. Technol.*, **41**, pp. 6216–6221.

13. Huang, G., Shi, X., Pinto, R. A., Petersen, E., and Weber, W. J., Jr. (2008). Tunable synthesis and immobilization of zero-valent iron nanoparticles for environmental applications, *Environ. Sci. Technol.*, **42**, pp. 8884–8889.

14. Kim, G., Jeong, W., and Choe, S. (2008). Dechlorination of atrazine using zero-valent iron(Fe0) under neutral pH conditions, *J. Hazard. Mater.*, **155**, pp. 502–506.

15. Wang, X. and Hsiao, B. S. (2016). Electrospun nanofiber membranes, *Curr. Opin. Chem. Eng.*, **12**, pp. 62–81.

16. Fang, X., Xiao, S., Shen, M., Guo, R., Wang, S., and Shi, X. (2011). Fabrication and characterization of water-stable electrospun polyethyleimine/polyvinyl alcohol nanofibers with super dye sorption capability, *New J. Chem.*, **35**, pp. 360–368.

17. Sehaqui, H., Mautner, A., de Larraya, U. P., Pfenninger, N., Tingaut, P., and Zimmermann, T. (2016). Cationic cellulose nanofibers from waste pulp residues and their nitrate, fluoride, sulphate and phosphate adsorption properties, *Carbohydr. Polym.*, **135**, pp. 334–340.

18. Perez-Page, M., Yu, E., Li, J., Rahman, M., Dryden, D. M., Vidu, R., and Stroeve, P. (2016). Template-based syntheses for shape controlled nanostructures, *Adv. Colloid Interface Sci.*, **234**, pp. 51–79.

19. Zhang, H., Xin, X., Sun, J., Zhao, L., Shen, J., Song, Z., and Yuan, S. (2016). Self-assembled chiral helical nanofibers by amphiphilic dipeptide derived from D- or L-threonine and application as a template for the synthesis of Au and Ag, *J. Colloid Interf. Sci.*, **484**, pp. 97–106.

20. Lu, Y., Li, Y., Zhang, S., Xu, G., Fu, K., Lee, H., and Zhang, X. (2013). Parameter study and characterization for polyacrylonitrile nanofibers fabricated via centrifugal spinning process, *Eur. Polym. J.*, 49, pp. 3834–3845.

21. Beachley, V. and Wen, X. (2010). Polymer nanofibrous structures: Fabrication, biofunctionalization, and cell interactions, *Prog. Polym. Sci.*, **35**, pp. 868–892.

22. Peng, S., Li, L., Lee, J. K. Y., Tian, L., Srinivasan, M., Adams, S., and Ramakrishna, S. (2016). Electrospun carbon nanofibers and their hybrid composites as advanced materials for energy conversion and storage, *Nano Energy*, **22**, pp. 361–395.

23. Xu, J.-W., Wang, Y., Yang, Y.-F., Ye, X.-Y., Yao, K., Ji, J., and Xu, Z.-K. (2015). Effects of quaternization on the morphological stability and antibacterial activity of electrospun poly(DMAEMA-co-AMA) nanofibers, *Colloids Surf., B.*, **133**, pp. 148–155.

24. Ranjbar-Mohammadi, M., Rabbani, S., Bahrami, S. H., Joghataei, M. T., and Moayer, F. (2016). Antibacterial performance and in vivo diabetic wound healing of curcumin loaded gum tragacanth/poly(e-caprolactone) electrospun nanofibers, *Mater. Sci. Eng. C.*, **69**, pp. 1183–1191.

25. Feng, C., Khulbe, K. C., Matsuura, T., Tabe, S., and Ismail, A. F. (2013). Preparation and characterization of electro-spun nanofiber membranes and their possible applications in water treatment, *Sep. Purif. Technol.*, **102**, pp. 118–135.

26. Wang, C.-B. and Zhang, W.-X. (1997). Synthesizing nanoscale iron particles for rapid and complete dechlorination of TCE and PCBs, *Environ. Sci. Technol.*, **31**, pp. 2154–2156.

27. Li, X.-Q., Elliott, D. W., and Zhang, W.-X. (2006). Zero-valent iron nanoparticles for abatement of environmental pollutants: Materials and engineering aspects, *Crit. Rev. Solid State Mat. Sci.*, **31**, pp. 111–122.

28. Sun, Y.-P., Li, X., Cao, J., Zhang, W., and Wang, H. P. (2006). Characterization of zero-valent nanoparticles, *Adv. Colloid Interface Sci.*, **120**, pp. 47–56.

29. Stefaniuk, M., Oleszczuk, P., and Ok, Y. S. (2016). Review on nano zerovalent iron (nZVI): From synthesis to environmental applications, *Chem. Eng. J.*, **287**, pp. 618–632.

30. Luo, F., Yang, D., Chen, Z., Megharaj, M., and Naidu, R. (2016). One-step green synthesis of bimetallic Fe/Pd nanoparticles used to degrade Orange II, *J. Hazard. Mater.*, **303**, pp. 145–153.

31. Gao, Y., Wang, F., Wu, Y., Naidu, R., and Chen, Z. (2016). Comparison of degradation mechanisms of microcystin-LR using nanoscale zero-valent iron (nZVI) and bimetallic Fe/Ni and Fe/Pb nanoparticles, *Chem. Eng. J.*, **285**, pp. 459–466.

32. Wang, Y., Zhao, H., and Zhao, G. (2015). Iron-copper bimetallic nanoparticles embedded within ordered mesoporous carbon as effective and stable heterogeneous Fenton catalyst for the degradation of organic contaminants, *Appl. Catal. B*, **164**, pp. 396–406.

33. Gautam, R. K., Rawat, V., Banerjee, S., Sanroman, M. A., Soni, S., Singh, S. K., and Chattopadhyaya, M. C. (2015). Synthesis of bimetallic Fe-Zn nanoparticles and its application towards adsorptive removal of carcinogenic dye malachite green and Congo red in water, *J. Mol. Liq.*, **212**, pp. 227–236.

34. Yu, Y., Jung, H. J., Je, M., Choi, H. C., and Choi, M. Y. (2016). Enhanced dechlorination of m-DCB using iron@graphite/palladium (Fe@C/Pd) nanoparticles produced by pulsed laser ablation in liquid, *Chemosphere*, **155**, pp. 250–256.

35. Chan, L. C., Donald, R. B., Dean, W. M., James, E. A., and Penn, R. L. (2010). Characterization and reactivity of iron nanoparticles prepared with added Cu, Pd, and Ni, *Environ. Sci. Technol.*, **44**, pp. 5079–5085.

36. Ren, Y. and Lai, B. (2016). Comparative study on the characteristics, operational life and reactivity of Fe/Cu bimetallic particles prepared by electroless and displacement plating process, *RSC Adv.*, **6**, pp. 58302–58314.

37. Sun, Y.-P., Li, X.-Q., Zhang, W.-X., and Wang, H. P. (2007). A method for the preparation of stable dispersion of zero-valent iron nanoparticles, *Colloids Surf. A Physicochem. Eng. Asp.*, **308**, pp. 60–66.

38. Madhavi, V., Prasad, T. N. V. K. V., Reddy, B. R., Reddy, A. V. B., and Gajulapalle, M. (2014). Conjunctive effect of CMC-zerovalent iron nanoparticles and FYM in the remediation of chromium-contaminated soils, *Appl. Nanosci.*, **4**, pp. 477–484.

39. He, F. and Zhao, D. (2005). Preparation and characterization of a new class of starch-stabilized bimetallic nanoparticles for degradation of chlorinated hydrocarbons in water, *Environ. Sci. Technol.*, **39**, pp. 3314–3320.

40. Wang, W., Jin, Z.-H., Li, T.-L., Zhang, H., and Gao, S. (2006). Preparation of spherical iron nanoclusters in ethanol-water solution for nitrate removal, *Chemosphere*, **65**, pp. 1396–1404.

41. Laumann, S., Micic, V., and Hofmann, T. (2014). Mobility enhancement of nanoscale zero-valent iron in carbonate porous media through co-injection of polyelectrolytes, *Water Res.*, **50**, pp. 70–79.

42. Liu, J., Liu, A., and Zhang, W. (2016). The influence of polyelectrolyte modification on nanoscale zero-valent iron (nZVI): Aggregation, sedimentation, and reactivity with Ni(II) in water, *Chem. Eng. J.*, **303**, pp. 268–274.

43. Liu, C., Li, X., Ma, B., Qin, A., and He, C. (2014). Removal of water contaminants by nanoscale zero-valent iron immobilized in PAN-based oxidized membrane, *Appl. Surf. Sci.*, **321**, pp. 58–165.

44. Zhou, Z., Ruan, W., Huang, H., Shen, C., Yuan, B., and Huang, C.-H. (2016). Fabrication and characterization of Fe/Ni nanoparticles supported by polystyrene resin for trichloroethylene degradation, *Chem. Eng. J.*, **282**, pp. 730–739.
45. Sikhwivhilu, K. and Moutloali, R. M. (2015). Functionalized PVDF membrane-immobilized Fe/Ni bimetallic nanoparticles for catalytic degradation of methyl orange dye: A comparative study, *Mater. Today: Proceedings*, **2**, pp. 4070–4080.
46. Wang, X., Chen, C., Liu, H., and Ma, J. (2008). Preparation and characterization of PAA/PVDF membrane-immobilized Pd/Fe nanoparticles for dechlorination of trichloroacetic acid, *Water Res.*, **42**, pp. 4656–4664.
47. Tong, M., Yuan, S., Long, H., Zheng, M., Wang, L., and Chen, J. (2011). Reduction of nitrobenzene in groundwater by iron nanoparticles immobilized in PEG/nylon membrane, *J. Contam. Hydrol.*, **122**, pp. 16–25.
48. Formalas. (1934). Process and apparatus for preparing artificial threads, *US patent*, pp. 1,975,504.
49. Thompson, C. J., Chase, G. G., Yarin, A. L., and Reneker, D. H. (2007). Effects of parameters on nanofiber diameter determined from electrospinning model, *Polymer*, **48**, pp. 6913–6922.
50. Rutledge, G. C. and Fridrikh, S. V. (2007). Formation of fibers by electrospinning, *Adv. Drug Delivery Rev.*, **59**, pp. 1384–1391.
51. Stepanyan, R., Subbotin, A. V., Cuperus, L., Boonen, P., Dorschu, M., Oosterlinck, F., and Bulters, M. J. H. (2016). Nanofiber diameter in electrospinning of polymer solutions: Model and experiment, *Polymer*, **97**, pp. 428–439.
52. Han, T., Reneker, D. H., and Yarin, A. L. (2007). Buckling of jets in electrospinning, *Polymer*, **48**, pp. 6064–6076.
53. Nguyen, T.-A., Park, S., Kim, J. B., Kim, T. K., Seong, G. H., Choo, J., and Kim, Y. S. (2011). Polycrystalline tungsten oxide nanofibers for gas-sensing applications, *Sens. Actuators B Chem.*, **160**, pp. 549–554.
54. Holzwarth, J. M. and Ma, P. X. (2011). Biomimetic nanofibrous scaffolds for bone tissue engineering, *Biomaterials*, **32**, pp. 9622–9629.
55. Gorji, M., Bagherzadeh, R., and Fashandi, H. (2017). Electrospun nanofibers in protective clothing. In: *Elecrospun Nanofibers*, Afshari, M. (ed.), pp. 571–598 (Woodhead Publishing).
56. Xiao, S., Shen, M., Guo, R., Wang, S., and Shi, X. (2009). Immobilization of zerovalent iron nanoparticles into electrospun polymer nanofibers:

Synthesis, characterization, and potential environmental applications, *J. Phys. Chem. C*, **113**, pp. 18062–18068.

57. Xiao, S., Shen, M., Ma, H., Fang, X., Huang, Q., Weber, W. J., and Shi, X. (2011). Manipulation of the loading and size of zero-valent iron nanoparticles immobilized in electrospun polymer nanofibers, *J. Nanosci. Nanotechnol.*, **11**, pp. 1–9.
58. Xiao, S., Shen, M., Guo, R., Huang, Q., Wang, S., and Shi, X. (2010). Fabrication of multiwalled carbon nanotube-reinforced electrospun polymer nanofibers containing zero-valent iron nanoparticles for environmental applications, *J. Mater. Chem.*, **20**, pp. 5700–5708.
59. Xiao, S., Wu, S., Shen, M., Guo, R., Huang, Q., Wang, S., and Shi, X. (2009). Polyelectrolyte multilayer-assisted immobilization of zero-valent iron nanoparticles onto polymer nanofibers for potential environmental applications, *ACS Appl. Mater. Interfaces*, **1**, pp. 2848–2855.
60. Ma, H., Huang, Y., Shen, M., Guo, R., Cao, X., and Shi, X. (2012). Enhanced dechlorination of trichloroethylene using electrospun polymer nanofibrous mats immobilized with iron/palladium bimetallic nanoparticles, *J. Hazard. Mater.*, **211–212**, pp. 349–356.
61. Wang, X., Chen, C., Chang, Y., and Liu, H. (2009). Dechlorination of chlorinated methanes by Pd/Fe bimetallic nanoparticles, *J. Hazard. Mater.*, **161**, pp. 815–823.
62. Ma, H., Huang, Y., Shen, M., Hu, D., Yang, H., Zhu, M., Yang, S., and Shi, X. (2013). Enhanced decoloration efficacy of electrospun polymer nanofibers immobilized with Fe/Ni bimetallic nanoparticles, *RSC Adv.*, **3**, pp. 6455–6465.
63. Xu, T., Ni, D., Chen, X., Wu, F., Ge, P., Lu, W., Hu, H., Zhu, Z., and Chen, W. (2016). Self-floating graphitic carbon nitride/zinc phthalocyanine nanofibers for photocatalytic degradation of contaminants, *J. Hazard. Mater.*, **317**, pp. 17–26.
64. Xiao, S., Ma, H., Shen, M., Wang, S., Huang, Q., and Shi, X. (2011). Excellent copper(II) removal using zero-valent iron nanoparticle-immobilized hybrid electrospun polymer nanofibrous mats, *Colloids Surf. A Physicochem. Eng. Asp.*, **381**, pp. 48–54.
65. Schrick, B., Blough, J. L., Jones, A. D., and Mallouk, T. E. (2002). Hydrodechlorination of trichloroethylene to hydrocarbons using bimetallic nickel-iron nanoparticles, *Chem. Mater.*, **14**, pp. 5140–5147.
66. Zhang, W., Wang, B., and Lien, H. L. (1998). Treatment of chlorinated organic contaminants with nanoscale bimetallic particles, *Catal. Today*, **40**, pp. 387–395.

67. Xu, J. and Bhattacharyya, D. (2005). Membrane-based bimetallic nanoparticles for environmental remediation: Synthesis and reactive properties, *Environ. Prog.*, **24**, pp. 358–366.

68. Tans, S. J., Devoret, M. H., Dai, H., Thess, A., Smalley, R. E., Geerligs, L. J., and Dekker, C. (1997). Individual single-wall carbon nanotubes as quantum wires, *Nature*, **386**, pp. 474–477.

69. White, C. T. and Todorov, T. N. (2001). Nanotubes go ballistic, *Nature*, **411**, pp. 649–651.

70. Hou, Y., Tang, J., Zhang, H., Qian, C., Feng, Y., and Liu, J. (2009). Functionalized few-walled carbon nanotubes for mechanical reinforcement of polymeric composites, *ACS Nano*, **3**, pp. 1059–1062.

71. Sui, X. and Wagner, H. D. (2009). Tough nanocomposites: The role of carbon nanotube type, *Nano Lett.*, **9**, pp. 1423–1426.

# Chapter 12

# Environmental Effects of the Application of Iron Nanoparticles for Site Remediation

**Ekain Cagigal,[a] Marta Ocejo,[a] José Luis R. Gallego,[b] Ana I. Peláez,[b] and Eduardo Rodríguez-Valdés[b]**
[a]*Tecnalia Research & Innovation, Bizkaia Science Technology Park, 48160 Derio, Spain*
[b]*Environmental Technology, Biotechnology and Geochemistry Group University of Oviedo, C/ Gonzalo Gutiérrez Quirós s/n, 33600 Mieres, Spain*
ekain.cagigal@tecnalia.com

## 12.1 Introduction

As long as the injection of nanoparticles (NPs, mainly nanozerovalent iron, nZVI) has become a common technology for soil and groundwater remediation, a concern has simultaneously grown regarding the side effects and potential impact on ecosystems, as occurring with other engineered nanometric scale materials. Based on precautionary principle, the technology is still pending approval by regulators in several countries until uncertainties regarding risk

*Iron Nanomaterials for Water and Soil Treatment*
Edited by Marta I. Litter, Natalia Quici, and Martín Meichtry

ISBN 978-981-4774-67-3 (Hardcover), 978-981-4669-49-8 (eBook)
www.panstanford.com

analysis are clarified, representing a major challenge for further technology market implementation.

In order to perform a proper assessment of the environmental risk associated with the technological deployment, a number of issues should be comprehensively evaluated and understood: degree of toxicity caused by NPs, toxicity mechanisms involved, fate and transport of NPs (i.e., physicochemical interaction with the media), impact on microbial communities, etc. These features will be discussed in this chapter, reviewing both methodological approaches and results of related research studies performed in this field.

The scope of this section will be particularly limited to risk assessments on ecosystems, excluding potential risks toward human health. In fact, injection of NPs in the subsurface involves an application area where humans are not the main receptors exposed to potential hazards (except workers in charge of manufacturing, injection, and, to a less extent, monitoring), but the microbial communities, indigenous organisms and natural media, as parts of the whole ecosystem. Nevertheless, human exposure should not be ignored, but addressed from a health and safety perspective.

The types of NPs designed and produced for remediation of contaminated sites have evolved and increased over the last years. Although metallic zerovalent iron (ZVI) and iron oxides have been most extensively employed, other developments have incorporated particles with a different nature, such as bimetallic iron particles, incorporating catalytic amounts of palladium, platinum, nickel, copper, and other metals; composite NPs, for example, carbo-iron or iron-zeolites; silver NPs, gold NPs, different metallic combinations, bio-nanomagnetite, and others. According to the highest number of research and field cases dealing with nZVI and iron oxides, most of the toxicity studies have been focused on this type of products, i.e., the findings referred to alternative NPs are scarce and quite often limited to particular laboratory tests and essays, and thus it is difficult to obtain a comprehensive (eco)toxicological behavior of the product(s) [1]. On this basis, this review will be mainly aimed at nZVI, which causes a more stressful environment in terms of biological response, and secondarily at iron oxides, delivered or originated as a result of the oxidation of metallic iron. However, it should be carefully considered that the use of other metals—despite involving catalytic amounts—might introduce new elements of inherently higher toxicity. It should be also pointed out that bare nZVI have proved to achieve poor success on remediation strategies, due to strong adsorption to soil particles, which reduces mobility in the subsurface. Accordingly, most of the produced nZVI incorporate a coating to enhance colloidal stability. In terms of toxicity, most of the coatings are usually biopolymers not affecting the indigenous soil biota; however, nanoparticle surface modification causes differences in their interaction with soil organisms, commonly leading to reduced biological damage.

## 12.2 Geochemistry, Fate, and Transport of Iron Nanoparticles in the Subsurface

### 12.2.1 Iron: Geochemical Features of a Remediation Tool

Iron is the fourth most abundant element in the earth's crust. It exists mainly in two oxidation states: the moderately water-soluble Fe(II) (ferrous iron) and the highly insoluble Fe(III)(ferric iron), stable in oxidant and neutral or alkaline conditions. Iron primary and secondary minerals such as sulfides (mainly pyrite, $FeS_2$), oxides (hematite, $Fe_2O_3$; magnetite, $Fe_3O_4$), hydrated oxides (FeOOH, goethite, limonite), and carbonates (siderite, $FeCO_3$) are within the most habitual minerals in the surficial geosphere; also amorphous forms of iron are important both in natural and industrial processes. However, the native or metal form of iron, i.e., zerovalent iron (ZVI), is rare in natural environments.

The variable oxidation state of iron facilitates its main role in many biogeochemical systems and cycles. Iron plays a significant role in contaminant mobility, sorption, and breakdown due to its activity as electron donor (oxidation of ferrous to ferric iron); in this regard, note that ZVI is an excellent electron donor, acting as a precursor for iron oxides because of the spontaneous oxidation of Fe(0) in aqueous medium to Fe(II) and Fe(III). Because the stability domain of Fe(III) is narrower than that of Fe(II) in most environmental conditions, Fe(II) will be preferentially released when ZVI is added to the subsurface, although both Fe(II) and Fe(III) will be finally oxidized to insoluble species (magnetite, maghemite, lepidocrocite, and goethite). Remarkably, Fe(II) oxidation in the presence of $H_2O_2$ occurs in biological cells resulting in the release of cytotoxic, reactive oxygen species [2].

In terms of heavy metal(loid)s remediation, mobilizing agents can be used to enhance the removal of heavy metal(loid) s through plant uptake (phytoextraction) and soil washing. On the contrary, immobilizing agents can be used to reduce the transfer of metal(loid)s to food chains via plant uptake (phytostabilization) and leaching to groundwater [3]. In this context, iron-based technologies in soil or groundwater remediation can be roughly divided into two (overlapping) groups based on the chemistry involved in the remediation process: (i) technologies that use iron as a sorbent, (co-)precipitant or contaminant immobilizing agent (sorptive/ stabilization technologies); (ii) those which use iron as an electron donor to break down or to convert contaminants into a less toxic or mobile form (reductive technologies). It should be noted, however, that many technologies utilize both processes to a greater or lesser degree (see Table 12.1).

As indicated in Table 12.1, ZVI was decades ago documented as a good electron donor in order to release electrons in aquatic environments. In fact, it has been used as a reactive material in subsurface permeable reactive barriers to degrade groundwater pollutants since the early 1990s [5]. ZVI is very active in transforming halogenated compounds, polychlorinated hydrocarbon pesticides, and other organics [6], and in the immobilization of As [7].

However, nZVI introduce a significantly increased available reactive surface areas (very high surface-area-to-weight ratio) compared to larger sized ZVI particles, thereby resulting in higher reactivity rates than micron-scale ZVI when normalized to mass [8–10], which consequently enhances contaminant degradation reactions. Therefore, the use of nZVI to remediate soil and groundwater has increased within the last years, and the use of nZVI has been effective for degrading organic pollutants such as chlorinated organic compounds, and for immobilizing metal(loid)s specially in groundwater [9, 11–13], but also in soils [14, 15].

**Table 12.1** Selected iron-based remediation technologies [3, 4]

| Technology | Iron-compound implied | Type | Remedial mechanism | Target contaminants |
|---|---|---|---|---|
| Chemical oxidation | Fenton's reagent | In situ/ex situ soil/groundwater | Chemical oxidation | Organics in general |
| Assisted natural remediation | Goethite, ZVI, others | In situ soil | Immobilization | Heavy metal(loid)s |
| Iron-based As removal | Fe oxides, Fe waste | In situ soil | Immobilization and ulterior phytoremediation | As and other heavy metal(loid)s |
| Chemical reduction | Fe(II)-containing solutions | In situ soil/groundwater | Reductive precipitation | Cr (VI), chlorinated hydrocarbons |
| Nanoremediation | nZVI, nano-oxides | In situ soil/groundwater | Dechlorination reduction/immobilization | Heavy metal(loid)s, chlorinated hydrocarbons |
| Permeable reactive barriers | Commercial systems available (many types of iron compounds) | In situ groundwater | Sorption or degradation in barriers | Heavy metal(loid)s, chlorinated hydrocarbons |

### 12.2.2 nZVI in the Subsurface

Naturally occurring NPs appear in the form of colloids, including clay, oxides, and organics. Therefore, it has been suggested that many manufactured metal oxide and other inorganic NPs will exhibit cluster-forming behavior similar to that of natural NPs [16]. In this sense, synthetic NPs released into the environment only represent a minuscule fraction of the total nanoparticulate matter [17].

Key characteristics determining the fate of iron NPs are related to their mobility and reactivity. Mobility and reactivity display an overall inverse relationship so that NPs with high reactivity often have low mobility, because of their propensity to agglomerate and sorb to surfaces [18]. For instance, nZVI have strong attractive interparticle forces; therefore, they tend to agglomerate to micron-sized particles, which have limited mobility in porous media [9, 19]. Under natural environmental conditions, NPs are affected the most significantly by pH, ionic strength, and content of organic matter and clay minerals [20]. It is also well known that natural organic substances are easily adsorbed on the surface of NPs, which may cause a change in the surface electrochemistry of nZVI [21].

To avoid these difficulties (agglomeration, interaction with organic matter, etc.), numerous polymers and other organic coatings have been used to stabilize the nZVI suspension [9, 11], thereby providing a number of potential benefits in the subsoil action of the nZVI particles [19]: (i) decreased filtration losses, (ii) decreased aggregation, (iii) enhanced sweep efficiencies, and (iv) external carbon source for enhanced biodegradation. Nevertheless, in groundwater there may be strong interactions of the stabilizers with humic acids, causing an increase in agglomeration, followed by sedimentation of NPs [21]. On the whole, the modification of nZVI, depending on the kind of target contaminants or the geochemical conditions, can behave differently and exert diverse effects on the environment [22].

nZVI within the subsurface are generally believed to be transformed (aging) to various iron oxyhydroxides, similar in composition to naturally occurring iron-based minerals, which, even with an inner core of ZVI, would be expected to essentially behave as iron hydroxides. Depending on general groundwater redox conditions, being either oxic or anoxic, iron hydroxides potentially form passivating layers on the nanoiron particle [18]. Depending on the matrix, increased agglomeration of particles may lead to the clogging of soil pores and to the inhibition of transport of the nZVI [23] through the accumulation of nZVI and of the nZVI-contaminant complex in the soil or on the bottom of water reservoirs. On the other hand, all modifications leading to a reduction in aggregation are basically favorable to nZVI scattering in the environment (renegade NPs), thus affecting other areas not directly connected with the source of the contaminant [23]. In addition, some authors have indicated that nZVI could remain active for longer periods, in the order of 6 months [24–26]. All things together, there is much debate over whether or not nZVI can be effectively transported by groundwater to impact an entire contamination plume, and if increasing mobility too much will add concerns about particles moving offsite and becoming a toxicological issue (renegade NPs). Therefore, a detailed conceptual site model (CSM) is needed in order to address the possible risk from renegade NPs, in addition to the CSM, which should already have been developed for the contamination problem at the site.

As a conclusion, the behavior of nZVI within the subsoil is site specific because it is highly affected by local geochemistry. Some of the issues affecting fate and transport include the contaminant concentration, particle agglomeration, age of the nZVI particles, soil matrix, ionic strength of the groundwater, hydraulic properties of the aquifer, depth to the water table, presence of organic matter, and other geochemical properties such as pH, DO, and oxidation reduction potential (ORP).

### 12.2.3 Fate and Transport of nZVI: Evaluation Methods

As described above, many diverse factors and related restrictions mainly depending on raw properties (e.g., particle size, surface charge, bulk composition) affect nZVI behavior. These features change with time due to the reactivity of the nZVI and their interaction with the environment. Therefore, it is not easy to evaluate the behavior of the nZVI in a site-specific application. In addition, whenever a new product is used, it seems necessary to thoroughly compare the characterization results for commercial products with the manufacturers' stipulations.

**Table 12.2** Direct and indirect characterization methods of the presence and activity of nZVIs in the subsoil [19]

| Methods | | Remarks |
|---|---|---|
| Direct | Microscopy and spectroscopy | Mainly electronic microscopy and X-ray methods [27] |
| | Color and absorptivity | UV-Vis absorbance at 800 nm recommended [28] |
| | Total iron | Usually overestimates delivered nZVI |
| | Total dissolved iron | Fe (II)–Fe (III) considered insoluble, Recommended coupled to total iron (problem: nZVI usually smaller than filters' pore size) |
| Indirect | pH | pH grows 2–3 units in the vicinity of injection wells while show little changes in monitoring wells [29] |
| | Dissolved oxygen | Complete depletion of DO usually observed near injection wells |
| | Conservative tracers | Indication of hydrogeological response to fluid injection<br>Classic tracers may interact with nZVI<br>Carrier materials could be good tracers |
| | Dissolved hydrogen | Not habitual, useful to evaluate ZVI content in commercial products |
| | Redox active probes | Under development [30] |
| | ORP | Highly negative ORP values expected although affected by deposition of nZVI particles and $H_2$ concentration<br>Widely used, complementary to iron analysis<br>ORP conventional electrodes need cleaning efforts prior and during the measurements |
| | Geophysical methods | Under development [31]<br>Options: magnetic, electrical, electromagnetic methods |

**Note**: Direct methods can directly detect ZVI or its oxidation derivatives; indirect methods are based on changes of water chemistry or electrometric properties of impacted environments.

In this context, sometimes the characterization of a specific formulation of nZVIs by themselves is the main object of research [27], as this has an obvious impact on the difficult matter of the prediction of NPs fate. However, in the background of this chapter, a review by Shi et al. [19] has recently introduced a useful sorting of direct and indirect characterization methods (Table 12.2).

Together with the various possibilities for monitoring purposes, it should be considered the reduction of the concentration of contaminants by itself as a complementary monitoring method, i.e., both an increased yield in the reductive dehalogenation of chlorinated organics or the sequestration of heavy metal(loid)s. Moreover, collateral effects as, for instance, the increased dissolved $H_2$ and $Fe^{2+}$, and the microbial metabolism of organic amendments can be observed and also favor bioremediation [19]. These effects open new possibilities to combine different technologies.

## 12.3 (Eco)toxicity and Methods of Characterization

In the recent years, the efforts on appraisal of the ecological relevance of using engineered NPs have been focused on two different research lines. On one hand, attempts for a standardized methodological approach have been introduced aiming at a common scheme for the toxicological characterization of any engineered material, where iron-based NPs are encompassed. The main goal of these protocols is to be useful as a predictive tool for pre-assessing the toxic response of a certain nanomaterial, prior to real application. On the other hand, from a more scientific standpoint, different research projects have been developed for the detailed study of the toxicity related to the application of nZVI in groundwater remediation processes. In this case, the methods have been more heterogeneous, accommodated to the study of particular interests of the research: response of different species (bacteria, fungi, algae, crustaceans, etc.), processes causing toxicity, resistance mechanisms developed by organisms, parameters influencing toxicity, impact(s) on microbial community structure, or any other target.

From the standardization perspective, international scientific committees have been organized with the intention of stating general guidelines for the direct toxicological assessment and physicochemical characterization (supporting toxicology studies) of any kind of engineered nanoscale materials [32, 33]. It is a wide-ranging approach covering many different products and materials, and based on classical screening methods for toxicology assessment: algae growth inhibition test, crustacean (*Daphnia sp.)* acute immobilization test, water quality determination of the toxic effect of sediment and soil samples on growth, fertility, and reproduction of nematoda (*Caenorhabditis elegans*). However, these guidelines have been compiled merely to aid in the preliminary stages of appraisal for nanomaterials, and not as a methodology for the (legal) approval of commercialization, which should be fatherly demonstrated according to in-depth tests and conclusions. Moreover, it has sometimes been questioned whether classical procedures lead to proper hazard identification when testing NPs, considering that physical effects, such as aggregation, agglomeration, sedimentation, shading and others, might mask the actual exposure and response [34].

Since nanoscale size might lead to new feature and properties, and subsequently environmental impacts in the short to long term, these products, nZVI in this case, require individual and customized studies for an extensive toxicological characterization. (Eco)toxicology of nZVI has been widely investigated for the last years through many different laboratory in vitro tests and studies, providing an exhaustive knowledge and elucidating its potential effects on the surrounding media [2, 35]. In terms of degree of organisms complexity, different (eco)toxicity studies have been performed. As stated below, in vitro tests have shown a bactericidal effect of nZVI to a number of species, for instance, certain aqueous cultures of *Escherichia coli* or *Bacillus nealsonii* [36–38]. Aquatic invertebrates are mainly assessed through *Daphnia sp* tests. Not many studies on the effect of nZVI to *Daphnia magna* have been published, but it is highlighted the significant relevance of this exposure, even leading in some cases to the death of the organism. As far as studied, many different effects have been measured depending on the commercial product [39]. Although soil ecotoxicity (earthworms' tests) has not been very extensively studied, some authors have published relevant conclusions, remarking potential negative effects on soil invertebrates [40]. Some phytotoxicity studies have also been released, pointing out not significant effect except for exposures at very high concentration of iron, and in this case, a clear dependence on aging stage of the nZVI is demonstrated [41].

### 12.3.1 Effects on Microorganisms

Microorganisms, including algae, are at the bottom of the aquatic food chain. Results and conclusions from research in this field have drawn attention to several issues, as described below. Measurements of cell viability, cell growth, cell integrity, or biological activity in various microbial species exposed to nZVI—ranging from 1 to 10,000 mg/L, and 5 min to 42 days—showed strong to severe negative effect in most of the cases, while only a few studies reported no bactericidal effects [2, and references therein]. On the contrary, fungi species presented a significantly larger tolerance [42].

Most of the studies agreed on addressing cell membrane disruption and oxidative stress through the generation of divalent iron, and reactive oxygen species as main mechanisms involved in cytotoxicity. Denaturalization of lipopolysaccharides and electron and ionic membrane transport proteins due to reducing properties of nZVI can affect membrane permeability and facilitate $Fe^{2+}$ entrance into the cell [43]. Ferrous iron could interact with mitochondrial $H_2O_2$ originating highly reactive oxygen species and causing cell destruction by oxidative stress. Other involved mechanisms could be complexation of iron with lipoteichoic acids conforming cell wall in Gram+ bacteria, or the precipitation of iron oxides caused by some anionic structures present in certain bacteria. Both processes would interfere in the cellular uptake of nutrients [42, 44].

The extent of iron oxide NPs sorption on the surface of the microorganisms appears to be related to electrostatic interaction forces (i.e., the zeta potential) and determines the killing effect. Interestingly, bacteria seem to be more sensitive to the surface-covering effect than eukaryotic microorganisms, as showed in *Escherichia coli* and *Saccharomyces cerevisiae* (Fig. 12.1) [45]. It has been suggested that the NPs could be internalized by endocytosis in the yeast cells, to protect this microorganism from the toxic effects of the NPs [45].

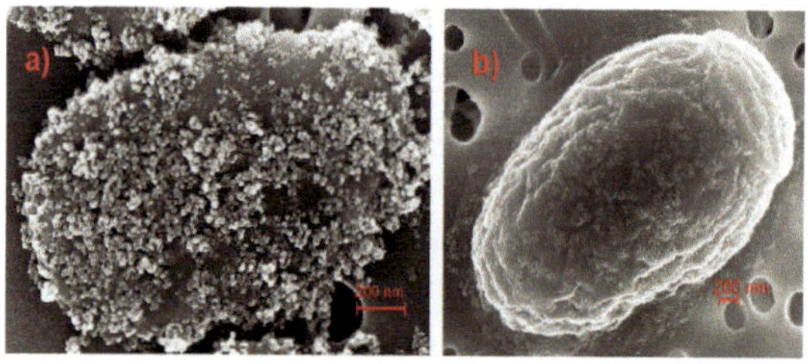

**Figure 12.1** Scanning electron microscopy observation of attached FeOx NPs to the surface of *E. coli* (a) and *S. cerevisiae* (b) cells. Reprinted from Ref. [45], Copyright 2010, with permission from Elsevier.

The rigidness of the fungi cell wall seems to act also as a protection against nZVI adverse effects [42, 46]. Toxicity mechanisms associated to the formation of highly reactive oxygen species have also been demonstrated, which triggers damage on lipids, proteins, and nucleic acids, and accordingly, resulting in cellular death [36]. These mechanisms have even been characterized kinetically, as a first-order reaction defined by a constant, which involves processes of transport, attachment, and reaction [47].

The sensitivity or tolerance to nZVI differs between the bacterial genera or even between different species. Several *Klebsiella* species are resistant to high doses of nZVI [37, 48] and the higher resistance of *Bacillus subtilis* to $Fe^{2+}$ than *E. coli* or *Pseudomonas fluorescens* in in-vitro experiments is conceivably due to its higher negative charge that gives rise to a higher nZVI electrostatic repulsion (see above) and as a consequence to a reduced toxicity [42, 49]. However, $Fe^{3+}$ was shown to have stronger inactivation effects on *B. subtilis* than $Fe^{2+}$, especially at higher dosages [50]. As well as the mechanisms of cellular damage are being elucidated, the resistance responses are also revealed. Some results suggest that one of the protective strategies developed by the organisms is the formation of spores, preventing bacteria from direct contact to iron NPs [51]. Another biological response might be the downregulation of membrane, and periplasmic proteins (porins, iron ABC transporter, siderophore receptor) involve in iron uptake [52, 53]. Remarkably, bacterial cells in stationary growth are less sensitive to nZVI compared with those in lag or exponential phase, being this likely related to the higher expression of several genes involved in cell stress response during stationary growth [2, 54]. Otherwise, schemes based on the production of extracellular polymeric substances have been suggested, aiming at bonding to nZVI and thus constraining interaction between iron and bacteria and, therefore, its harmful effects [55, 56]. Besides the aforementioned mechanisms preventing iron uptake by cells, other responses are developed for minimizing oxidative stress. The high tolerance to nZVI of some bacteria as *Klebsiella oxytoca* seems to be related with adaptative stress response, involving tryptophanase and indole as a signal molecule of environmental disturbances [48]. In other cases, the defensive response has been linked to the maintenance of cellular homeostasis and the upregulation of proteins involved in reducing intracellular oxidative stress, as catalase, a key component for detoxification of reactive oxygen species as $H_2O_2$ [36, 51–53].

## 12.3.2 Parameters Affecting Toxicity

Literature shows a comprehensive set of studies dealing with the investigation of parameters affecting toxicity caused by nZVI. Increasing NPs dosage seems to be related to an increasing toxic response [36, 37, 55, 56], unless a high concentration of NPs induces iron aggregation leading to less colloidal stability, higher sedimentation rates, and consequently, milder toxic effects [57–59]. Exposure time is also another key factor on toxic effects, observing in most of the cases an unmistakable response in the early stages of the test, and minimizing—or even disappearing—in the long term, likely related to the oxidation of nZVI [52, 60]. Therefore, (an)aerobic conditions strongly influence nZVI oxidation rate and pathways, controlling the proportion of mineralogical forms of iron oxides, and thus the related harmful effects. Thus, the bactericidal effect in the presence of oxygen is significantly lower than under de-aerated conditions, and complete oxidation of nZVI virtually eliminates the toxic effects [43, 61]. In deep underground aquifers where oxygen is limited, Fe(0) can be converted to magnetite ($Fe_3O_4$) as the end product of nZVI oxidation, which has also much lesser bactericidal effect [36, 62, 63]. As previously stated, nZVI are commonly modified with surface coatings in order to enhance colloidal stability and thus migration in the subsurface. Laboratory tests have suggested that these surface modifiers constraint the adhesion of iron particles to the cells, likely by electrosteric repulsions, reducing the exposure rate and toxic response [61]. In some cases, the polymeric coatings not only minimize physical contact but also act as hydroxyl radical scavengers, mitigating oxidative stress [64]. Similar effects have showed natural organic matter and humic acids present in the soil, causing an attenuation of toxicity by adsorption to nZVI and competing with nZVI–cell interaction [49, 61]. These attenuation effects have interesting potential applications, as discussed below.

### 12.3.3 Effects on Environmental Microbial Communities

An interesting environmental aspect related with the nZVI chemistry concerns to the fast release of $H_2$ by nZVI under anaerobic conditions. This could potentially favor the growth of some microorganisms that degrade chlorinated solvents, as hydrogenotrophic methanogens, which convert $H_2$ and $CO_2$ to $CH_4$, or homoacetogenic bacteria [2, and references therein]. Sulfate-reducing bacteria that eliminate heavy metals through the formation of metal-sulfide precipitates or the denitrifying bacteria that eliminate nitrogen are also favored [2, 38]. Positive effects of nZVI have been also described for anammox bacteria and have been attributed to the removal of dissolved oxygen and the concomitant nitrate reduction to ammonium, and to the stimulation of the secretion of EPS by anammox bacteria, which favors in this way their stability within the bioreactor through their aggregation [65]. Based on these findings, combining the microbial activities with nZVI treatments (either concurrently or sequentially) might be a more efficient alternative in certain cases to remediate contaminated sites [49, 66]. In addition, the attenuation of the undesirable effects of nZVI on bacteria with nZVI coated with some compounds such as polyaspartate or other natural organic compounds such as humic acid, mentioned above, does not significantly alter $H_2$ production and, therefore, stimulation of methanogenic and sulfate-reducing populations persists [38, 49]. Clearly, one of the critical issues needed to be analyzed in the future in order to effectively implement those findings in situ is to investigate how the syntrophic or competitive relationships conceivably present within the microbial community in natural environments influence the potential stimulatory effects of nZVI described above [2].

Whereas the in vitro tests on individual bacteria have shown clear inhibitory effects, when trying to examine the effect of nZVI on microbial populations in microcosm and/or polluted environment, the bactericidal effects may be not so obvious and would be mainly detected as changes in the taxonomic and, therefore, functional composition of the indigenous microbial communities [2, 38]. How this affect the critical microbial functions involved in global nutrient cycling is a critical objective, as the bacterial species are functionally redundant, and an apparent shift in composition does not necessarily means functional loss [2]. Moreover, the possibility that the microbial communities affected by nZVI could revert to the initial status after the termination of the process [38, 66] must also be analyzed in field studies. It would be desirable to analyze the biogeochemical cycles of interest in the particular environment studied, before, during, and after the nZVI treatment. In this context, the combination of traditional microbial ecology methods and the new efficient and affordable molecular methodologies, including OMICS approaches [67–70], will provide suitable tools to address these questions. Also because the physicochemical environmental conditions modulate nZVI toxicity, a better understanding of them is of primary importance for evaluating the ecological consequences of the use of nZVI in the specific site. It is clear that the analysis of the interaction of nZVI with microbial populations as a whole and with the environmental matrix is a demanding challenge that must be addressed through exhaustive studies in the future, before being able to predict, from the effects observed in the laboratory, the effects of nZVI on in situ environmental remediation projects.

The currently available information reviewed in this chapter shows a growing interest in the recent years in analyzing the effect of nZVI on microbial populations that may be considered useful toxicological indicators; however, additional efforts are needed in the analysis of invertebrates and, in general, in the study of multicellular eukaryotic organisms, where information, despite the advances made, does not present the same equilibrium at present. In any case, the methods and studies already available represent an incipient useful predictive tool to assess the toxicological response of the injectable NPs.

## 12.4 Ecological Risk Assessment

Due to the uncertainties about environmental behavior of nZVI pointed above, appraisals based on an integral assessment of the ecological stress on the entire ecosystems, which demands large sets of toxicity data (e.g., the potentially affected fraction of species), cannot be still properly applied. Alternatively, the classical methodology for environmental risk assessment can be employed, on the basis of a conceptual site model (i.e., the scenario where the injection of NPs takes place) defined and characterized by the sources, pathways, and potential receptors of nZVI.

Based on this approach, a case-by-case scheme should be outlined, compiling information about different aspects: type and nature of NPs selected for injection, injection method (gravity, direct push, or controlled fracking), presence of surface modifiers of NPs ((bio)degradability and biocompatibility), physicochemical features of NPs (reductive potential, solubility, particle size, etc.), properties of the aquifer (presence of fractures, groundwater flow, permeability, porosity, ionic background, etc.), toxic behavior, pathways, main receptors of the ecosystem (bacteria, nematodes, plants, invertebrates, groundwater, surface water, etc.), distance to potential receptors, and many other site-specific issues that should be taken into consideration.

Multiple factors are involved in the environmental impact of using iron NPs, within a complex and intricate biogeochemical system, with many potential roles depending on the particular conditions of the remediation site and plan. Nevertheless, some general ideas can be outlined in terms of ecological impact related to the deployment of iron NPs. nZVI are potentially more stressful than iron oxides due to their intrinsic higher reactivity, and hence these might trigger more severe (bio)chemical reactions regarding ecological impact. Most of the commonly used surface modifiers are biologically compatible, not affecting the ecosystems, or even decreasing the effect of the iron itself. Although the aquifer characteristics include a large number of parameters influencing NPs dispersion, it seems to be demonstrated that their mobility is pretty limited and they only migrate a few meters from the injection point; thus meaning the area of influence in terms of risk, related to potential receptors, is rarely larger than 20 m and commonly circumscribed to a couple of meters.

However, because of the ambiguity inherent in the general concepts outlined here, a case-based approach as proposed above will definitely be needed for a proper risk assessment of the technology application.

## 12.5 Conclusion

Comparatively, the use of iron-based NPs with remedial purposes involves, a priori, a lesser environmental risk rather than other engineered NPs, for instance titanium, silver, or gold. This statement is based on the low inherent toxicity of iron species, the quick tendency to aggregate in the application media, and the small ability for mobilization and dispersion in porous aquifers. Nonetheless, different studies have pointed to toxicological effects on organisms or terrestrial communities, which suggest the potential existence of some ecological effects, although their relevance should be more carefully assessed.

Even though many efforts have been conducted during the last years for a better understanding of the chemical properties of iron NPs, there is still a lack of knowledge regarding some aspects, particularly related to their environmental impact. Most of the studies have been focused on the examination of toxic effects for particular organisms or microbial communities so far, omitting the potential stress of the ecosystem as a whole and the disturbance of its functionalities and services. Accordingly, more research is still required aiming a wider picture of the ecological significance of this technology. At the same time, it is also demanded the development, enhancement, or adaptation of existing and/or new methodologies and tools for an integral ecological diagnosis of the ecosystems exposed to nZVI-based treatment processes.

Improvements in the environmental assessment of nZVI in situ injection will be helpful in two different levels. First, from a methodological point of view, a more comprehensive understanding of the biogeochemical mechanisms involved in NPs–media interaction will lead to a better remediation planning, minimizing undesired collateral effects; for instance, through the development of predictive models. On the other hand, extensive knowledge of the technology and the potential consequences of its deployment will help to overcome the predictable barriers related to environmental objections of regulators and public bodies. As a result, a broader social audience will be better able to accept these new remediation practices.

## References


1. Zhao, X., Liu, W., Cai, Z., Han, B., Qian, T., and Zhao, D. (2016). An overview of preparation and applications of stabilized zero-valent iron nanoparticles for soil and groundwater remediation, *Water Res.*, **100**, pp. 245–266.
2. Lefevre, E., Bossa, N., Wiesner, M. R., and Gunsch, C. K. (2016). A review of the environmental implications of in situ remediation by nanoscale zero valent iron (nZVI): Behavior, transport and impacts on microbial communities, *Sci. Total Environ.*, **565**, pp. 889–901.
3. Bolan, N. S., Kunhikrishnan, A., Thangarajan, R., Kumpiene, J., Park, J., Makino, T., Kirkham, M. B., and Scheckel, K. (2014). Remediation of heavy metal(loid)s contaminated soils: To mobilize or to immobilize? *J. Hazard. Mater.*, **266** , pp. 141–166.
4. Cundy, A. B., Hopkinson, L., and Whitby, R. L. (2008). Use of iron-based technologies in contaminated land and groundwater remediation: A review, *Sci. Total Environ.*, **400**, pp. 42–51.
5. Henderson, A. D. and Demond, A. H. (2007). Long-term performance of zero-valent iron permeable reactive barriers: A critical review, *Environ. Eng. Sci.*, **24**, pp. 401–423.
6. Fu, F., Dionysiou, D. D., and Liu, H. (2014). The use of zero-valent iron for groundwater remediation and wastewater treatment: A review, *J. Hazard. Mater.*, **267**, pp. 194–205.
7. Ungureanu, G., Santos, S., Boaventura, R., and Botelho, C. (2015). Arsenic and antimony in water and wastewater: Overview of removal techniques with special reference to latest advances in adsorption, *J. Environ. Manag.*, **151**, pp. 326–342.
8. Kanel, S. R., Manning, B., Charlet, L., and Choi, H. (2005). Removal of arsenic (III) from groundwater by nanoscale zero-valent iron, *Environ. Sci. Technol.*, **39**, pp. 1291–1298.
9. O'Carroll, D., Sleep, B., Krol, M., Boparai, H., and Kocur, C. (2013). Nanoscale zero valent iron and bimetallic particles for contaminated site remediation, *Adv. Water Resour.*, **51**, pp. 104–122.

10. Statham, T. M., Mason, L. R., Mumford, K. A., and Stevens, G. W. (2015). The specific reactive surface area of granular zero-valent iron in metal contaminant removal: Column experiments and modelling, *Water Res.*, **77**, pp. 24–34.

11. Karn, B., Kuiken, T., and Otto, M. (2009). Nanotechnology and in situ remediation: A review of the benefits and potential risks, *Environ. Health Persp.*, **117**, pp. 1823–1831.

12. Li, X., Elliot, D. W., and Zhang, W. (2006). Zero-valent iron nanoparticles for abatement of environmental pollutants: Materials and engineering aspects, *Crit. Rev. Solid State*, **31**, pp. 111–122.

13. Li, X. and Zhang, W. (2007). Sequestration of metal cations with zerovalent iron nanoparticles. A study with high resolution X-ray photoelectron spectroscopy (HRXPS), *J. Phys. Chem. C*, **111**, pp. 6939–6946.

14. Gil-Díaz, M., Alonso, J., Rodríguez-Valdés, E., Pinilla, P., and Lobo, M. C. (2014). Reducing the mobility of arsenic in brownfield soil using stabilised zero-valent iron nanoparticles, *J. Environ. Sci. Heal. A Tox. Hazard. Subst. Environ. Eng.*, **49**, pp. 1361–1369.

15. Gil-Díaz, M., Diez-Pascual, S., González, A., Alonso, J., Rodríguez-Valdés, E., Gallego, J. R., and Lobo, M. C. (2016). A nanoremediation strategy for the recovery of an as-polluted soil, *Chemosphere*, **149**, pp. 137–145.

16. Gilbert, B., Lu, G., and Kim, C. S. (2007). Stable cluster formation in aqueous suspensions of iron oxyhydroxide nanoparticles, *J. Colloid Interf. Sci.*, **313**, pp. 152–159.

17. Wiggington, N. S., Haus, K. L., and Hochella, M. F. Jr. (2007). Aquatic environmental nanoparticles, *J. Environ. Monitor*, **9**, pp. 1306–1316.

18. Bardos, P., Bone, B., Elliott, D. W., Hartog, N., Henstock, J. E., and Nathanail, C. P. (2011). *CL:AIRE: A Risk/Benefit Approach to the Application of Iron Nanoparticles*, Final Report. Published by DEFRA, ISBN: DefraCB440.

19. Shi, Z., Fan, D., Johnson, R. L., Tratnyek, P. G., Nurmi, J. T., Wu, Y., and Williams, K. H. (2015). Methods for characterizing the fate and effects of nano zerovalent iron during groundwater remediation, *J. Contam. Hydrol.*, **181**, pp. 17–35.

20. Jung, B., O'Carroll, D., and Sleep, B. (2014). The influence of humic acid and clay content on the transport of polymer-coated iron nanoparticles through sand, *Sci. Total. Environ.*, **496**, pp. 155–164.

21. Dong, H. and Lo, I. M. (2013). Influence of humic acid on the colloidal stability of surface-modified nano zero-valent iron, *Water Res.*, **47**, pp. 419–427.

22. Stefaniuk, M., Oleszczu, P., and Yong, S. (2016). Review on nano zerovalent iron (nZVI): From synthesis to environmental applications, *Chem. Eng. J.*, **287**, pp. 618–632.

23. Keane, E. (2010). Fate, transport and toxicity of nanoscale zero-valent iron (nZVI) used during superfund remediation (U.S. Environmental Protection Agency, Washington DC).

24. Adeleye, A. S., Keller, A. A., Miller, R. J., and Lenihan, H. S. (2013). Persistence of commercial nanoscaled zero-valent iron (nZVI) and by-products, *J. Nanopart. Res.*, **15**, pp. 1–18.

25. Liu, Y. Q. and Lowry, G. V. (2006). Effect of particle age (Fe-O content) and solution pH on NZVI reactivity: H-2 evolution and TCE dechlorination, *Environ. Sci. Technol.*, **40**, pp. 6085–6090.

26. Phenrat, T., Long, T. C., Lowry, G. V., and Veronise, B. (2009). Partial oxidation (aging) and surface modification decrease the toxicity of nanosized zerovalent iron, *Environ. Sci. Technol.*, **43**, pp. 195–200.

27. Chekli, L., Bayatsarmadi, B., Sekine, R., Sarkar, B., Shen, A. M., Scheckel, K. G., Scheckel, K. G., Skinner, W., Naidu, R., Shon, H. K., Lombi, E., and Donner, E. (2016). Analytical characterisation of nanoscale zero-valent iron: A methodological review, *Anal. Chim. Acta*, **903**, pp. 13–35.

28. Johnson, R. L., Nurmi, J. T., Johnson, G. S. O., Fan, D. M., Johnson, R. L. O., Shi, Z. Q., Salter-Blanc, A. J., Tratnyek, P. G., and Lowry, G. V. (2013). Field-scale transport and transformation of carboxymethylcellulose-stabilized nano zero-valent iron, *Environ. Sci. Technol.*, **47**, pp. 1573–1580.

29. Wei, Y. T., Wu, S. C., Chou, C. M., Che, C. H., Tsai, S. M., and Lien, H. L. (2010). Influence of nanoscale zero-valent iron on geochemical properties of groundwater and vinyl chloride degradation: A field case study, *Water Res.*, **44**, pp. 131–140.

30. Jones, B. D. and Ingle Jr., J. D. (2005). Evaluation of redox indicators for determining sulfate-reducing and dechlorinating conditions, *Water Res.*, **39**, pp. 4343–4354.

31. Joyce, R. A., Glaser, D. R., Werkema, D. D., and Atekwana, E. A. (2012). Spectral induced polarization response to nanoparticles in a saturated sand matrix, *J. Appl. Geophys.*, **77**, pp. 63–71.

32. ISO/TR 13014. (2012). Nanotechnologies: Guidance on physico-chemical characterization of engineered nanoscale materials for toxicologic assessment.

33. ISO/TR 16197. (2014). Nanotechnologies: Compilation and description of toxicological screening methods for manufactured nanomaterials.

34. Hjorth, R., Coutris, C., Nguyen, N., Sevcu, A., Baun, A., and Joner, E. (2015). Ecotoxicity testing of nanoparticles for remediation of contaminated soil and groundwater, *Proc. 13th International UFZ-Deltares Conference on Sustainable Use and Management of Soil, Sediment and Water Resources*, Copenhagen, Denmark.

35. Ahamed, N. (2014). Ecotoxicity concert of nano zero-valent iron particles: A review, *J. Crit. Rev.*, **1**, pp. 36–39.

36. Auffan, M., Achouak, W., Rose, J., Roncato, M.-A., Chanéac, C., Waite, D. T., Masion, A., Woicik, J. C., Wiesner, M. R., and Bottero, J.-Y. (2008). Relation between the redox state of iron-based nanoparticles and their cytotoxicity toward *Escherichia coli*, *Environ. Sci. Technol.*, **42**, pp. 6730–6735.

37. Fajardo, C., Ortiz, L. T., Rodriguez-Membibre, M. L., Nande, M., Lobo, M. C., and Martin, M. (2012). Assessing the impact of zero-valent iron (ZVI) nanotechnology on soil microbial structure and functionality: A molecular approach, *Chemosphere*, **86**, pp. 802–808.

38. Kirschling, T. L., Gregory, K. B., Minkley, E. G., Lowry, G. V., and Tilton, R. D. (2010). Impact of nanoscale zero valent iron on geochemistry and microbial populations in trichloroethylene contaminated aquifer materials, *Environ. Sci. Technol.*, **44**, pp. 3474–3480.

39. Keller, A. A., Garner, K., Miller, R. J., and Lenihan, H. S. (2012). Toxicity of nano-zero valent iron to freshwater and marine organisms, *PLoS ONE*, **7**(8), e43983.

40. El-Temsah, Y. S. and Joner, E. J. (2012). Ecotoxicological effects on earthworms of fresh and aged nano-sized zero-valent iron (nZVI) in soil, *Chemosphere*, **89**, pp. 76–82.

41. Wang, J., Fang, Z., Cheng, W., Tsang, P. E., and Zhao, D. (2016). Ageing decreases the phytotoxicity of zero-valent iron nanoparticles in soil cultivated with Oryza sativa, *Ecotoxicology*, **25**(6), pp. 1202–1210.

42. Diao, M. and Yao, M. (2009). Use of zero-valent iron nanoparticles in inactivating microbes, *Water Res.*, **43**, pp. 5243–5251.

43. Lee, C., Kim, J. Y., Lee, W. I., Nelson, K. L., Yoon, J., and Sedlak, D. L. (2008). Bactericidal effect of zero-valent iron nanoparticles on *Escherichia coli*, *Environ. Sci. Technol.*, **42**, pp. 4927–4933.

44. Chen, Q., Gao, M., Li, J., Shen, F., Wu, Y., Xu, Z., and Yao, M. (2012). Inactivation and magnetic separation of bacteria from liquid suspensions using electrosprayed and nonelectrosprayed nZVI

particles: Observations and mechanisms, *Environ. Sci. Technol.*, **46**, pp. 2360–2367.

45. Schwegmann, H., Feitz, A. J., and Frimmel, F. H. (2010). Influence of the zeta potential on the sorption and toxicity of iron oxide nanoparticles on *S. cerevisiae* and *E. coli*, *J. Colloid Interf. Sci.*, **347**, pp. 43–48.
46. Otero-González, L., García-Saucedo, C., Field, J. A., and Sierra-Álvarez, R., (2013). Toxicity of $TiO_2$, $ZrO_2$, $Fe^0$, $Fe_2O_3$, and $Mn_2O_3$ nanoparticles to the yeast, *Saccharomyces cerevisiae*, *Chemosphere*, **93**, pp. 1201–1206.
47. Barton, L. E., Auffan, M., Olivi, L., Bottero, J.-Y., and Wiesner, M. R. (2015). Heteroaggregation, transformation and fate of $CeO_2$ nanoparticles in wastewater treatment, *Environ. Pollut.*, **203**, pp. 122–129.
48. Saccà, M. L., Fajardo, C., Nande, M., and Martín, M. (2013). Effects of nano zero-valent iron on *Klebsiella oxytoca* and stress response, *Microb. Ecol.*, **66**, pp. 806–812.
49. Chen, J., Xiu, Z., Lowry, G. V., and Alvarez, P. J. J. (2011). Effect of natural organic matter on toxicity and reactivity of nano-scale zero-valent iron, *Water Res.*, **45**, pp. 1995–2001.
50. Chen, Q., Li, J., Wu, Y., Shen, F., and Yao, M. (2013). Biological responses of Gram-positive and Gram-negative bacteria to nZVI ($Fe^0$), $Fe^{2+}$ and $Fe^{3+}$, *RSC Adv.*, **3**, pp. 13835–13842.
51. Fajardo, C., Saccà, M. L., Martinez-Gomariz, M., Costa, G., Nande, M., and Martin, M. (2013). Transcriptional and proteomic stress responses of a soil bacterium *Bacillus cereus* to nanosized zero-valent iron (nZVI) particles, *Chemosphere*, **93**, pp. 1077–1083.
52. Saccà, M. L., Fajardo, C., Costa, G., Lobo, C., Nande, M., and Martin, M. (2014). Integrating classical and molecular approaches to evaluate the impact of nanosized zero-valent iron (nZVI) on soil organisms, *Chemosphere*, **104**, pp. 184–189.
53. Saccà, M. L., Fajardo, C., Martinez-Gomariz, M., Costa, G., and Nande, M. (2014). Molecular stress responses to nano-sized zero-valent iron (nZVI) particles in the soil bacterium *Pseudomonas stutzeri*, *PLoS ONE*, **9**(2), e89677.
54. Chaithawiwat, K., Vangnai, A., McEvoy, J. M., Pruess, B., Krajangpan, S., and Khan, E. (2016). Impact of nanoscale zero valent iron on bacteria is growth phase dependent, *Chemosphere*, **144**, pp. 352–359.
55. Kim, H. S., Kim, T., Ahn, J.-Y., Hwang, K.-Y., Park, J.-Y., Lim, T.-T., and Hwang, I. (2012). Aging characteristics and reactivity of two types of nanoscale

zero-valent iron particles (FeBH and $FeH_2$) in nitrate reduction, *Chem. Eng. J.*, **197**, pp. 16–23.

56. Le, T. T., Murugesan, K., Kim, E.-J., and Chang, Y.-S. (2014). Effects of inorganic nanoparticles on viability and catabolic activities of *Agrobacterium* sp. PH-08 during biodegradation of dibenzofuran, *Biodegradation*, **25**, pp. 655–668.

57. Lowry, G. V. and Casman, E. A. (2009). *Nanomaterials: Risks and Benefits*, eds. Linkov, I., Steevens, J., "Nanomaterial transport, transformation, and fate in the environment", (Springer Netherlands, Dordrecht), pp. 125–137.

58. Phenrat, T., Saleh, N., Sirk, K., Tilton, R. D., and Lowry, G. V. (2007). Aggregation and sedimentation of aqueous nanoscale zerovalent iron dispersions, *Environ. Sci. Technol.*, **41**, pp. 284–290.

59. Tratnyek, P. G. and Johnson, R. L. (2006). Nanotechnologies for environmental cleanup, *Nano Today*, **1**, pp. 44–48.

60. An, Y., Li, T., Jin, Z., Dong, M., Xia, H., and Wang, X. (2010). Effect of bimetallic and polymer-coated Fe nanoparticles on biological denitrification, *Bioresource Technol.*, **101**, pp. 9825–9828.

61. Li, Z., Greden, K., Alvarez, P. J. J., Gregory, K. B., and Lowry, G. V. (2010). Adsorbed polymer and NOM limits adhesion and toxicity of nanoscale zerovalent iron to *E. coli*, *Environ. Sci. Technol.*, **44**, pp. 3462–3467.

62. Auffan, M., Rose, J., Wiesner, M. R., and Bottero, J.-Y. (2009). Chemical stability of metallic nanoparticles: A parameter controlling their potential cellular toxicity in vitro, *Environ. Pollut.*, **157**, pp. 1127–1133.

63. Grieger, K. D., Fjordøge, A., Hartmann, N. B., Eriksson, E., Bjerg, P. L., and Baun, A. (2010). Environmental benefits and risks of zero-valent iron particles (nZVI) for in situ remediation: Risk mitigation or trade-off? *J. Contam. Hydrol.*, **118**, pp. 165–183.

64. Zhou, L., Thanh, T. L., Gong, J., Kim, J.-H., Kim, E.-J., and Chang, Y.-S. (2014). Carboxymethyl cellulose coating decreases toxicity and oxidizing capacity of nanoscale zerovalent iron, *Chemosphere*, **104**, pp. 155–161.

65. Ren, L.-F., Ni, S.-Q., Liu, C., Liang, S., Zhang, B., Kong, Q., and Guo, N. (2015). Effect of zero-valent iron on the start-up performance of anaerobic ammonium oxidation (anammox) process, *Environ. Sci. Pollut. R.*, **22**, pp. 2925–2934.

66. Xiu, Z.-M., Jin, Z.-H., Li, T.-L., Mahendra, S., Lowry, G. V., and Alvarez, P. J. J. (2010). Effects of nano-scale zero-valent iron particles on a mixed

culture dechlorinating trichloroethylene, *Bioresource Technol.*, **101**, pp. 1141–1146.

67. Imhoff, J. F. (2016). New dimensions in microbial ecology-functional genes in studies to unravel the biodiversity and role of functional microbial groups in the environment, *Microorganisms*, **4**, pp. 19.

68. Méndez-García, C., Mesa, V., Sprenger, R. R., Richter, M., Diez, M. S., Solano, J., Bargiela, R., Golyshina, O. V., Manteca, Á., Ramos, J. L., Gallego, J. R., Llorente, I., Martins dos Santos, V. A., Jensen, O. N., Peláez, A. I., Sánchez, J., and Ferrer, M. (2014). Microbial stratification in low pH oxic and suboxic macroscopic growths along an acid mine drainage, *ISME J.*, **8**, pp. 1259–1274.

69. Pérez-García, O., Lear, G., and Singhal, N. (2016). Metabolic network modeling of microbial interactions in natural and engineered environmental systems, *Front. Microbiol.*, **7**, pp. 673.

70. Tobalina, L., Bargiela, R., Pey, J., Herbst, F. A., Lores, I., Rojo, D., Barbas, C., Peláez, A. I., Sánchez, J., von Bergen, M., Seifert, J., Ferrer, M., and Planes, F. J. (2015). Context-specific metabolic network reconstruction of a naphthalene-degrading bacterial community guided by metaproteomic data, *Bioinformatics*, **31**, pp. 1771–1779.

## Chapter 13

# Future and Perspectives of the Use of Iron Nanoparticles for Water and Soil Remediation

**Marta I. Litter**
*Gerencia Química, Comisión Nacional de Energía Atómica,*
*1650 San Martín, Prov. de Buenos Aires, Argentina*
marta.litter@gmail.com

## 13.1 Introduction

A careful examination of the chapters in this book indicates that the actual number of applications of zerovalent iron nanoparticles (nZVI) in various sites and effluents is constantly growing. Many evidences have also proven that nanoscale Fe(0) is a highly efficient and affordable material. As also highlighted in some chapters, nanoparticulated iron oxides can also be applied in some cases, presenting some advantages compared with nZVI, such as stability toward oxidation or more efficient in removing certain pollutants, such as arsenic. Our analysis in this chapter will be centered on nZVI, although it can be extended to all nanoparticulated iron materials.

*Iron Nanomaterials for Water and Soil Treatment*
Edited by Marta I. Litter, Natalia Quici, and Martín Meichtry

ISBN 978-981-4774-67-3 (Hardcover), 978-981-4669-49-8 (eBook)
www.panstanford.com

In the last 10 years, the adaptation of nZVI to remove contaminants has received increasing attention due to its inherent extraordinary properties due to the chemical and physical structure of the material such as high specific surface area and high reactivity.

nZVI represent a potentially efficient alternative to current materials used for water treatment and have been thoroughly studied for remediation purposes considering the numerous reviews on the subject. At least 11 review papers on the nZVI remediation technology have been published in the last 5 years [1–10], summarizing different aspects in this field (see, e.g., Refs. [11, 12] and all references therein). The technology has been shown efficient to remediate an impressive range of contaminants at much greater reaction rates than other reagents. nZVI can be used as an effective and versatile tool for the purification of waters and soils.

To be used as solid catalysts, iron nanoparticles must fulfill certain requirements, for example, high activity, marginal leaching of active cations, and stability over a wide range of pH and temperature. Last but not least, the materials should be available at a reasonable cost.

The remediation technology with nZVI involves a series of steps:

1. Transport, in the aqueous phase (or other delivery fluid) to the contaminated zone;
2. Attachment to soils in the contaminated zone or partitioning to the non-aqueous phase liquid (NAPL)/aqueous phase;
3. Reaction with the target contaminant to form less toxic or less mobile products.

The application of nanoparticles (NPs) in the remediation of the environment not only reduces the concentration of potential noxious substances, but also decreases the costs of large-scale remediation and of the duration of the process. Possible applications vary from small-scale (domestic point-of-use treatments), environmental in situ methods, and much larger batch or flow through industrial applications. For groundwater remediation, the injection of nanoscale iron-based particles delivered in the subsurface as aqueous slurries is a promising, effective technology.

## 13.2 Limitations of the Use of Zerovalent Iron Nanoparticles

Although significant potential has been demonstrated for the application of nZVI to remediate a wide range of priority pollutants in groundwater or wastewater, several important limitations constraint the proper deployment of the technology and the widespread commercial application. The most important restrictions are due to the following problems:

- Lack of stability and easy leaching;
- Rapid and easy aggregation and settling of nZVI, due to strong attractive interparticle forces, primarily magnetic, leading to agglomeration to micron-sized particles of limited mobility in porous media;
- Not enough reactive lifetime of the NPs, especially when stabilizers are not used;
- Difficulty for the separation of nZVI from the treated solution;
- Low selectivity in reaction with chemical substances;
- Limited reusability due to rapid oxidation of iron nanoparticles under ambient conditions;
- Reaction of nZVI with natural groundwater constituents, reducing the available active material for reaction with the target contaminants;
- Displacement of dilute contaminant plumes upon nZVI injection at some sites, suggesting that nZVI technology is more suitable to source zones rather than to dilute plumes;
- Stabilizers added to nZVI or supports for the NPs such as carbon, mesoporous silica, and colloidal clays have an additional cost, and some of them may be regarded as possible contaminants;
- Lack of comparable studies for different nZVI materials and deployment strategies;
- Possibility of remobilization of heavy metals and radionuclides over extended periods;
- The transport distance is still an undefined and unpredictable factor for the effective application of the technology, especially in soils of low permeability;

- The effect of the delivered NPs on the hydraulic conductivity has not yet been sufficiently investigated in field, especially in the long term.

## 13.3 Environmental Concerns

There are a number of potentially serious issues regarding the environmental fate of engineered NPs and their potential impacts on human health [13] derived from their small size and high reactivity, and the potential mobility of engineered NPs in both environmental and biological systems. For example, environmental exposure to NPs may allow their deep penetration into the lungs via inhalation or the passage of NPs across cell membranes directly into cells or tissues under controlled exposure conditions. In general, the transport mechanisms of engineered NPs through the environment and into plants and animals, and the associated risks, remain poorly understood (e.g., [14]), and this may significantly limit their widespread application as remediation tools.

In addition, toxic effects of nZVI on biota and on the biogeochemical conditions appear as potential threats. Long-term environmental impacts of nZVI either on living organisms or on the whole environment remain rudimentary. There are big concerns over the long-term fate transformation and ecotoxicity of nZVI in environmental systems.

## 13.4 Recent Advances in the Use of Iron Nanoparticles

In view of the interest in nZVI, numerous studies have been undertaken in recent years, basically aimed at searching innovative production techniques, modification of the physicochemical functionality, and enhancement of the stability and mobility of the particles.

Immobilization of nZVI onto suitable supports is one of the most studied fields, as it provides an easy operation; however, an excellent nZVI reduction ability has to be preserved. Different solid porous materials such as carbons, exchange resins, bentonite, kaolinite, zeolites, or chitosan beads have been tested.

To overcome the aggregation problem, various polymers and other coatings have been used to stabilize nZVI particles.

Doping of ZVI with other metals, i.e., the preparation of bimetallic Fe particles with transition metals such as palladium (Pd), copper (Cu), nickel (Ni), or platinum (Pt), on the iron surface has been effective in the improvement of the removal of contaminants, especially for halogenated hydrocarbons [15].

The particle stabilization technique using polyelectrolytes is a valuable alternative for preparing soil deliverable nZVI and represents a great step toward field application of the in situ remediation technology [9]. Depending on the site characteristics and properties of the target contaminants, neutral or charged polysaccharides may be used to yield stabilized nZVI of desired particle size and transportability. The pre-agglomeration stabilization technique is not only thermodynamically more favorable but also more effective in fabricating smaller NPs and facilitating size control. In addition to much improved soil deliverability, stabilized nZVI show greater reactivity and much lower adverse cytotoxicity. The effectiveness and feasibility of stabilized nZVI for remediation have been widely tested or demonstrated in the laboratory, pilot, and field scales.

## 13.5 Technical Constraints and Future Research

Like for any other new technologies, there are some critical constraints and knowledge gaps for the nZVI-based in situ technology, for which urgent further research is needed. However, despite the unquestionable effectiveness of nZVI and the simplicity of their delivery into the ground as a dry powder or slurry for a direct treatment, there are multiple disadvantages of using the free NPs. The dispersion of nZVI in an aquifer will be limited by geochemical/geophysical characteristics, mineral sorption, microbiological activity, aggregation and formation of voluminous corrosion products, etc., which are variables difficult to predict. For this reason, a unique tailoring of nZVI for each situation would be required. Also changes in the groundwater system (natural or otherwise) may also cause contaminants adsorbed to the nZVI surface to remobilize.

Additional research is needed at all scales to improve the understanding of the nZVI potential for remediation. At the nanoscale, an improved understanding of the governing chemical reactions, the kinetic behavior and the physical mechanisms governing transport in waters and soils will yield a potential transfer of the technology to real application. More field applications are required with detailed characterization before and after nZVI injection to assess nZVI mobility and the extent of contaminant destruction. All these information will lead to the development and validation of numerical models that can be used to predict remediation at a wide range of field sites. Despite a number of successful large-scale applications of chemical injection methods, further research is still needed on the limitations and feasibilities at the field scale, particularly regarding the interaction of iron with the intermediate products of degradation of the contaminant, the role of organic matter present in the soils in the efficiency of the treatment process, and the behavior of other, non-target, site contaminants during the treatment [16]. The following considerations can be taken into account:

- NPs should have the ability to remove a large range of contaminants.
- Surface area and reactivity of NPs have to be maximized.
- NPs should present excellent mechanical properties to allow optimal flow rates.
- NPs need to be supported in the form of membranes, mats, beads, etc.; in this way, they can be used in fixed bed reactors, filter columns, permeable reactive barriers, or domestic filters (even simple household jars).
- NPs need to have strong adhesion to the support to ensure no release into the environment.
- NPs should have low production costs to ensure realistic deployment.
- New advances in bimetallic particles, improved stabilizers, and improved formulations have to be investigated for enhancing partitioning of nZVI to NAPL, including field mobility and performance at the field scale.
- The particle stabilization technique should be modified to yield NPs more dispersible in soils and to facilitate the deliverability of stabilized nZVI to contaminated source zones.

- The reactive lifetime of nZVI should be increased by a combination of stabilizers of various properties to minimize the corrosion reactions and maintaining a high reactivity.
- The long-term stability of metals/radionuclides immobilized by nZVI at the field scale is a critical factor to assess the effectiveness of the technology and optimize the process design.
- The effect of the delivered NPs on the hydraulic conductivity should be confirmed at the field scale and over extended periods of time.
- Mechanistic transport models considering the chemical transformation of nZVI are needed to predict the transport and fate of stabilized nZVI in soils.
- Toxic effects of nZVI on biota and on the biogeochemical conditions, especially for bimetallic and immobilized particles should be reinforced.
- Long-term environmental impacts should be investigated more, studying various groups of organisms at various stages of development, and at various degrees of transformation of nZVI, including the potential stress of the ecosystem.

To be commercially viable, sustainability of the technology needs to be assured. The nanomaterials, in any form they are used, should gather the following properties:

- Should not be very much expensive
- Should be recyclable
- Should be reusable and for this, adsorbed contaminants need to be removed by simple methods
- Methods of disposal of used materials and treatment of wastes are needed, especially if they are hazardous wastes
- Removed metals need to be recovered for further application, with the consequent economical return

Standardization is also needed, by elaboration of testing procedures, indicating the optimal amount of reactive NP material.

As said, for groundwater remediation, injection of NPs as slurries is a promising, effective technology. For a successful full-scale application, generated reactive zones should be wide and homogeneous. For this, the development of reliable transport models to predict the mobility of the particles during the injection and to estimate the radius of influence and final distribution of the reactive material is needed; these studies should include numerical models to predict the long-term fate of the injected particles after the remediation.

Summarizing, big efforts should be made to improve and adapt this technology for water and soil treatment, combining the studies with fundamental work for thorough understanding of the mechanisms and kinetic aspects related to the systems. In spite of the above problems, nZVI look extremely promising not only for water treatment but also for many other important global industries.

## References

1. Fu, F., Dionysiou, D. D., and Liu, H. (2014). The use of zero-valent iron for groundwater remediation and wastewater treatment: A review, *J. Hazard. Mater.*, **267**, pp. 194–205.
2. Xie, Y., Dong, H., Zeng, G., Tang, L., Jiang, Z., Zhang, C., Deng, J., Zhang, L., and Zhang, Y. (2017). The interactions between nanoscale zero-valent iron and microbes in the subsurface environment: A review, *J. Hazard. Mater.*, **321**, pp. 390–407.
3. Kharisov, B. I., Rasika Dias, H. V., Kharissova, O. V., Jiménez-Pérez, V. M., Olvera Pérez, B., and Muñoz Flores, B. (2012). Iron-containing nanomaterials: Synthesis, properties, and environmental applications, *RSC Adv.*, **2**, pp. 9325–9358.
4. Kharisov, B. I., Kharissova, O. V., Rasika Dias, H. V., Ortiz Méndez, U., Gómez de la Fuente, I., Peña, Y., and Vázquez Dimas, A. Iron-based nanomaterials in the catalysis, http://dx.doi.org/10.5772/61862.
5. Li, S., Wang, W., Liang, F., and Zhang, W.-X. (2017). Heavy metal removal using nanoscale zero-valent iron (nZVI): Theory and application, *J. Hazard. Mater.*, **322**, pp. 163–171.
6. Stefaniuk, M., Oleszczuk, P., and Sik, Y. (2016). Review on nano zerovalent iron (nZVI): From synthesis to environmental applications, *Chem. Eng. J.*, **287**, pp. 618–632.
7. Tesh, S. J. and Scott, T. B. (2014). Nano-composites for water remediation: A review, *Adv. Mater.*, **26**, pp. 6056–6068.
8. Zhao, X., Liu, W., Cai, Z., Han, B., Qian, T., and Zhao, D. (2016). An overview of preparation and applications of stabilized zero-valent iron

nanoparticles for soil and ground water remediation, *Water Res.*, **100**, pp. 245–266.

9. Zou, Y., Wang, X., Khan, A., Wang, P., Liu, Y., Alsaedi, A., Hayat, T., and Wang, X. (2016). Environmental remediation and application of nanoscale zerovalent iron and its composites for the removal of heavy metal ions: A review, *Environ. Sci. Technol.*, **50**, pp. 7290–7304.
10. Thomé, A., Reddy, K. R., Reginatto, C., and Cecchin, I. (2015). Review of nanotechnology for soil and groundwater remediation: Brazilian perspectives, *Water Air Soil Pollut.,* **226:121**, DOI 10.1007/s11270-014-2243-z.
11. Crane, R. and Scott, T. (2012). Nanoscale zero-valent iron: Future prospects for an emerging water treatment technology, *J. Hazard. Mater.*, **211–212**, pp. 112–125.
12. Zhang, W. (2003). Nanoscale iron particles for environmental remediation: An overview, *J. Nanoparticle Res.*, **5**, pp. 323–332.
13. Colvin, V. (2003). The potential environmental impact of engineered nanoparticles, *Nature Biotechnol.*, **21**, pp. 1166–1170.
14. Nowack, B. and Bucheli, T. D. (2007). Occurrence, behavior and effects of nanoparticles in the environment, *Environ. Pollut.*, 150, pp. 5–22.
15. O'Carroll, D., Sleep, B., Krol, M., Boparai, H., and Kocur, C. (2013). Nanoscale zero valent iron and bimetallic particles for contaminated site remediation, *Adv. Water Res.*, **51**, pp. 104–122.
16. Cundy, A. B., Hopkinson, L., and Whitby, R. L. D. (2008). Use of iron-based technologies in contaminated land and groundwater remediation: A review, *Sci. Total Environ.*, **400**, pp. 42–51.

# Index